Electricity and Electronics Fundamentals for Industrial Maintenance

Thomas Kissell

Terra Community College
Fremont, Ohio

PEARSON

Prentice
Hall

Upper Saddle River, New Jersey
Columbus, Ohio

Library of Congress Cataloging in Publication Data

Kissell, Thomas E.
  Electricity and electronics fundamentals for industrial maintenance / Thomas Kissell.
     p. cm.
   ISBN 0-13-117598-X (pbk.)
      1. Electric machinery—Maintenance and repair—Textbooks. 2. Industrial
equipment—Maintenance and repair—Textbooks. I. Title

TK454.K52 2005
621.31'042'0288—dc22

2004053159

Executive Editor: Ed Francis
Production Editor: Christine Buckendahl
Production Coordination: Carlisle Publishers Services
Design Coordinator: Diane Ernsberger
Cover Designer: Jason Moore
Production Manager: Deidra Schwartz
Marketing Manager: Mark Marsden

Pearson Education Ltd.
Pearson Education Singapore Pte. Ltd.
Pearson Education Canada, Ltd.
Pearson Education—Japan

Pearson Education Australia Pty. Limited
Pearson Education North Asia Ltd.
Pearson Educación de Mexico, S.A. de C.V.
Pearson Education Malaysia Pte. Ltd.

0-13-117598-X

PREFACE

The latest trend in modern industry is for one person to troubleshoot and repair all the systems on a machine. This requires a technician or maintenance repair person to fully understand the electrical, electronic control, and PLC systems that are found in modern industrial equipment. *Electricity and Electronics Fundamentals for Industrial Maintenance* is written to help students readily learn the operational theory, installation, and troubleshooting of these major systems.

This book focuses specifically on the interactions and interdependence of electrical and electronic components and systems. Traditionally, these topics would be covered one at a time in two or more separate textbooks. Obviously, this artificially isolates the topics and ignores the critical interrelationships among the components and systems. In contrast, this text is a comprehensive resource that enables students to learn not only the individual subjects but also the way these components and circuits intertwine to make a simple or complex machine operate. Thus, one technician can learn to troubleshoot effectively and repair any component or circuit found on modern automated machines regardless if it is a simple control circuit or a complex PLC-controlled system with electronic variable-frequency drives.

A troubleshooting technician who is responsible for ensuring that simple and complex automation runs smoothly in a factory today needs to learn only the *most essential* theories and troubleshooting tips to become effective. This book uses a unique proven curriculum change from traditional electrical and electronic programs where electrical and electronic technicians do not need months of calculating series and parallel resistive and reactive circuits that are used for design; rather they need a detailed understanding of how series and parallel circuits operate in AC, DC, and electronic circuits and what new test instruments are used to troubleshoot these components and circuits. They also need to be able to identify electrical and electronic components and traditional circuits and use their knowledge of these to quickly and efficiently troubleshoot them. The author has spent numerous hours over the past 20 years working directly with these technicians right on the factory floor of some of the most modern automated factories teaching the technicians how to efficiently an-

alyze these systems when they become faulty and how to quickly provide the correct repairs. The author has brought this knowledge back into the college classroom and has redefined the traditional electrical programs and courses at Terra Community College with the help of the electrical faculty, by integrating the material in this text into all the courses in the electrical curriculum. These changes over the past 10 years have allowed the electrical program to have retention and completion success as high as 90% and dramatically improved student performance while on the job. This program has become one of the first two-year college programs in North America to be awarded a prestigious Society of Manufacturing Engineers MEP grant in 2002. The curriculums for the electrical courses covered in this book are available from the author.

Lowering the Cost of Textbooks for Students

One of the most important problems for students today is the high cost of textbooks. It is very difficult to produce a quality textbook at a lower price year after year. The author is approaching this problem in a different way. This text is designed to be used in a complete series of electrical and electronic courses for the electrical and maintenance technician. The first chapters can be used in the basic or fundamentals electricity course, and remaining chapters can be used in courses that cover motor-control circuits and concepts, AC and DC motors and transformers, programmable controllers, and industrial electronics. This means a student needs to purchase only one book for as many as three or four courses, thus lowering the total cost of purchasing texts. The book is also an excellent investment for the student's technical library, as many of the diagrams and components are among the remaining technical archives of older circuits and diagrams that are still in use in modern automated equipment.

Features of the Book

The book uses a unique approach to ensure students fully understand the theory of operation of each component covered. This includes providing a picture of the component as it is normally found in the electrical system. Then an exploded-view picture or diagram and electrical symbol are presented along with a detailed explanation of the scientific and electrical theory that the component uses for normal operation. In addition, the component is shown in the type of circuit in which it will traditionally be found, and the text explains how this circuit operates as part of the complex machine automation. Finally, a detailed process of troubleshooting procedures is covered, including the best test equipment to use to test the component in or out of the circuit. The text also stresses the need for students to be able to move seamlessly between wiring diagrams and ladder diagrams. Chapter 21 in this text shows, step by step, how to convert a wiring diagram to a ladder diagram and how to convert a ladder diagram into a wiring diagram. This information is used throughout the text to help students understand which diagram is best used in each troubleshooting situation. Each chapter includes many photographs and diagrams to explain the theory of operation and show test points for troubleshooting individual components and complete systems. This text is designed to be simple enough for the beginner, yet comprehensive enough to be a complete reference for the technician to take on the job and use to review a technique or theory while troubleshooting.

The first chapter is devoted to all the aspects of safety that a student needs to understand to work safely in the laboratory and on the job. The remainder of the book is split into three parts to cover basic electrical fundamentals, motor controls and motors, and electronics and PLCs. Since basic electricity fundamentals tend to be the most difficult and diverse of these three topics, the fundamentals are presented in several subsections. Chapters 2 through 6 provide an introduction to basic theory, series circuits, and parallel circuits using meters. Unlike traditional texts that delve deeply into classical calculations, these chapters show how the student must apply the basics of electricity to become a well-trained troubleshooting technician. For example, the first circuits the student is introduced to use lamps for loads rather than traditional resistive circuits. The reason for this is that students can easily see when the lamp is illuminated and thus know when current is flowing, whereas it is difficult and confusing for students to analyze resistive circuits if they do not know whether or not the circuits are working.

Chapters 7 through 12 explain in a comprehensive manner the operational theory for a wide variety of electrical control components such as relays and motors. Chapter 7 reviews magnetic theory, Chapter 8 addresses AC fundamentals, and Chapter 9 covers basic step-up and step-down transformer fundamentals as well as delta and wye transformers that provide industrial voltage sources. Chapters 10 and 11 cover relays, contactors, and motor starters. Complete instructions are also included to help students learn how to troubleshoot these circuits and devices.

Chapter 12 concentrates on motor-control devices and circuits and helps students begin to learn how individual components work together as systems. Also, students are exposed to traditional start-stop circuits and many other traditional motor-control circuits.

Chapters 13 and 14 deal with single-phase and three-phase motors so that students can fully understand how to wire, install, and troubleshoot the most widely used load in electrical circuits. Chapter 15 covers DC motors, and Chapters 16 and 17 cover DC and AC generators.

Chapter 18 is a comprehensive chapter where students will learn to program and troubleshoot programmable logic controllers (PLCs). Chapter 19 focuses on electronic theory and on the operation of the wide variety of components and circuits that are used in industrial circuits and variable-frequency drives.

Chapter 20 introduces the procedures used in lockout, tag-out safety. This chapter is placed near the end of the text, but it can be integrated into any of the earlier chapters where it may be introduced as part of overall safety instruction.

The final chapter, Chapter 21, is unique to this textbook in that it is one of the only sources available that teach students a step-by-step procedure to convert a wiring diagram to a ladder diagram and to convert a ladder diagram to a wiring diagram. These conversion procedures are very important since it is essential for students to fully understand how to use both ladder and wiring diagrams while they troubleshoot a complete industrial machine.

This book is designed to provide as little or as much coverage to electrical theory through complex troubleshooting. It may be used in an introductory electrical course or as a student progresses to learn all aspects of electricity, electronics, and programmable controllers. The author has crafted this text to provide a single source that allows students to learn all these aspects of an exciting career.

ACKNOWLEDGMENTS

I would like to thank my wife, Kathleen Kissell, for all her help as graphic artist and project manager of this book and for keeping the figures straight. I also thank her for all her support and for putting up with all the activities that are associated with writing a book. I thank Ed Francis for his encouragement and help at all stages of this book.

I would also like to say a special thank you to the following teacher and students in the Senior Writing Class at Put-in-Bay High School, which is located on South Bass Island, in Lake Erie in Ohio: Ms. Katie Schneider, Caleigh Eriksen, Ericca Hirt, Nicole Karr, Caroline Koehler, Shaena Kowalski, and Sarah Seaberg. These students helped develop the glossary words for this text as part of a writing project.

Thanks also to the reviewers of this edition for their helpful comments and suggestions: Harold Brinkley, Ivy Technical State College; and Larry Chastain, Athens Technical College.

Finally, I would like to personally thank the following persons and their companies who have helped locate photographs and diagrams for this book and graciously provided permission to use these materials:

Kathy McCoin, Acme Electric Corporation, Power Distribution Products Division

Dennis Kennedy, Cooper Bussmann

Christine Wagner, Danaher Industrial Controls

Timothy Kovach, Eaton Electrical

Beverly Summers and Larry Wilson, Fluke Corporation

Lawrence Wegner and John McDevitt, Klein Tools, Inc.

Jennifer Ray, Leonard Safety Equipment

Peter Hayward, Ridge Tool Company

Richard Kirwin, Snap-on Tools

Susan Christensen, Rockwell Automation's Allen-Bradley Business

Steve Wehrle and Pam Cupp, Greenlee Textron

Terri Reid, Romy Corliss, and David Cohen, Honeywell Sensing and Control, Honeywell International, Inc.

Dennis Berry, National Fire Protection Association

Rae Hamilton, National Electrical Manufacturers Association

Julie Cohen, Square D / Schneider Electric

Kip Larson, SymCom Inc.

Ted Suever, Triplett Corporation

To Kathy
and her little angels;
Amber, Cassidy, Shelby, Christina, Robby and George

CONTENTS

CHAPTER 10

Relays, Contactors, and Solenoids   119

CHAPTER 11

Motor Starters and Over-Current Controls   131

CHAPTER 12

Motor-Control Devices and Circuits   143

CHAPTER 13

Single-Phase AC Motors   169

CHAPTER 19
Electronics for Maintenance Personnel   261

CHAPTER 20
Lockout, Tag-out   283

CHAPTER 1

Shop Safety and Shop Practices

OBJECTIVES

After reading this chapter you will be able to:

1. List typical safety clothing and equipment you should use on the job.

2. Inspect machines to ensure safety guards are in place and operating.

3. Identify problems with housekeeping to avoid accidents.

4. Develop a plan and implement it for fire safety.

1.0

INTRODUCTION

The most important part of working on the factory floor is personal safety. You are responsible for your own safety and the safety of others working with you. In this chapter you will learn about safety measures that will help you protect yourself and protect others, and you will learn about basic machine and shop safety, all of which will ensure a safe workplace. Your instructor will use many of the same safety rules and practices in your lab at school to ensure your safety in that work environment as well as to help you learn them.

1.1

SAFETY GLASSES, PROTECTIVE CLOTHING, AND EQUIPMENT

When you are working on the factory floor, you will be required to wear a variety of safety gear, such as safety glasses and earplugs, to protect yourself. Safety glasses with shatterproof glass lenses and side shields should be worn in your lab and the factory at all times to protect your eyes from debris. Problems can also occur with electrical equipment if there is an electrical short, when metal from electrical contacts can melt at high temperatures and spray into your face and eyes. Safety glasses with side shields will protect your eyes against these problems too. Make sure your safety glasses are comfortable to wear over extended periods. You can get safety glasses with prescription lenses to match your regular glasses.

Figure 1–1 shows several types of eye-protection equipment: safety glasses, safety goggles, and safety shields. Safety goggles with colored lenses are required when you are using torches or welding equipment. Safety shields must be used when you are mixing or pouring dangerous liquids that might splatter and get in your eyes or on your face.

In most factory settings you will also need to wear ear protection, which may include earplugs or earmuff protectors or both. Figure 1–2 shows typical earplugs and earmuffs. The conditions in your factory can be tested to determine the decibel level of noise and the type of ear protection that will be best over a long period of time. Ear damage occurs very slowly over time, and it is almost impossible to detect without sophisticated testing equipment. On any given day, you will not realize the noise is causing any damage to your ears, but after a number of years you may find that you cannot understand certain conversations around you or that you may need to turn up the volume on the television or radio. All excessive noise in a factory is dangerous and will eventually cause permanent damage to your hearing. Certain frequencies of noise will cause the deterioration to occur more quickly. You should begin to wear hearing protection as soon as you enter areas where you are exposed to noise, and you should continue to wear it at all times as long as you are in the exposed area.

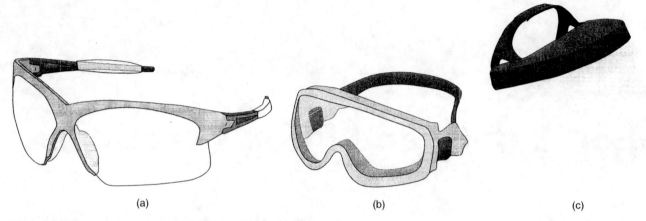

Figure 1–1   (a) Safety glasses; (b) safety goggles; (c) face shield.

You should also be aware of loose clothing and jewelry when you are working around equipment that has moving parts or when you work in electrical cabinets. Jewelry such as watches, finger rings, and other body rings need to be removed since they can be snagged on equipment and cause a laceration to the skin or broken fingers. In some cases, the jewelry may become entangled in equipment such as gears or belts, and workers may be pulled deeper into the machine where extreme injury or death may result. You should also be aware that you can receive an extreme electrical shock if metal jewelry comes into contact with elected terminals.

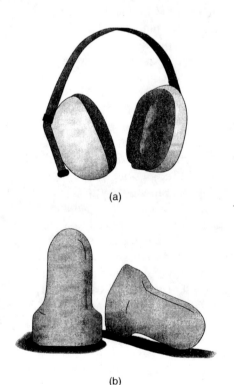

Figure 1–2   (a) Earmuffs; (b) earplugs.

Protective clothing may be required where welding or cutting occurs. Other protective clothing and respirators may be required when spray painting or doing other similar activities. You will receive a briefing before beginning any job where protective clothing is required, and you should always follow the instructions provided.

1.1.1
Back Brace for Lifting and Carpal Tunnel Braces

Certain activities in industry require repeated activity that aggravates specific muscle groups. For example, if you must lift loads over ten pounds repeatedly, you should wear a back brace that is specifically designed to help prevent back muscle strains. Other problems may occur if your job requires repetitive motions, such as tightening nuts and bolts continually. The muscles in your hands and wrist will begin to weaken, and they may begin to cramp. This condition can be corrected or prevented by special braces that provide support for the muscles. In some instances, the muscles are so deteriorated that the person must be removed from the job and given a different job. Some employers recognize the activities that cause these types of injuries and either provide protection or replace the job with some type of automation. You can help prevent these types of injuries by wearing the appropriate braces. You should also request help in lifting large weights or determine the maximum weight a person should lift without equipment.

1.1.2
Steel-Toed Shoes and Hard Hats

Another safety concern in most factories involves protecting your head against falling objects or bumping into solid objects with your head. A hard hat will protect

Figure 1–3 Hard hat.

Figure 1–4 Steel-toed boots.

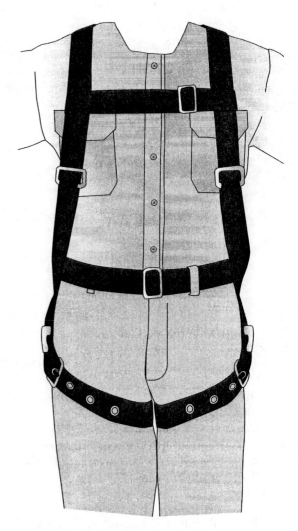

Figure 1–5 Safety harness.

you against these hazards. Figure 1–3 shows an example of a typical hard hat.

In some factories, you will be required to wear steel-toed shoes to prevent injuries due to dropping heavy objects on your feet. Steel toes are available in a wide variety of shoes and boots, allowing you to wear shoes or boots that are both comfortable and safe. Figure 1–4 shows steel-toed boots. It is important to wear steel-toed shoes at all times in areas where they are required and there is a danger of foot injuries. Some shoes are also specially fitted with steel soles to prevent punctures from walking on sharp objects.

1.1.3
Safety Harness

At times you may need to work in areas above the floor. In most factories, electrical technicians may need to check fuses in an overhead bus duct or change lightbulbs in overhead fixtures. Any time you are working above the floor you should wear a safety harness. Figure 1–5 shows a typical safety harness that will protect you from falling. You must attach one end of the safety harness to your safety belt, and the other ends

must be connected to the equipment you are working on. If you slip and fall, the safety harness will prevent you from falling to the floor.

1.2
SAFETY MATS AND EQUIPMENT TO PREVENT FATIGUE

Some jobs in the factory require standing on concrete floors for long periods of time. In these conditions a *cushion safety mat may be used to help absorb some of the shock* and *limit fatigue.* Other mats are provided to keep personnel from coming into contact with wet or damp surfaces. Some automation applications that have robots or presses with moving parts provide mats that have special sensors in them to ensure that personnel are not in the area of the machine's moving parts when the machine starts. These mats may include safety light curtains that will automatically stop the machine if you move off the mat or break the light curtain.

1.3
SAFETY LIGHT CURTAINS AND TWO-HAND START BUTTONS

Safety light curtains are used to ensure that personnel are not in the work area of automation and machines. These safety devices are specifically designed to prevent personnel from entering areas where they might be injured when equipment is in operation. If someone moves through the light curtain, the machine will be stopped.

Some machines have a safety circuit built into their start push buttons. These machines have two push buttons that are located approximately 2 feet apart. Both push buttons must be depressed at the same time to start the machine. Since the push buttons are mounted 2 feet apart, the operator must have one hand on each switch to start it. This ensures that both of the operator's hands are clear of the machine, so the operator will not be injured. These starting controls are added to machines that have moving parts that may injure hands if they are not clear when the machine is in operation. The two-handed push buttons are designed so they cannot be held down or bypassed in some manner. Since there are two switches, operators may try to hold one push button down with a screwdriver or other tool to bypass the safety feature so they can start the machine with one hand. This is usually done to speed up loading and unloading the machine. The two-handed push-button circuit is specifically designed so that if one button is held down, the machine will not restart. Do not try to disable these types of safety circuits under any circumstances since the operator can be seriously injured.

1.4
SAFETY GATES AND SHIELDS

Another type of safety device that is added to machines in a factory consists of a safety gate and a safety fence placed around large-scale automation such as robot cells. In some factories one or more robots are used to stack boxes onto pallets or do other jobs where personnel may be in close proximity. In these applications the equipment, robots, and machines are enclosed behind fencing. When personnel *must work on the equipment inside the fence, they must enter through a safety gate* that is wired so that all motion inside the fence is stopped any time the gate has been opened. When personnel leave the fenced-in area, the gate is closed and the machines are reset.

1.4.1
Machine Guards

Machine guards are provided to keep hands and clothing out of rotating parts on machines, such as shafts, gears, and belts. These guards are designed to be removed easily for maintenance and repairs, but they should remain in place any time the machine is operating. If a guard is removed during maintenance, do not try to restart the machine until the guard is back in place.

1.4.2
Glass Machine Guards

Some machines have guards that are see-through, such as glass guards on grinders and glass doors. Glass allows you to see into the moving part of the machine without removing the guard completely. For example, the glass guards on grinding machines protect operators while allowing them to view the grinding operation through the guard. Over time the glass may become dirty or get scratched to the extent that you cannot see through it. If the safety glass is dirty or damaged, it should be cleaned or replaced so that employees can utilize it as intended.

1.5
CHECKING GUARDS AND ACTIVE SAFETY DEVICES

One essential activity that needs to be completed every day is a walk-around check to ensure that safety guards are in place and that all active safety devices are operational. This type of check is called a daily preventative maintenance (PM) check. This check should be made each day to assure the safety of all personnel as they work around equipment or in dangerous areas of the factory. A more complete test and cleaning of these safety devices should be scheduled approximately twice a year. During this more in-depth test, the safety guards should be inspected for wear or deterioration of mounting mechanisms to ensure that the guards will withstand normal use and wear over the next six months.

A formal procedure should be provided that allows employees to write up any piece of equipment that has a missing guard or a piece of safety equipment that is not operating correctly. The procedure should include a means to inspect and repair the problem as quickly as possible. A complete record should be kept on each piece of equipment to track these types of safety problems so that inspections and repairs can be made on similar equipment in the factory. This is very important where a factory has more than one of any machine, such as multiple plastic presses. If a safety guard fails on one machine, it is a good bet that the same problem will occur on the other machines, especially if they are identical in make and model.

1.6
CLEANLINESS AND
SHOP HOUSEKEEPING

One of the major conditions that affect shop safety is the cleanliness of the shop. Many accidents occur when things are not put away or cleaned up. The shop should be cleaned on a regular basis, and there should be proper locations to store everything that is used around machines on the shop floor. When you locate oil spills or water spills, you should make sure that they are cleaned up thoroughly and as quickly as possible. Water can produce conditions that can cause someone to slip and fall, and it also may allow an electrical shock to occur if the power cords from hand tools, such as drills, are allowed to lie in the water.

Another problem related to cleanliness is the disposal of oily rags. The proper way to dispose of all oil rags is to use fire-rated storage cans. These containers have a fire-rated lid so that a fire cannot start in the container. It is also important to empty these containers on a regular basis.

1.7
LEAKS

Oil leaks and water leaks should be cleaned up and repaired as soon as possible to limit the amount of liquid that is released. The leaks can cause exposure to hazardous material as well as provide a potential source for slips and falls. In most factories, periodic maintenance schedules identify specific parts of each machine that should be checked for leaks on a daily basis. When a leak is first identified, it should be fixed immediately before it becomes excessive.

1.8
MSDS

In a factory or industry, you may work with any number of chemicals and other types of materials that are hazardous. In 1983, the Occupational Safety and Health Administration (OSHA) required manufacturing companies to maintain and distribute Material Safety Data Sheets (MSDS) on any material used in the factory. MSDS information may be stored in notebooks or on computer programs somewhere that is totally accessible to all personnel. Additional information is available by contacting the manufacturer of any material. When you are working where you may be exposed to a hazardous material, you should wear the proper safety equipment and clothing and locate the MSDS information and follow the instructions provided therein.

1.9
FIRST AID, CPR AND
PORTABLE DEFIBRILLATORS

As many people as possible at every work site should have training for first aid, cardiopulmonary resuscitation (CPR) and the proper use of portable defibrillators. If someone's heart stops beating they will need CPR or a defibrillator to help revive them until they can be moved to a hospital. Local Red Cross agencies usually provide first aid and CPR courses for workers. If someone at the plant receives a severe electrical shock or collapses and stops breathing, he or she will need the aid of someone who is certified in CPR.

First-aid safety includes first response to problems such as severe cuts and other lacerations or broken bones. If people are injured by machines, it is important to turn off power to the machine as quickly as possible, determine if you can safely move the injured person, and call for help immediately. First-aid courses will teach you how to apply pressure to deep wounds to stop bleeding. These courses will also tell you when it is unsafe to move someone with a head or neck injury.

1.10
FIRE SAFETY

Fire safety is very important in industrial settings, and it can be broken into two major categories: preventing fires and fighting fires with fire extinguishers. To ensure fire safety, it is essential to understand the basic principles of fire. Fire requires fuel, oxygen, heat, and a chain reaction (source of ignition) to start and sustain it. When you are trying to prevent or fight a fire, you must separate these elements. Fuels may include paper, rubber, oils, gases, or any other material that can burn.

The National Electric Code (NEC) classifies the locations and types of materials that have the possibility of burning as Class I, Class ll, and Class III:

Class I Petroleum-Refining Facilities
Petroleum distribution points
Petrochemical plants
Dip tanks containing flammable or combustible liquids
Dry-cleaning plants
Plants manufacturing organic coatings
Spray-finishing areas (residue must be considered)
Solvent extract plants
Locations where inhalation anesthetics are used
Utility gas plants, operations involving storage and handling of liquefied petroleum and natural gas
Aircraft hangars and fuel servicing areas

Class II Grain Elevators and Bulk Handling Facilities
Manufacturing and storage of magnesium
Manufacturing and storage of starch
Fireworks manufacturing and storage
Flour and feed mills
Areas for packaging and handling of pulverized sugar and cocoa
Facilities for the manufacture of magnesium and aluminum powder

Class III Rayon, Cotton, and Other Textile Mills
Combustible fiber manufacturing and processing plants
Cotton gins and cottonseed mills
Flax-processing plants
Clothing manufacturing plants
Sawmills and other woodworking locations
Coal preparation plants
Spice-grinding plants
Confectionary manufacturing plants

1.10.1
Sources of Ignition

One way to prevent a fire is to remove the source of ignition from the fuel. Some sources of ignition that you should be aware of include open flames from torches and heaters, matches, electrical equipment and heating elements, friction, high-intensity lighting, combustion sparks from grinding and welding, hot surfaces, static electricity, and spontaneous ignition, caused by improper storage of oily rags. You must also store flammable liquids in proper storage containers and keep them away from open flames and sources of ignition at all times.

Many companies implement proactive programs when any of these sources are exposed. For example, when you must use welding equipment or torches, safety covers and fire extinguishers are moved into place and additional personnel are used to watch for fire problems. Potential fire problems are also prevented by the use of safety guards to keep fuels away from hot surfaces.

1.10.2
Fire Extinguishers

Fire extinguishers are rated by classification. Class A fire extinguishers can be used on ordinary combustible material such as paper, wood, or clothing. This type of fire can easily be extinguished by water. A Class B fire extinguisher is designed for fires in flammable liquids, grease, and other material that can be extinguished by smothering or removing air (oxygen) from the chain reaction. Class C fire extinguishers are used on fires from live electrical equipment. The extinguishers for these types of fires must use material that is nonconducting, so that the person using the extinguisher is not exposed to additional hazard from electrical shock when fighting the fire. Class D fire extinguishers are used on fires in combustible metals such as sodium, magnesium, or lithium. Special extinguishers must be used to lower the temperature and remove oxygen from these fires.

Color codes for these fire extinguishers have been developed, and it is important that you recognize the type of fire and use the proper extinguisher. Class A fire extinguishers are green, Class B are red, Class C are blue, and Class D are yellow. It is also important that you recognize the types of materials found in your shop area and have the proper fire extinguisher for each material.

1.10.3
Fire Extinguisher Safety

Fire extinguishers should be inspected on an annual or semiannual basis. During the inspection, each fire extinguisher should be checked to ensure that it is fully charged and ready for use. All fire extinguishers should be clearly marked for the type of fire they are specified for and clearly displayed where all employees can locate them when they are needed.

1.10.4
Personnel Safety during a Fire

When a fire occurs in your building, it is important that several things occur. First and foremost, the proper authorities must be notified. This step should include setting off all fire alarms in the building to warn the other personnel so that they can evacuate the area. Next be sure to notify the plant fire brigade if you have one, and notify the local fire company so that it can begin its response. If the fire is small, the local plant fire brigade may begin to fight the fire with fire extinguishers. If the fire is larger, the most important thing to do is to get everyone safely out of the building. Be sure you learn the proper exit routes for all areas of the factory that you work in. It is also important to have a procedure in place to account for all personnel once they have evacuated the building.

1.11
ELECTRICAL SAFETY

When you are working in a factory, it is vital to understand how to work safely around electrical cabinets and circuits. This includes inspecting electrical safety grounds to ensure that they are connected correctly and are operational and knowing how to work safely around circuits that are under power. The safest way to work around electrical circuits is to turn the power off and lock out and tag out the electrical circuit. Lockout and tag-out procedures are required by OSHA and are explained in detail in Chapter 20.

In some cases you may be required to test voltage and current in circuits that have electrical power applied. When you must work on circuits that are under power, be sure to wear heavy rubber-sole shoes and try to work with only one hand exposed. If you work with both hands in an electrical panel, you may conduct electrical current through one arm, through your body (near your heart), and out the other arm. This is the most dangerous type of shock; it may cause severe damage to you or stop your heart. If you are working with only one hand at a time in the electrical panel, the chances of conducting electrical current directly through your body during an electrical shock is minimal.

An important new innovation in electrical safety is shown in Fig. 1–6. The Triplett Corporation has designed a new type of voltmeter display called a *heads-up display*. It consists of a glass lens attached to a visor

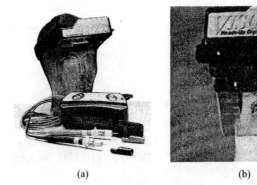

(a)   (b)

Figure 1–6 (a) Voltmeter with heads-up display so technician can work safely while making voltage measurements; (b) close-up of voltage displayed on heads-up display. (Courtesy of Triplett Corporation)

that is worn on your head. The heads-up display attaches to a digital voltmeter and displays the voltmeter reading directly on the glass lens that is located directly in front of your eye. This allows you to safely make voltage measurements in cabinets that have a large number of exposed electrical terminals. When you are working in an electrical cabinet, there is always a danger that your skin will accidentally come into contact with an exposed electrical connection and you will receive a severe electrical shock; or you might place the meter probes across two terminals and cause an arc. This is extremely dangerous when you have your hands inside the cabinet on voltmeter leads, and you are looking away from the meter leads and the exposed electrical terminals to read the meter face. The heads-up display allows you to look at your meter leads and concentrate on tile placement of tile leads in the cabinet, while you see the meter reading on the display.

If you do not have a heads-up display on your voltmeter, you should be careful to place the meter where you can easily read the face while you have your hands in the cabinet. In some instances, you may need a helper to read the meter as you concentrate on placing the meter leads on exposed electrical terminals.

1.11.1
Ground Fault Interruption Circuit Breakers

If you are using electrical hand tools or if you must work in areas that are damp or have standing water, it is important that all circuits have ground fault interruption circuit breakers (GFIs). GFIs are available as single-pole, two-pole, and three-pole devices. The GFI-type breaker measures all the electrical current that is supplied to a circuit or power tool and compares it with the amount of current returning from the tool. If there is no electrical short circuit to ground and the power tool is working correctly, the amount of current returning from the circuit will be the same. If less current is returning, the GFI assumes that a short circuit is occurring and a dangerous condition will result, and so it will trip to the open position. This means you could be using a power tool that has a faulty power cord and as soon as a short circuit occurs and you start to get an electrical shock, the GFI will interrupt all current in the circuit and prevent you from being shocked. GFI protection is also available as a built-in feature in some extension or power cords for portable electrical power tools.

Another type of GFI has an adjustable detection setting. When you are working with GFIs, you should never increase their trip point or tamper with their mechanism if they are tripping. If you tamper with the GFI, someone may receive a severe electrical shock. When GFIs are properly installed and maintained, they will prevent injury and death due to electrical shocks. It is important that you test all GFI circuit breakers at their predetermined test intervals.

1.11.2
Ground Wires

All electrical circuits and metal power tools require an electrical ground wire. The ground wire is identified by its green color and is used to provide a direct path to ground, which will cause excessive electrical current to flow through fuses or circuit breakers and will cause them to trip. If a power tool requires a ground circuit, it will have a third prong on its plug end. You should use these tools only in an electrical outlet receptacle that also has a ground terminal. If you have a power tool with a three-prong plug, you should never cut the ground prong off the plug so that it will fit a two-prong (non-grounded) receptacle. You should also never use a two-prong to three-prong adapter with a power tool that requires a three-prong plug. Some power tools have a plastic case; these tools will have a two-prong plug that does not need to be grounded when it is used. You will learn more about short circuits, grounding circuits, fuses, and circuit breakers in a later chapter. It is important that you inspect the power cord of any electrical power tool that has a metal case to ensure that it has a ground wire.

1.11.3
Ground Indicators on Electrical Panels

Some electrical panels have a set of two white lights on the front that are used to indicate that the cabinet or equipment is properly grounded. If the equipment ground wire is faulty, the lights will not be illuminated. If you work on an electrical cabinet where the ground-indicating lights are not illuminated, it may indicate that the ground wire is now faulty, and you should not touch the cabinet until you can determine if the ground wire is effective.

1.11.4
Electrical Enclosures That Provide Electrical Safety

At times you will encounter a variety of electrical enclosures that are specifically designed for safety. These enclosures have safety covers or seals that are designed protect them. When machinery is first installed, you may be requested to select the proper enclosure for the application that will provide safety. You may also be required to ensure the proper enclosure is selected if it is being replaced. If you are working on or troubleshooting an existing system and need to open the enclosure, you must be sure that the seals on the enclosure is not damaged when you replace the cover. You should also inspect the seal when the enclosure is opened and replace it if is damaged.

1.11.5
Hazardous Locations

You should become familiar with the different types of hazardous locations that exist where you work. When you arrive at your new work site, you should ask about any hazardous locations and any safety equipment you must use when working in these locations.

1.12
WORKING SAFELY WITH HAND TOOLS AND POWER TOOLS

At times you will be required to work with hand tools and power tools, which increase the chances of having an accident. You should learn the safe operation of these tools and practice safe working habits when you use them. Always read and follow the directions for safe operation before using tools. It is also important to have proper safety equipment, such as safety glasses, when you are operating hand tools. Do not use tools if you have not received the proper instructions to use them safely.

1.13

SAFETY BRIEFINGS AND SAFETY MEETINGS

Many companies use safety meetings to present basic safety information on a weekly basis. This provides a forum to identify and discuss safety problems throughout the plant. Other plants use a safety committee to identify safety areas of concern and to post safety warnings and posters throughout the plant to keep everyone safety conscious.

Some companies have designated personnel who inspect the various shop areas for safety compliance, whereas other companies empower all employees to keep an eye on safety conditions throughout the plant. Remember that you are responsible for your own safety at all times. Make working safely a habit.

1.14

SAFETY STANDARDS ORGANIZATIONS IN THE UNITED STATES, CANADA, AND EUROPE

Figure 1–7 shows the different safety standards organizations found throughout the United States, Canada, and Europe. The most familiar organizations are the National Electrical Manufacturers Association (NEMA), Underwriters Laboratories (UL), the National Fire Pro-

EEMAC
Electrical Equipment Manufacturers Advisory Council (Canada)

UL
Underwriters Laboratories Inc. (US)

IEC
International Electrotechnical Commission (Europe)

CSA
Canadian Standards Associations

ANSI
American National Standards Institute (US)

DIN
Deutsches Institut fur Normung (Germany)

NEMA
National Electrical Manufacturers Association (US)

ISO
International Organization for Standardization (Europe)

VDE
Verband Deutscher Elektrotechnicker (Germany)

NEC
National Electric Code (US)

NFPA
National Fire Protecion Agency (US)

CENELEC
European Committee for Electotechnical Standardization

Figure 1–7 Standards organizations in the United States, Canada, and Europe.

tection Agency (NFPA), and the National Electric Code (NEC). You will see the logos of these safety organizations stamped on the electrical equipment you use on the job. To learn more about the standards for each of these organizations, you can use the Internet to check out each organization's web site.

QUESTIONS

Short Answer

1. Explain why you should wear safety glasses with side shields at all times while you are working in a factory.

2. Explain why you should always wear earplugs or other types of ear protection while you are working in a factory.

3. List two safety items you can wear to protect yourself against falling debris in a factory.

4. Explain what MSDS is.

5. Identify a type of fire for each of the three classifications.

True or False

1. Class I fires include petroleum and solvent-type fires.

2. The reason a two-hand push-button station is used on a stamping press is so that you can operate it with either switch.

3. You can use a Class A fire extinguisher on an electrical fire.

4. A Class B fire extinguisher is painted red.

5. A GFI circuit breaker should be used with an extension cord if it is used in a damp or wet location.

Multiple Choice

1. The Class A fire extinguisher is used for paper or wood fires, and it is painted _____.
   a. red
   b. green
   c. yellow

2. A GFI (ground fault interruption) receptacle is _____.
   a. a special receptacle that has a third terminal for ground wire
   b. a special receptacle for power tools that allows more than one power tool to be plugged into it at the same time
   c. a special circuit breaker that can determine if a short circuit occurs and automatically opens the circuit breaker to protect the circuit

3. A two-handed push-button station is _____.
   a. a safety circuit that requires an operator of a stamping press to have both hands on a button for the press to begin its cycle
   b. a convenience circuit that allows an operator to operate the press from two different locations

   c. a single push button that has an oversized spring that requires the operator to use both hands to depress it

4. Ear protection should be used _____.
   a. at all times even in locations with moderate or small amounts of noise because this noise is damaging over a long period of time
   b. only in high-noise areas
   c. only if a person has been tested and found to have some hearing loss

5. If an electrical power tool has a power cord with a three-prong plug and if the only electrical outlet has two prongs, you should _____.
   a. cut the third prong (ground lug) off the power cord so that you can use the outlet
   b. find an adapter that allows you to plug a three-prong plug into a two-prong outlet
   c. not use this outlet because the power tool needs to be grounded for safe operation, and the two-prong outlet is not grounded

PROBLEMS

1. Make a fire safety plan for *your lab area and classroom* that includes an evacuation plan.

2. Make a safety plan that includes a method of accounting for *all personnel* if evacuation must occur.

3. Make a safety plan that lists all the potential fire safety problems in your lab.

4. Make an electrical safety plan that includes a weekly inspection of all the power cords used for power tools to ensure that they are properly grounded.

5. Make a safety plan that includes the proper periodic inspection of all safety guards for all equipment in your lab.

CHAPTER 2

Tools for Electrical Technicians

OBJECTIVES

After reading this chapter you will be able to:

1. Identify the proper electrical meters to use for troubleshooting.

2. Select the proper electrical tools for punching or cutting holes in an electrical cabinet.

3. Select the proper electrical tools for cutting and bending electrical conduit.

4. Select the proper pliers for cutting and stripping wire.

5. Select the proper screwdriver for installing wiring in an electrical cabinet.

6. Select the proper wrenches to use in electrical maintenance.

7. Select the proper tool pouch for carrying electrical tools.

2.0
OVERVIEW OF ELECTRICAL TOOLS

An electrical technician must be able to properly use a wide variety of meters, construction tools, wire cutters, screwdrivers, wrenches, and other tools. This chapter introduces the tools you will be using on the job in installing, troubleshooting, and repairing electrical equipment.

2.1
ELECTRICAL METERS

As an electrical technician or person who will work on electrical systems in maintenance, you will need to be able to select the proper type of meter and use it correctly. Figure 2–1 shows a variety of meters you will use. Figure 2–1a shows a Klein Wiggy voltage tester, which is also called a solenoid-type voltage tester. This type of tester is available from a number of meter manufacturers, and it operates in a similar way to a voltmeter in that it is able to determine the presence of voltage in a circuit. The advantage of a solenoid-type tester over a traditional digital meter is that the solenoid in the Wiggy requires a substantial amount of voltage and current to cause the meter to activate. When the meter leads are placed across a 110-VAC power source, the solenoid will activate and the meter indicator will "jump" to indicate the amount of voltage the meter is measuring. If the voltage is a "stray" voltage or a voltage that would cause a digital voltmeter to create a "phantom" voltage and it does not have current with it, it will not cause the solenoid to energize on the Wiggy. The other feature that makes the Wiggy different from a traditional analog or digital voltmeter is that it indicates the presence of voltage in three distinct ways. First, the meter has a needle that moves to indicate the amount of voltage numerically; second, the meter vibrates when its solenoid is energized by the presence of voltage; and third, it will emit an audible hum when voltage is present.

It is important to remember that the Wiggy and all voltmeters must have their leads placed across a voltage differential such as the voltage supply lines, line 1 L1 to neutral or line 1 to line 2, to indicate the presence of voltage. It is possible to put the meter on two separate wires that are connected to terminal line 1 (L_1) in the circuit. Since both wires come from the same source, there is no difference of potential, and the meter will not indicate a voltage. This may give you a false sense of security, and you will assume the wires are not energized. But if you cut through one of these wires and let your wire cutters touch metal conduit that is grounded, you could receive a severe electrical shock. This is true of all digital and analog voltmeters. You should always test across the two lines, and if you do not measure voltage,

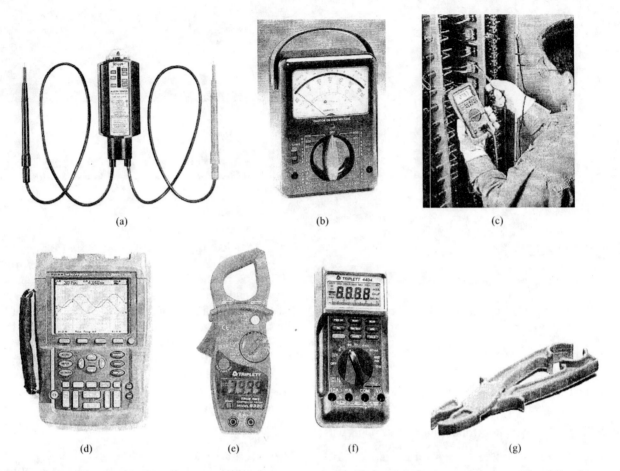

Figure 2–1 (a) Klein Wiggy solenoid tester (Copyright © 2004 Klein Tools, Inc); (b) Triplett analog volt, ohm, and milliamp (YOM) (Courtesy of Triplett Corporation); (c) technician using a Fluke digital voltmeter (Reproduced with permission of Fluke Corporation); (d) Fluke scope meter (Reproduced with permission of Fluke Corporation); (e) Fluke clamp meter (Courtesy of Triplett Corporation); (f) Triplett digital voltmeter (DVM) (Courtesy of Triplett Corporation); (g) Snap-on nylon fuse puller (Courtesy of Snap-on Tools); (h) Triplett digital voltmeter (DVM) (Courtesy of Triplett Corporation).

you should then test for voltage between each line and a known good ground wire or ground terminal.

It is also important to understand that the solenoid-type tester does not have high impedance like an analog or digital voltmeter. This means that a small amount of current will flow through the meter when you are making a voltage measurement. For this reason you must be careful using this type of meter around switches and other electronic circuits where the small amount of current may cause erroneous circuit or component operation. You should always use an analog or digital voltmeter that has high impedance when you are making measurements in electronic components or electronic circuits.

An electronic tester is available that will emit a buzz or tone when it is placed near any single wire that is energized with voltage, regardless of the presence of a neutral or ground wire. This tester will allow you to determine if a wire is energized with voltage or not.

Figure 2–1b shows a meter that indicates the amount of voltage and resistance with a meter movement over a numerical face. The amount of voltage is determined by the location of the needle over the face of the meter. You must be aware of the polarity of the leads if you are using this type of meter to measure DC voltage. If you have the polarity wrong (the positive lead on the negative voltage terminal), the needle on the meter will try to move backward. If this occurs, you can simply exchange the positive and negative leads or change the polarity switch and the meter needle will read upscale. A range switch allows you to select a range that is higher than the voltage you are trying to measure. This is important because if you select too low a range, the needle will try to move too far and the meter could be damaged. Thus you should always set the meter on the highest range and switch it to a lower range if the reading is too low. If you use the meter to measure AC voltage, the polarity does not matter, but you will still need to set the meter to the highest reading and set the switch lower if needed.

Figure 2–1c shows a technician making a measurement with a digital voltmeter. Notice that the technician places the meter in a location where he can put the meter leads on the electrical terminals and look at the meter display to read the amount of voltage the meter is indicating. Experienced technicians tend to place the meter so that they can comfortably read the meter display and keep an eye on the terminals where the leads are touching. This helps to prevent accidental electrical shocks that may occur if you place the meter so that you must look away from the cabinet to read the meter.

The electrical meter in Fig. 2–1d is called a Fluke scope meter. The scope meter can read voltage in a digital readout just like a digital voltmeter can, it can produce a copy of the waveform of the voltage like an oscilloscope can, and it can display the frequency as a number on its display. The scope meter is battery-powered, and it has extremely high impedance, and so it can be placed across any terminals in any electronic or electrical circuit and safely provide a measurement of voltage and frequency. Some models of the scope meter have the ability to measure two separate signals and display the phase shift or time difference between the two. It is also possible to measure extremely small amounts of voltage and display the waveform it creates to determine if any "noise" or unwanted signals are present. This is very useful in measuring computer-level signals that are present in an electrical cabinet.

Figure 2–1e shows a clamp-on–type ammeter. The meter is clamped around the wire to measure AC current. This type of meter can also be used to measure voltage and resistance with probes, just like a digital voltmeter. The clamp-on meter primarily measures AC current and some models have special circuits that allow DC current to be measured too. The clamp-on ammeter operates on the principle of electromagnetic induction, which is similar to the operation of a transformer. You can read more about transformers in Chapter 7.

Figure 2–1f shows a typical digital voltmeter, sometimes called a DVM. The digital voltmeter displays the amount of voltage the meter is measuring. Some DVMs, known as digital multimeters (DMMs), measure resistance and small currents up to 1,000 milliamps (1 amp). The digital meter has a range-selection button that allows you to set the range for the voltage or current you are measuring. Some of these meters also have an autoranging function, which means they will automatically adjust to the proper range. In addition, the newest models have automatic mode selection so that the meter and circuit are protected if you set the meter for resistance and mistakenly place the probes across a voltage. When any meter is set to measure resistance, its internal circuitry creates a low-impedance circuit that will act like a short circuit if the meter leads are placed across a voltage. If this were to occur, the meter would be severely damaged, and a large arc could be produced where the meter leads touch the electrical voltage source. The lesson here is that you should always set the meter for voltage if you

are not absolutely sure the circuit does not have voltage present.

Figure 2–1g shows a fiberglass fuse puller. You may encounter a socket with a fuse that you suspect is faulty. If you want to remove the fuse from the socket, you will need to use a fuse puller made of insulating material so that you do not receive an electrical shock. The fuse puller allows you to grasp the fuse securely and pull it from its socket without exposing you to additional shock hazards in the electrical panel. You can also use the fuse puller to put a new fuse back into the socket.

2.2
HOLE PUNCHES AND HOLE-CUTTING TOOLS

If you are requested to install electrical conduit or switches into an electrical panel, you will need a hole punch or hole saw to safely cut a hole in the metal cabinet. Figure 2–2 pictures a number of tools specifically designed to punch or cut holes safely and efficiently in metal cabinets. The hole punch will make an exact-size hole in the cabinet , and it is generally used for making holes that are larger than $1/2$ inch in diameter. The reason a hole punch is preferred over a drill for making larger holes is that a very large drill motor would be required to turn a drill bit big enough to make holes larger than $1/2$ inch. A hole saw is used when the metal is thinner. The hole punch shown in Fig. 2–2a is manufactured by the Greenlee Textron Company and is sometimes referred to as a "Greenlee punch." Figure 2–2b shows how the hole punch operates.

If you are using the hole punch to cut a $3/4$-inch hole in a metal electrical panel, you will need to drill a small $1/4$-inch hole in the metal cabinet where you want to locate the larger hole for a switch. The $1/4$-inch hole allows the bolt for the hole punch to be inserted through the metal plate of the panel. The hole punch is disassembled so that the two parts of the punch are mounted on each side of the metal plate you are piercing. The bolt will be used to pull the two sides of the punch together to pierce the metal plate. Figure 2–2e shows a hydraulic version of the hole punch. The hydraulic hole punch allows larger punches to be used when you need to make larger holes or when material is thicker. The hole punch die is assembled on the two sides of the metal plate that is to be pierced, and the hydraulic hand pump is used to supply hydraulic pressure through a hose to the die. Some larger hydraulic hole punches have a foot pedal for easier use.

Figure 2–2c shows a set of hole saws. These circular metal saws are mounted in a drill and are rotated at high speeds. The teeth in the saw cut the metal just like a wood saw.

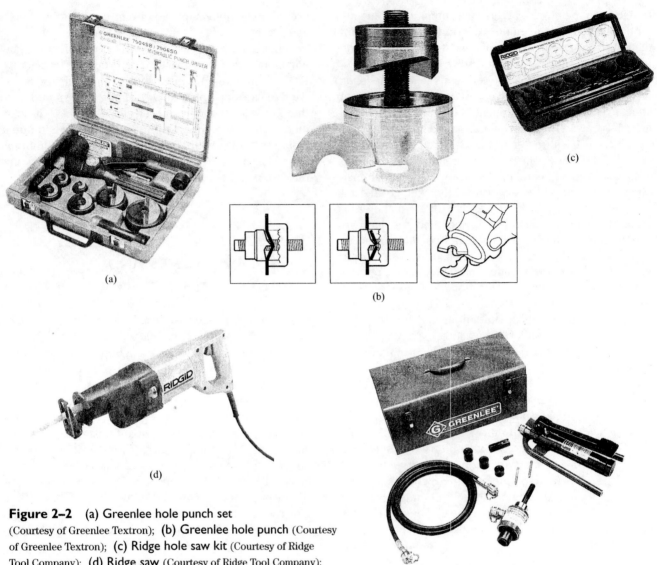

Figure 2–2 (a) Greenlee hole punch set
(Courtesy of Greenlee Textron); (b) Greenlee hole punch (Courtesy
of Greenlee Textron); (c) Ridge hole saw kit (Courtesy of Ridge
Tool Company); (d) Ridge saw (Courtesy of Ridge Tool Company);
(e) Greenlee hydraulic punch set (Courtesy of Greenlee Textron).

Figure 2–2d shows a reciprocating saw that can use a metal cutting blade to cut conduit or other metal. This saw can be used as a cutoff saw, or it can be used to cut freehand holes into panels. It can also be used to cut off bolts or other metal shafts.

2.3

ELECTRICAL CONDUIT BENDING, CUTTING, AND THREADING TOOLS

The conduit bender pictured in Fig. 2–3a is used to make measured bends in metal conduit. To operate this bender, you place your foot on the baseplate and pull on the handle. The bender has a set of indicators that show the amount of the bend in degrees. A hydraulic bender is available for bending larger conduits.

Figure 2–3b shows a flexible metal fish tape used to pull wire through a conduit. The fish tape is stored in a plastic case when it is wound up, and it is extended into the empty conduit until the hook on its end extends out of the conduit. You can attach the wire you are pulling into the conduit onto the hook of the fish tape. The fish tape is strong enough to pull multiple wires into the conduit. You must be very careful not to let the metal fish tape come into contact with exposed electrical wires or terminals, as the metal tape will conduct current that can cause severe electrical shock.

Figure 2–2c presents a cutting tool for cutting metal conduit. Figure 2–2d illustrates a tool for reaming the

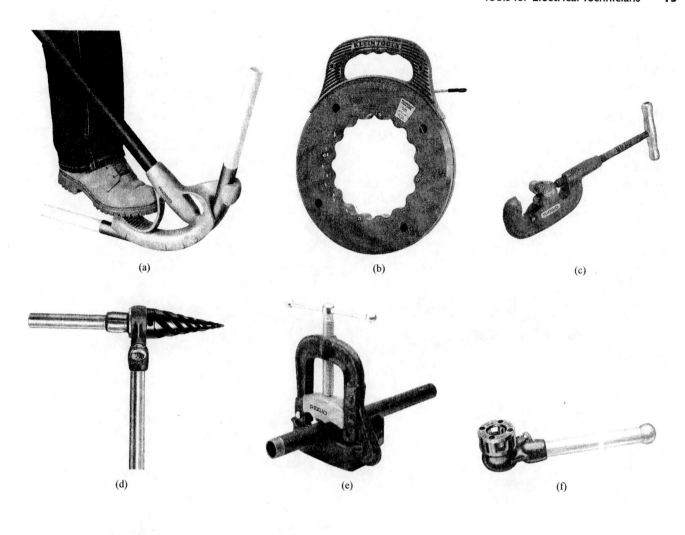

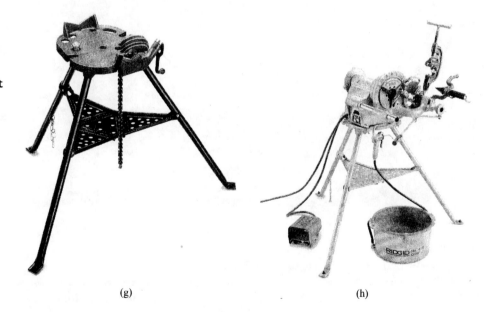

Figure 2–3

(a) Ridge tubing bender (Courtesy of Ridge Tool Company); (b) Klein flexible fish tape with grip handle (Copyright © 2004 Klein Tools, Inc.); (c) Ridge pipe conduit cutter (Courtesy of Ridge Tool Company); (d) Ridge pipe reamer (Courtesy of Ridge Tool Company); (e) Ridge portable pipe vise (Courtesy of Ridge Tool Company); (f) Ridge pipe (conduit) threader (Courtesy of Ridge Tool Company); (g) Ridge pipe tripod stand with vise (Courtesy of Ridge Tool Company); (h) Ridge tripod stand with electric pipe threader and pipe cutter (Courtesy of Ridge Tool Company).

pipe and removing rough edges after the pipe cutter is used to cut it. This is an important step since the rough edges of the pipe can cut or damage the insulation on wire that is pulled into the pipe or conduit.

Figure 2–2e shows a portable pipe vise, and Fig. 2–2g and h shows models of vice clamps that are mounted on a tripod. The tripod provides a stable work surface to hold conduit or pipe while you are cutting or threading it. Figure 2–2f shows a pipe-threading tool. This tool is used to put pipe threads on metal conduit so that it can be threaded together with threaded fixtures.

2.4
ELECTRICAL PLIERS, WIRE CUTTERS, AND WIRE STRIPPERS

Electrical technicians and electricians must use a wide variety of pliers, wire cutters, and wire strippers specifically designed for the electrical trade. Figure 2–4 shows a variety of wire cutters and strippers you may use on the job. The wire stripper is designed so that only the insulation on the wire is cut, leaving the wire un-

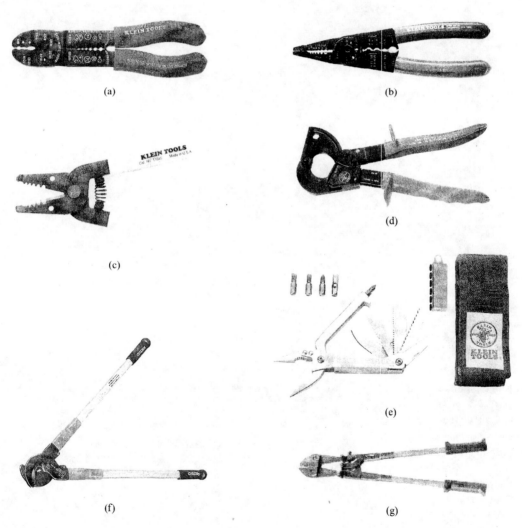

Figure 2–4   (a) Klein five-in-one tool that cuts bolts, cuts and strips wire, measures stud sizes, and crimps insulated terminals (Copyright © 2004 Klein Tools, Inc.);   (b) Klein combination wire- stripping and wire-cutting tool (Copyright © 2004 Klein Tools, Inc.); (c) Klein wire stripper and cutter (Copyright © 2004 Klein Tools, Inc.);   (d) Klein ratcheting cable cutter (Copyright © 2004 Klein Tools, Inc.);   (e) Klein electrical maintenance multitool (Copyright © 2004 Klein Tools, Inc.);   (f) Ridge cable cutter (Courtesy of Ridge Tool Company);   (g) Klein bolt cutter (Copyright © 2004 Klein Tools, Inc.).

Figure 2–5   (a) Klein slip joint pliers; (b) Klein side-cutting pliers with smooth streamlined nose; (c) Klein side-cutting pliers with crimping die; (d) Klein electronic diagonal-cutting pliers; (e) Klein pump pliers; (f) Klein long-nose, side-cutting pliers; (g) Klein curved-nose pliers. (Copyright © 2004 Klein Tools, Inc.)

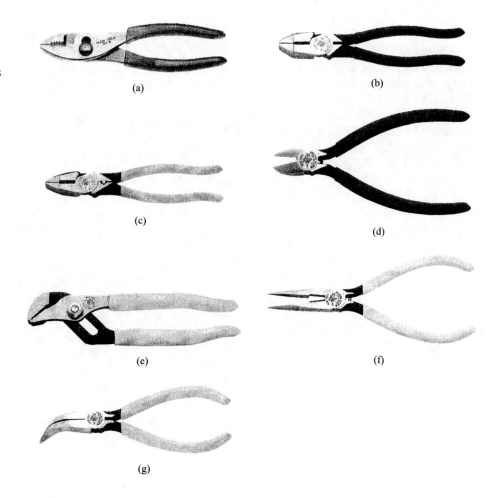

(a)

(b)

(c)

(d)

(e)

(f)

(g)

damaged when the jaws are closed tightly. The stripper has a variety of holes that fit wire gauges from size 8 to size 22. It is also possible to purchase wire strippers designed for smaller-size holes, and larger cutters are used to cut large-size wires.

Figure 2–5 illustrated a variety of slip joint, side-cutting, needle-nose, and clamping pliers that electrician will use on the job. The reason a wide variety of pliers are provided is to give you access in places that are hard to reach.

2.5
SCREWDRIVERS FOR ELECTRICAL INSTALLATION AND MAINTENANCE

Figure 2–6 presents a variety of regular and Phillips types of screwdrivers. Some of the screwdrivers in this

figure are called self-starting screwdrivers because they hold a screw securely so that it can be started into a threaded hole with one hand. Other screwdrivers are specifically offset to provide access to very tight spaces so that you can turn screws slightly to tighten or loosen them.

2.6
WRENCHES FOR ELECTRICAL MAINTENANCE

Figure 2–7 shows a variety of wrenches electricians may use on the job. Figure 2–7a pictures combination box and open-end wrenches that are used to tighten or loosen nuts and bolts. These types of wrenches are available in metric and standard sizes. Figure 2–7b shows a set of socket wrenches. The sockets snap onto a ratchet drive, which allows nuts and bolts to be

Figure 2–6   (a) Vaco #2 Phillips screwdriver; (b) Vaco 6-piece screwdriver set; (c) Vaco 4-in-1 screwdriver; (d) Klein starting screwdriver (Copyright © 2004 Klein Tools, Inc.);   (e) Klein starting screwdriver (Copyright © 2004 Klein Tools, Inc.);   (f) Snap-on offset screwdrivers (Courtesy of Snap-on Tools).

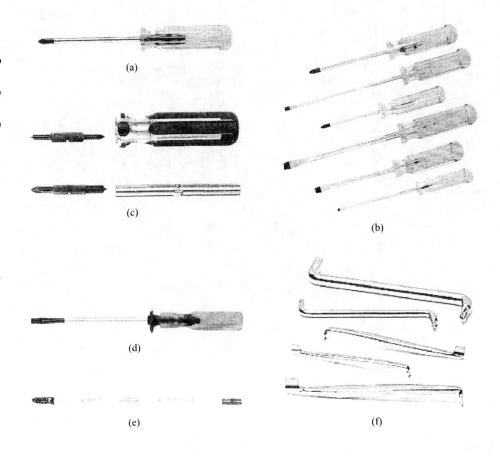

tightened or loosened quickly. The ratchet is also very useful in tight spaces.

The box end wrench shown in Fig. 2–7c has a ratchet feature that allows it to be used in tight spaces with a minimum of travel to tighten and loosen nuts and bolts. Figure 2–7d shows a set of nut drivers—special wrenches that have an end that looks like a socket wrench and a handle that is similar to a screwdriver. The nut driver is very useful for removing and replacing small sizes of screws or bolts that have hex heads. Figure 2–7e shows a set of hex key wrenches, sometimes called Allen wrenches. Figure 2–7f shows hex key wrenches with a T-handle. The T-handle allows more torque to be placed on the wrench so that larger hex head bolts can be tightened or loosened.

Figure 2–7g shows an adjustable wrench. One of the best-known brands of adjustable wrenches is a Crescent wrench. An adjustable wrench is extremely useful when you can only carry one wrench instead of a complete set of wrenches. It can be used for a wide variety of bolt and nut sizes by simply changing the opening of its jaws.

The pipe wrench, shown in Fig. 2–7h, has an adjustable jaw that grips the smooth surface of a pipe

or conduit so that it can be rotated to tighten or loosen its threads. Figure 2–7i shows a chain wrench, and Figure 2–7j shows a strap wrench. Both of these wrenches are used like a pipe wrench. They are adjustable and are used to loosen or tighten pipe and conduit.

2.7
TOOL POUCHES AND TOOL BELTS

The general tool set in Fig. 2–8a contains wrenches and screwdrivers. The tool set has the most common tools that an electrician may use for maintenance. Figure 2–8b shows a tap and die set. You use a tap to make interior threads inside a hole and a die to make exterior threads on a bolt. The die is used to "dress" or clean up threads on a bolt, and the tap can be used not only to make initial threads but also to dress up threads that are worn or damaged. You may need to drill a hole and tap threads into it to allow you to mount electrical

Figure 2–7   (a) Klein combination wrench set (Copyright © 2004 Klein Tools, Inc.);   (b) Klein socket wrench set (Copyright © 2004 Klein Tools, Inc.);   (c) Klein box reversing ratchet set (Copyright © 2004 Klein Tools, Inc.);   (d) Klein nut driver set (Copyright © 2004 Klein Tools, Inc.); (e) Klein hex key Allen wrench set (Copyright © 2004 Klein Tools, Inc.); (f) Klein T-handle hex key set (Copyright © 2004 Klein Tools, Inc.);   (g) Klein adjustable wrench (Copyright © 2004 Klein Tools, Inc.);   (h) Ridge pipe wrench (Courtesy of Ridge Tool Company);   (i) Ridge chain wrench (Courtesy of Ridge Tool Company);   (j) Ridge strap wrench (Courtesy of Ridge Tool Company).

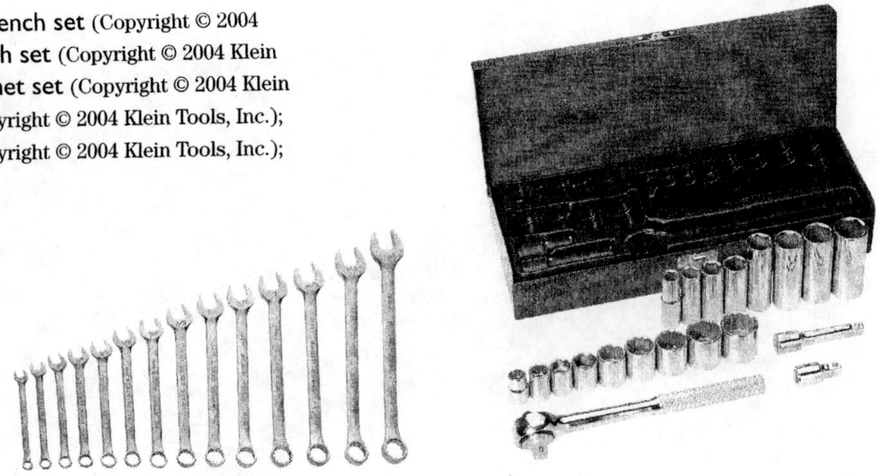

(a)

(b)

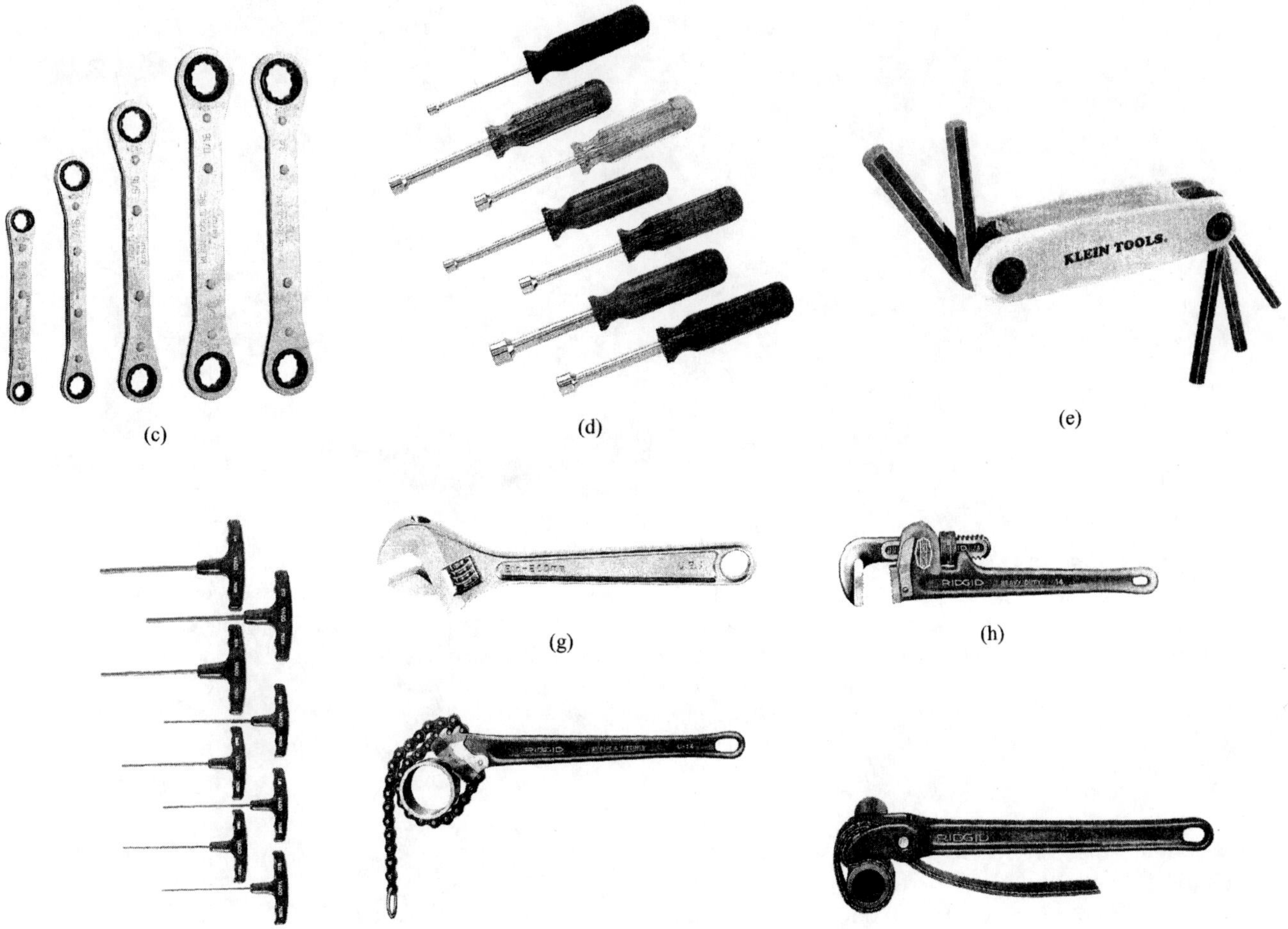

(c)

(d)

(e)

(f)

(g)

(h)

(i)

(j)

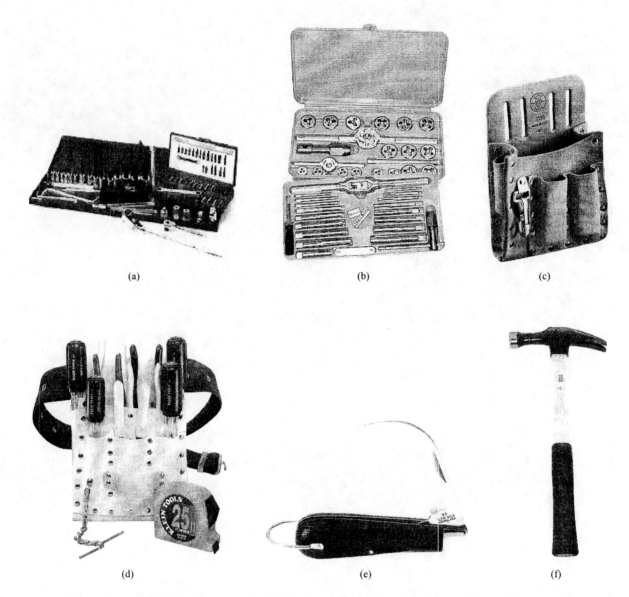

Figure 2–8   (a) Klein general tool set (Copyright © 2004 Klein Tool, Inc.);   (b) Snap-on tap and die set (Courtesy of Snap-on Tools);   (c) Klein tool pouch (Copyright © 2004 Klein Tool, Inc.);   (d) Klein electrician's tool pouch (Copyright © 2004 Klein Tool, Inc.);   (e) Klein electrician's pocket knife (Copyright © 2004 Klein Tool, Inc.);   (f) Klein hammer (Copyright © 2004 Klein Tool, Inc.).

components into an electrical panel with a screw or small bolt. The tapped hole takes the place of a nut, so you do not need a wrench on the back side of panel to hold the nut.

Figure 2–8c and d shows two different types of tool pouches. The tool pouch allows you to carry tools easily and safely so that they are always within reach.

The electrician's pocket knife in Fig. 2–8e is used to remove, or "skin," the insulation off a wire. The hammer in Fig. 2–8f is used for maintenance. It has a claw on one end to help remove nails or other fasteners.

2.8
PROPER USE OF TOOLS

You will be exposed to a wide variety of tools when you are on the job. It is important to understand that tools are designed for a specific job and that you should always use the proper tool for the right job. Many on-the-job injuries are caused by improperly using tools. If you do not have the correct tool to accomplish a job safely, you should request the proper tool.

QUESTIONS

Short Answer

1. Explain how a Wiggy-type voltage tester is different from a traditional voltmeter.

2. Identify four types of pliers.

3. Explain how a clamp-on ammeter measures AC current.

4. Explain how a hole punch is used to make a hole in a metal cabinet for a push-button switch.

5. Name two different types of screwdrivers.

True or False

1. A digital voltmeter displays the amount of voltage it is measuring.

2. A Wiggy-type voltmeter indicates the presence of voltage with a needle on a scale and with a solenoid that bounces or buzzes.

3. You can safely remove a fuse with a common pair of pliers instead of a recommended fiberglass fuse puller.

4. A hole punch is a better tool to use than a drill to put a hole in an electrical cabinet.

5. Wire cutters are available in a variety of sizes to allow all sizes of wire to be easily cut.

Multiple Choice

1. Which tool would you use to strip a wire to remove its insulation?
   a. Wire cutters
   b. Common pliers
   c. Wire strippers
   d. Side-cutting pliers

2. A scope meter can _____.
   a. display voltage as a number
   b. display the waveform of the voltage being measured
   c. display the frequency of the voltage being measured
   d. all of the above

3. Offset screwdrivers are used to _____.
   a. provide more torque than a regular screwdriver
   b. tighten or loosen hard-to-reach screws that have small clearance or are in tight spaces
   c. reach screws in deep recesses
   d. all of the above

4. A nut driver is _____.
   a. a type of pliers
   b. a type of wrench with a hex head
   c. a tool used to strip wires
   d. a special offset screwdriver

5. A tap and die set is used in this way:
   a. The tap makes interior threads, and the die makes exterior threads.
   b. The tap makes exterior threads, and the die makes interior threads.
   c. The tap can make both interior and exterior threads.
   d. The die can make both interior and exterior threads.
   e. Both c and d.

PROBLEMS

1. You are requested to mount three push-button switches in a new metal electrical panel. Select the tools you would use to do this job.

2. You are requested to cut three pieces of wire and pull them through a new section of conduit. Identify the tools that would help you with this job.

3. You are requested to mount a panel meter that is 2 inches in diameter in the front of the metal electrical panel and secure it with six $1/4$-20 screws. Identify the tools that would help you complete this project.

4. You are requested to cut a piece of conduit to length and thread it on each end. Identify the tools you would select to do this job.

5. You are requested to loosen several screws with Phillips heads that are in a very tight location. Identify any tools that would help you with this process.

CHAPTER 3

Fundamentals of Electricity

OBJECTIVES

After reading this chapter you will be able to:

1. Explain the term *electricity*.

2. Describe the term *voltage*.

3. Explain the term *current*.

4. Describe the term *resistance*.

5. Understand the sources of electricity.

6. Understand Ohm's law.

7. Explain the function of an electrical wiring diagram.

8. Explain the function of an electrical ladder diagram.

3.0
OVERVIEW OF ELECTRICITY

As a maintenance technician, you must thoroughly understand electricity so that you are able to troubleshoot the electrical components of systems, such as motors and controls. This chapter provides you with the basics of electricity, which will be the building blocks for more complex circuits that you will encounter in the field. You do not need any previous knowledge of electricity to understand this material. The simplest form of electricity to understand is *direct current* (DC) electricity, so many of the examples in this chapter use DC electricity. It is also important to understand that the majority of electrical components and circuits in field equipment use *alternating current* (AC) electricity, so some of the examples use AC electricity. This chapter begins by defining basic terminology such as voltage, current, and resistance and by showing simple relationships between them.

3.1
EXAMPLE OF A SIMPLE ELECTRICAL CIRCUIT

It is easier to understand the basic terms of electricity if they are connected with a working circuit with which you can identify. As an example, we will use the motor that moves a conveyor belt. The motor must have a source of voltage and current. *Current* is a flow of electrons, and *voltage* is the force that makes the electrons flow. This is just an introduction; much more detail about the terms is provided later in this chapter.

This circuit is shown in Fig. 3–1. It shows the source of voltage is 110 V, which is identified by L1 and N. The switch is a manually actuated switch for this circuit, so the switch provides a way to turn the conveyor motor on and off. When the switch is moved to the on position, or closed, voltage and current move through the wires to the motor. When voltage and current reach the motor, its shaft begins to turn. The wires in the circuit are called *conductors*. The conductors provide a path for voltage and current, and they provide a small amount of opposition to the current flow. This opposition is called *resistance*.

After the motor has run for a while, the switch is opened, and the voltage and current are stopped at the switch and the motor is turned off. When the switch is closed again, the voltage and current again reach the motor. Now that you have an idea of the basic terms *voltage*, *current*, and *resistance*, you are ready to learn more about them.

The source of energy that makes the shaft in the motor turn is called electricity. *Electricity* is defined as a flow of electrons. You should remember from the previous discussion that the definition of *current* is also a flow of electrons. In many cases the terms *electricity* and *current* are used interchangeably. *Electrons* are the negative parts of an atom, and *atoms* are the basic

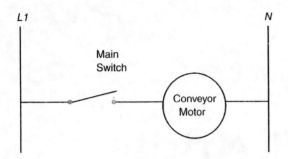

Figure 3–1   Electrical diagram of a manually operated switch controlling a conveyor motor.

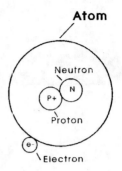

Figure 3–2   An atom with its electron, proton, and neutron identified.

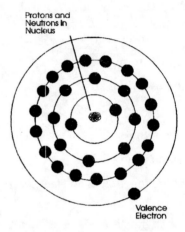

Figure 3–3   A copper atom has 29 electrons. This atom has 1 electron in its valence shell that can break free to become current flow.

building blocks of all matter. All matter can be broken into one of 118 materials called *elements*. Two of the elements that are commonly used in electrical components and circuits are copper, which is used for wire, and iron, which is used to make many of the parts, such as motors.

To understand electricity at this point, we need to understand only that the atom has three parts: electrons, which have a negative charge; protons, which have a positive charge; and neutrons, which have no charge (also called a neutral charge). Figure 3–2 shows a diagram of the simplest atom, which has one electron, one proton, and one neutron. From this diagram you can see that the proton and neutron are located in the center of the atom. The center of the atom is called the *nucleus*. The electron moves around the nucleus in a path called an *orbit*, in much the same way the earth revolves around the sun.

Because the electron is negatively charged and the proton is positively charged, an attraction between these two charges holds the electron in orbit around the nucleus until sufficient energy, such as heat, light, or magnetism, breaks it loose so that it can become free to flow. Any time an electron breaks out of its orbit around the proton and becomes free to flow, it is called electrical current. Electrical current can occur only when the electron is free from the atom so it can flow. The energy to create electricity can come from burning coal, nuclear power reactions, or hydroelectric dams.

The atom shown in Fig. 3–2 is a hydrogen atom; it has only one electron. In the circuit presented in Fig. 3–1, the copper wire is used to provide a path for the electricity to reach the conveyor motor. Because copper is the element most often used for the conductors in circuits, it is important to study the copper atom. The copper atom is different from the hydrogen atom in that the copper atom has 29 electrons, whereas the hydrogen atom has only 1. The copper wire that provides a path

for the electricity for the motor contains millions of copper atoms.

Figure 3–3 shows a diagram of the copper atom with its 29 electrons, which move about the nucleus in a series of orbits. Because the atom has so many electrons, they will not all fit in the same orbit. Instead, the copper atom has four orbits. These orbits are referred to as *shells*. There are also 29 protons in the nucleus of the copper atom, so the positive charges of the protons will attract the negative charges of the electrons and keep the electrons orbiting in their proper shells. The attraction for the electrons in the three shells closest to the nucleus is so strong that these electrons cannot break free to become current flow. The only electrons that can break free from the atom are the electrons in the outer shell. The outer shell is called the *valence shell*, and the electrons in the valence shell are called *valence electrons*. Notice that the copper atom only has one electron in the valence shell. All good conductors have a

Figure 3–4 Voltage is the force that knocks one electron out of its orbit. In this diagram the valence electron in the atom on the far right breaks free from its orbit and leaves a hole. The electrons in the atoms to the left of the first atom, move from hole to hole to create current flow.

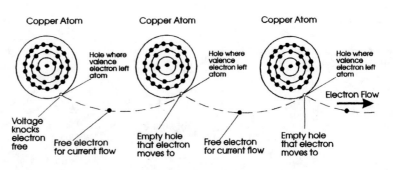

low number of valence electrons. For example, gold—which is one of the best conductors—has only one valence electron and silver—which is a good conductor—has three valence electrons.

In the motor circuit we studied earlier, a large amount of electrical energy, called *voltage*, is applied to the copper atoms in the copper wire. The voltage provides a force that causes a valence electron in each copper atom to break loose and begin to move freely. The free electrons actually move from one atom to the next. This movement is called *electron flow*; more precisely, it is called *electrical current*. Because the copper wire has millions of copper atoms, it will be easier to understand if we study the electron flow between three of them. Figure 3–4 shows a diagram of the electrons breaking free from three separate copper atoms.

When the electron breaks loose from one atom, it leaves a space, called a *hole*, where the electron was located. This hole will attract a free electron from the next nearest atom. As voltage causes the electrons to move, each electron moves only as far as the hole that was created in the adjacent atom. This means that electrons move through the conductor by moving from hole to hole in each atom. The movement will begin to look like cars that are bumper to bumper on a freeway.

3.2
EXAMPLE OF VOLTAGE IN A CIRCUIT

Voltage is the force that moves electrons through a circuit. Figure 3–5 shows the effects of voltage in a circuit where two identical lightbulbs are connected to separate voltage supplies. Notice that the bulb connected to the 12-V battery is much brighter than the bulb connected to the 6-V battery. The bulb connected to the 12-V battery is brighter because 12 V can exert more force on the electrons in the circuit than 6 V, and higher voltage will also cause more electrons to flow.

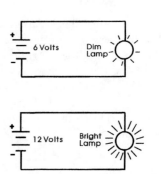

Figure 3–5 Two identical lightbulbs have different voltages applied to them. The lightbulb with 12 V applied to it is brighter than the one with 6 V applied to it.

In the original diagram of the motor, the amount of voltage is 110 V. The energy that produces this force comes from a generator. The shaft of the generator is turned by a source of energy such as steam. Steam is created from heating water to its boiling point by burning coal or from a nuclear reaction. The energy to turn the generator shaft can also come from water moving past a dam at a hydroelectric facility. In the lightbulb example, the voltage comes from a battery. The voltage in a battery was previously stored in a chemical form when the battery was originally charged. The larger the amount of voltage in a circuit, the larger the force that can be exerted on the electrons, which makes more of them flow. The scientific name for voltage is *electromotive force* (EMF).

3.3
EXAMPLE OF CURRENT IN A CIRCUIT

The definition of *current* is a flow of electrons. The actual number of electrons that flow at any time in a circuit can be counted, and their number is very large. The unit of current flow is the *ampere* (A). The word *ampere* is usually shortened to amp. One amp is equal

Figure 3–6 A number of electrons passing a point where they are counted. When the number reaches 6.24×10^{18}, 1 A is measured.

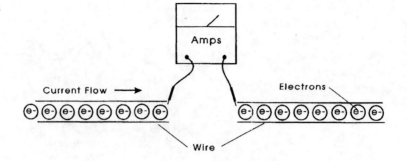

to 6.24×10^{18} electrons flowing past a point in 1 second (s). Figure 3–6 shows an example of the number of electrons flowing that equals 1 A. In this diagram you can see that each electron that is flowing is counted; when 6,240,000,000,000,000,000 electrons pass a point in 1 s, it is called an *ampere* (A). The ampere is used in all electrical equipment, such as motors, pumps, and switches, as the unit for electrical current.

3.4
EXAMPLE OF RESISTANCE IN A CIRCUIT

Resistance is defined as the opposition to current flow. Resistance is present in the wire that is used for conductors and in the insulation that covers the wires. The unit of resistance is the ohm, or Ω (the capital Greek letter omega, which represents the letter O). When the resistance of a material is very high, it is considered to be an insulator, and if its resistance is very low, it is considered to be a conductor.

It is important for conductors to have very low resistance so that electrons do not require a lot of force to move through them. It is also important for insulation that is used to cover wire to have very high resistance so that the electrons cannot move out of the wire into another wire that is nearby. The insulation also ensures that a wire on a hand tool such as a drill or saw can be touched without electrons traveling through it to shock the person using the tool. Examples of materials that have very high resistance and can be used as insulators are rubber, plastic, and air.

The amount of resistance in an electrical circuit can also be adjusted to provide a useful function. For example, the heating element of a furnace is manufactured to have a specific amount of resistance so that the electrons will heat it up when they try to flow through it. The same concept is used in determining the amount of resistance in the filament of a lightbulb. The lightbulb filament of a 100-W lightbulb has approximately 100 Ω

of resistance; 100 Ω is sufficient resistance to cause the electrons to work harder as they move through the filament, which causes the filament to heat up to a point where it glows. When the amount of resistance in a circuit is designed to convert energy, such as a heating element or motor, it is referred to as a *load*.

3.5
IDENTIFYING THE BASIC PARTS OF A CIRCUIT

It is important to understand that each electrical circuit must have a *supply of voltage, conductors* to provide a path for the electrons to flow, and at least one *load*. The circuit may also have one or more controls. Figure 3–7 shows an example of a typical electrical circuit with the voltage supply, conductors, control, and load identified. In this circuit, the voltage source is supplied through terminals L1 and N, the load is a motor, and the two wires connecting the motor to the voltage source are the conductors. The manual switch is a control, because it controls the current flow to the motor. In a large system in a factory, it is possible to have more than one

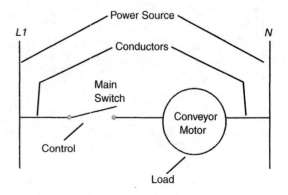

Figure 3–7 The basic parts of a circuit are identified in this example. They are the power source, conductors, control and the load.

load. For example, the system may have multiple conveyors or multiple hydraulic pumps. You will learn more about these components later.

3.6
WIRING DIAGRAM AND LADDER DIAGRAM

There are two basic ways to diagram electrical circuits. The components can be shown in a wiring diagram or a ladder diagram. The wiring diagram shows the components of a circuit as you would see them if you took a picture of the circuit. The ladder diagram shows the se-

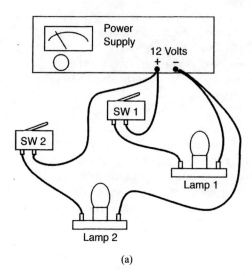

(a)

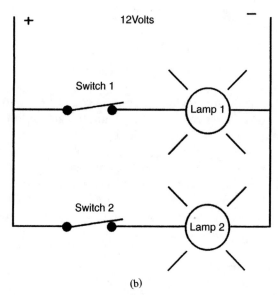

(b)

Figure 3–8   (a) Wiring diagram for two circuits in parallel; (b) ladder diagram of two circuits, each with one switch and one lamp.

quence of operation of all the components. Figure 3–8a illustrates a wiring diagram with two switches and a light. The switches and wires are shown where you would find them in a picture. The wiring diagram helps you know where the components are located in an electrical cabinet and where the terminals are located. This is important when you are installing components in a circuit. Figure 3–8b shows the same circuit, but this time drawn as a ladder diagram. The ladder diagram clearly shows that the two switches are wired in parallel with each other so either can turn on the light. This is called "showing the sequence of operation." In the wiring diagram it may be difficult to determine if the switches are connected in series or parallel, whereas in the ladder diagram it is very easy to see how the switches operate. You will learn that one of these diagrams is not necessarily better than the other; rather there is an appropriate time to use each. You will also see that it is easy to convert from either type of diagram to the other. Chapter 21 shows how to convert a wiring diagram to a ladder diagram, and how to convert a ladder diagram to a wiring diagram. In the remainder of this book, diagrams will be shown as both wiring diagrams and ladder diagrams to get you used to the value of each.

3.7
EQUATING ELECTRICITY TO A WATER SYSTEM

It may be easier to understand electricity if you think of it as a system such as a water system, with which you may be more familiar. If you examine water flowing through a hose, for example, the flow of water would be equivalent to the flow of current (amps), and the water pressure would be equal to voltage. If you stepped on the hose or bent the end over, it would create a resistance that would slow the flow, just as resistance in an electrical circuit slows the flow of current. Figure 3–9 provides a diagram that shows the similarities between the water system and an electrical circuit.

By definition, 1 V is the amount of force required to push the number of electrons equal to 1 A through a circuit that has 1 Ω of resistance. Also, 1 A is the number of electrons that flow through a circuit that has 1 Ω of resistance when a force of 1 V is applied, and 1 Ω is the amount of resistance that causes 1 A of current to flow when a force of 1 V is applied. You can see that voltage, current, and resistance can all be defined in terms of each other.

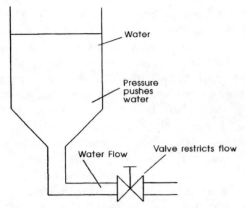

Water = Source of electrons
Water pressure = Voltage that pushes electrons
Water flow = Current flow
Valve restricts flow = Electrical resistance

Figure 3–9   A diagram of a water system that shows the similarities to an electrical system. The flow of water is similar to the flow of current (amperes), the pressure on the water is similar to voltage, and the resistance from the valve causes the flow of water to slow, just as resistance in an electrical system causes current flow to slow.

3.8

USING OHM'S LAW TO CALCULATE VOLTS, AMPERES, AND OHMS

There are times when you are working in a circuit that you need to calculate or estimate the volts, amperes, or ohms the circuit should have. In most cases when you are troubleshooting an electrical problem, you will also need to make measurements of the volts, amperes, and ohms to determine if the circuit is operating correctly. The problem with taking a measurement is that you will not be sure if the values are too high or too low or if they are just about right. A calculation can tell you if the measurements are within the estimated values the circuit should have.

A relationship exists between the volts, amperes, and ohms in every DC circuit that can be identified by Ohm's law. Ohm's law states simply that the amount of voltage in a DC circuit is always equal to the amperes multiplied by the resistance in that circuit. If a circuit has 2 A of current and 4 Ω of resistance, the total voltage is 8 V.

The formula can be changed to determine the amount of amperes or ohms. Because we know that $2 \times 4 = 8$, it stands to reason that 8 divided by 2 equals 4, and 8 divided by 4 equals 2. Thus, if you measure the volts in a circuit and find that you have 8 V and if you measured the amperes in a circuit and find that you

have 2 A, you can determine that you have 4 Ω by dividing 8 by 2. You can also determine the number of amperes you have in the circuit by measuring the volts and ohms and calculating; 8 V divided by 4 Ω is 2 A.

3.9

OHM'S LAW FORMULAS

When scientists, engineers, and technicians do calculations, they use letters of the alphabet to represent units such as volts, amps, and ohms. The letters are accepted by everyone, so when a calculation is passed from one person to another, everyone uses the same standard abbreviations.

In Ohm's law formulas, the letter E is used to represent voltage. The E is derived from first letter of the word *electromotive force*. The letter R is used to represent *resistance*. The letter I is used to represent current (amperes); it has been derived from the first letter of the word *intensity*.

Ohm's law is represented by the formula $E = IR$. When you use Ohm's law, the basic rule of thumb is that you should use E, I, and R to represent the unknown values in a formula when you begin calculations and you should use V for volts, A for amps, and Ω for ohms as units of measure when you have determined an answer or anytime the values are known or measured.

3.10

USING OHM'S LAW TO CALCULATE VOLTAGE

When you are solving for voltage, you need to know the amount of current (amperes) and the amount of resistance (ohms). If you do not know these two values, you cannot calculate voltage. For example, if you are asked to calculate the amount of voltage when the current is 20 A and the resistance is 40 Ω, you can multiply 20 by 40 to get an answer of 800 V. To solve a problem, start with the formula and substitute the values of I and R. Then you can continue the calculation. Notice that the units A (for amperes) and Ω (for ohms) are added when the values are known.

$$E = IR$$
$$E = 20 \text{ A} \times 40 \text{ }\Omega$$
$$E = 800 \text{ V}$$

3.11

USING OHM'S LAW TO CALCULATE CURRENT

The formula for calculating an unknown amount of current in a circuit is $I = \frac{E}{R}$. To calculate the current, you divide the voltage by the resistance. For example, if you have measured the circuit voltage and found that it is 50 V and if you have measured the resistance and found that it is 10 Ω, the current is found by dividing 50 V by 10 Ω, which equals 5 A.

$$I = \frac{E}{R}$$

$$I = \frac{50 \text{ V}}{10 \text{ Ω}}$$

$$I = 5 \text{ A}$$

3.12

USING OHM'S LAW TO CALCULATE RESISTANCE

The formula for calculating an unknown amount of resistance in a circuit is $R = \frac{E}{I}$. To calculate the resistance of a circuit, you divide the voltage by the current. For example, if you have measured voltage of 60 V and current of 12 A, the resistance is found by dividing the voltage by the current. Again, start with the formula and substitute values to get an answer:

$$R = \frac{E}{I}$$

$$R = \frac{60 \text{ V}}{12 \text{ A}}$$

$$R = 5 \text{ Ω}$$

3.13

USING THE OHM'S LAW WHEEL TO REMEMBER THE OHM'S LAW FORMULAS

A learning aid has been developed to help you remember all the Ohm's law formulas. Figure 3–10 shows this aid, and you can see that it is in the shape of a wheel, or pie. The top of the pie has the letter E, representing volt-

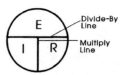

Figure 3–10   Ohm's law pie that shows a memory tool for remembering all the Ohm's law formulas.

age; the bottom left has the letter I, representing current (amperes); and the bottom right has the letter R, representing resistance (ohms).

If you want to remember the formula for voltage, put your finger over the E and the letters I and R show up side by side. The vertical line separating the I and R is called the *multiply line* because that is the math function you use with I and R. This represents the formula $E = IR$.

If you want to know the formula for current (I), put your finger over the I; the remaining letters are E over R. The horizontal line that separates the E and R is called the divide-by line, because to determine I you divide E by R. This represents the formula $I = \frac{E}{R}$.

If you want to know the formula for resistance (R), you put your finger over the R; the remaining letters are E over I. This represents the formula $R = \frac{E}{I}$.

3.14

CALCULATING ELECTRICAL POWER

Electrical power (P) is work that electricity can do. The units for electrical power are *watts* (W). Watts are determined by multiplying amps times volts ($P = IE$). Most electrical loads that are resistive in nature, such as heating elements, are rated in watts. For example if a heating element is rated for 5 A and 110 V, it uses 550 W of power. You can calculate the amount of power by using Watt's law, $P = IE$: 5 A × 110 V = 550 W.

This basic formula can be rearranged in the same way as the Ohm's law formula. You can calculate current by dividing the power (550 W) by the voltage (110 V), and you can calculate voltage by dividing the power (550 W) by the current (5 A). Figure 3–11 shows a pie that is a memory aid for remembering all of the formulas for Watts law. From this pie you can see that the original formula, $P = IE$, can be found by placing your finger over the P. The formula for solving for current ($I = \frac{P}{E}$) can be found by placing your finger over the I, and the formula to solve for voltage ($E = \frac{P}{I}$) can be found by placing your finger over the E.

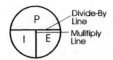

Figure 3–11 The formula pie for Watt's law formulas, which are used to calculate electrical power.

3.15
PRESENTING ALL THE FORMULAS

When you are troubleshooting, you may be able to get a voltage reading easily. If you can find a data plate that provides the wattage for a heating element or other load, you will be able to use one of the formulas that was presented to calculate (estimate) the current or the resistance. This estimate will give you an idea of what the values should be so that you can determine if the system is working correctly or not. When you are in the field making these measurements, it may be difficult to remember all the formulas, so a chart has been developed that provides all the formulas (Figure 3–12). From the figure you can see that the chart is in the shape of a wheel with spokes emanating from an inner ring. A cross is shown in the middle of the inner wheel, and it separates the inner wheel into four sections. These sections are represented by the letters P (power, in watts), I (amps), E (volts) and R (ohms).

The formulas in each section of the outer ring correspond to either P, I, E, or R. The formulas for wattage are shown emanating from the inner section marked with P. All the formulas for current are shown emanating from the inner section marked with I. All the formulas for voltage are shown emanating from the inner section marked with E, and all the formulas for resistance are shown emanating from the inner section marked with R.

If you are trying to calculate the power in a circuit or the power used by any individual component, you could use any of the three formulas shown on the wheel emanating from the section identified by the letter P. Thus, you could use $P = EI$, $P = I^2R$, or $P = E^2/R$. Your choice of formula would depend on the two values that are given.

3.16
USING PREFIXES AND EXPONENTS WITH NUMBERS

At times you have seen numbers that take up a lot of room to print out—such as 5,000,000 or 0.0000003. When you need to write a long number on the face of a meter or in a calculation, the number may need to be written in a form that does not take up so much room. Several standards have been adopted throughout the fields of mathematics and electricity for this purpose. The Chart in Fig. 3–13 shows several ways to present the number using less space. For example, if you wanted to shorten the number that represents the value for 5,000,000 (5 million) ohms you could use the prefix *mega* and the value would be expressed as 5 mega ohms or you could use the symbol 5 M Ω. If you wanted to work with this number in a calculator, you could use its exponent form, which would be 5^{e6}. The exponent may be expressed verbally as "5 times 10 to the 6th power."

If you need to shorten the value 7,000 (7 thousand) ohms you would use the prefix *kilo*, and the value would be expressed as 7 kilo ohms, or you could use the symbol 7 kΩ. If you used the exponent form in a calculator, you would use 7^{e3} Ω. When you express the exponent verbally, you would say "7 times 10 to the 3rd power."

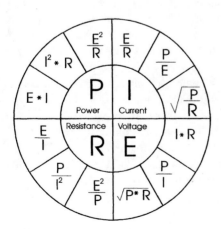

Figure 3–12 Formulas for calculating power, current, voltage, and resistance. Notice * means multiply.

Prefix	Symbol	Number	Exponent
Mega	M	1,000,000	10^6
Kilo	k	1,000	10^3
Milli	m	0.001	10^{-3}
Micro	μ	0.000001	10^{-6}

Figure 3–13 Prefixes, symbols, and exponents commonly found in electricity and on electrical equipment.

If you wanted to shorten the value 0.005 amp, you would use the prefix *milli*, and the value would be expressed as 5 milliamps, or you could use the symbol 5 mA. If you used the exponent in a calculation, you would use 5^{e-3} A. To express the exponent verbally, you would say "5 times 10 to the negative 3rd power."

If you wanted to shorten the value 0.000002 amp, you would use the prefix *micro*. This number would be expressed as 2 micro amps, or you could use the symbol 5 μA. If you used the exponent in a calculator, you would use 5^{e-6} A. To express the exponent verbally, you would say "5 times 10 to the negative 6th power."

Questions

Short Answer

1. Define the term *electricity*.
2. What is voltage?
3. What is current?
4. Explain what a valence electron is and why it is important to current flow.
5. What is resistance?
6. Define the term *power*.

True or False

1. The insulation coating on wire has high resistance.
2. Voltage is the flow of electrons.
3. Resistance is the opposition to current flow.
4. Watts are the units for current.
5. A good conductor has low resistance.
6. The negative part of an atom is an electron.

Multiple Choice

1. A valence electron is _____.
   a. the electron closest to the nucleus
   b. the electron in the outer shell
   c. an electron with a positive charge

2. Wattage is _____.
   a. the unit for power
   b. the unit for current
   c. the unit for voltage

3. Resistance is _____.
   a. the force that moves electrons
   b. wattage
   c. the opposition to current flow

4. Current will _____ when voltage increases and resistance stays the same.
   a. increase
   b. decrease
   c. stay the same

5. Wattage will _____ when voltage increases and current stays the same.
   a. increase
   b. decrease
   c. stay the same

Problems

1. Use Ohm's law to calculate the amount of voltage if current in a DC circuit is 5 A and resistance is 8 Ω.
2. Use Ohm's law to calculate the amount of current in a DC circuit if voltage is 25 V and resistance is 50 Ω.
3. Use Ohm's law to calculate the amount of resistance in a DC circuit if voltage is 40 V and current is 10 A.
4. Use the power formula to determine the number of watts for Problem 1.
5. Use the power formula to determine the number of watts for Problem 2.

CHAPTER 4

Voltmeters, Ammeters, and Ohmmeters

OBJECTIVES

After reading this chapter you will be able to:

1. Use a VOM meter to measure voltage.
2. Use a VOM meter to measure resistance.
3. Use a VOM meter to measure current in milliamperes.
4. Understand safety warnings for using ohmmeters and ammeters.
5. Explain the operation of a clamp-on ammeter.

4.0

MEASURING VOLTS, AMPS, AND OHMS

It is important to be able to measure the amount of voltage (volts), current (amperes), and resistance (ohms) in an industrial system to be able to determine when it is working correctly or if something is faulty. If you are checking a tire on your automobile, you can see if it is flat by simply looking at it. Because the flow of electrons through a wire is invisible, it is not possible to simply look at the wire and determine if it has voltage applied to it or if it has current flowing through it, so you must use a meter to make voltage, current, and resistance measurements.

4.1

MEASURING VOLTAGE

The meter used to measure voltage is called a voltmeter. Figures 4–1 and 4–2 show two types of voltmeters. The meter in Fig. 4–1 is called an *analog-type voltmeter* because it uses a needle and a scale to indicate the amount of voltage being measured. The meter in Fig. 4–2 is called a *digital-type voltmeter* because it displays numbers to indicate the amount of voltage being measured.

The voltmeter has very high internal resistance—approximately 20,000 ohms per volt—so its probes can be safely placed directly across the terminals of the power source without damaging the meter. The range-selector switch adjusts the internal resistance of the meter to set the maximum amount of voltage the meter can measure. The voltmeter can be used to make many different voltage measurements. For example, it can measure the amount of supply voltage in a system as well as the amount of voltage provided to each load. Figure 4–3 shows where the probes of a voltmeter should be placed to safely make voltage measurements in a circuit. In Fig. 4–3a, the voltmeter probes are shown across the terminals of a battery to measure the amount of battery voltage supplied. Figure 4–3b shows the proper location for placing the voltmeter probes to measure the voltage available to lamp 1, and Fig. 4–3c shows the proper location for placing the voltmeter probes to measure the voltage available to lamp 2. In each case the voltmeter probes are placed across the terminals where the voltage is being measured.

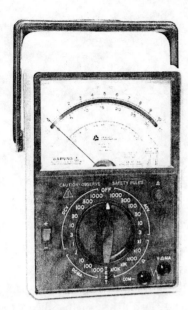

Figure 4–1 An analog voltmeter. (Courtesy of Triplett Corporation)

Figure 4–2 A digital voltmeter. (Courtesy of Triplett Corporation)

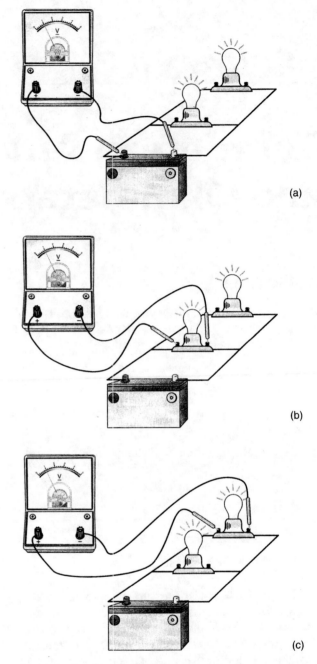

(a)

(b)

(c)

Figure 4–3 This diagram shows (a) the proper location for placing the probes of a voltmeter to measure the amount of voltage available at the battery, (b) the proper location for placing the voltmeter probes to measure the voltage available at lamp 1, and (c) the proper location for placing the voltmeter probes to measure the voltage available at lamp 2.

Typical meter ranges for digital voltmeters are 0–2, 0–20, 0–200, and 0–1,000 V. Special high-voltage probes must be used to measure voltages above 600 V. Typical voltages that are found in industrial systems are 24, 120, 208, 240, 277, and 480 V. The actual amount of voltage in your area may vary slightly; for example, 120 V may actually be measured as 110 V, and 480 V may be closer to 440 V.

The ranges for analog-type voltmeters are typically 0–2.5, 0–10, 0–50, 0–250, and 0–1,000 V. If you try to mea-

sure 120 V when the selector switch is set at 50 V, the needle will try to go past the maximum reading and the meter will be damaged. If you are using a digital voltmeter and you try to read a voltage that exceeds the setting of the range-selector switch, the display will indicate you are over the range but the meter will not be damaged.

4.2
MEASURING ELECTRICAL CURRENT

An ammeter is used to measure electrical current. Figure 4–4 shows two types of digital ammeters. (Note that ammeters can also be analog types.) The ammeter in Fig. 4–4a is part of a volt-ohm ammeter, and it is used to measure current that is less than 10 A. In many cases the ammeter is actually built in with the voltmeter, and the selector switch allows you to set the meter to be a voltmeter, ohmmeter, or ammeter. This kind of meter—a meter that can measure volts, ohms, or amperes—it is called a *VOM meter* (volt-ohm-milliampere meter). The ammeter is actually designed to read very small amounts of current, such as 1/1,000 A. The quantity 1/1,000 A is called a *milliamp* (mA), which can also be written 0.001 A. The prefix *milli*, as noted in Chapter 3, means *one-thousandth*. The VOM-type meter has a single meter movement that causes a needle to deflect when it senses current flow. This means that when a voltage is applied to the VOM meter, it must be routed through resistors that limit the amount of current that actually flows through the meter movement. When the VOM meter is set to measure current or resistance, the same meter movement is used. The only part of the meter that changes as it is switched between voltmeter, ammeter, and ohmmeter is the arrangement of resistors to limit the amount of current flow through the meter movement.

The meter shown in Fig. 4–4b is called a clamp-on ammeter, and it is used to measure current in larger AC circuits. As you can see in the figure, this type of meter has two *claws* at the top that form a circle when they are closed. A button on the side of the meter is depressed to cause the claws to open so that they can fit around a wire without the wire having to be disconnected at one end. The claws of the clamp-on ammeter are actually a transformer that measures the amount of current flowing through a wire by *induction*. When current flows through a wire, it creates a magnetic field that has flux lines. Because AC voltage reverses polarity once every 1/60 s, the flux lines collapse and cause a small current to flow through the transformer in the claw of the ammeter. This small current is measured by the meter movement, and the display indicates the actual current flowing in the conductor that the meter is clamped around. Induction is explained in more detail in Chapters 7 and 8.

4.3
MEASURING MILLIAMPS

At times you will need to measure the milliamperes in a circuit. It is important to remember, as noted previously, that *milli* is a prefix that means *one-thousandth*. For example, you may need to measure the amount of current that a solid-state control is using. Figure 4–5 shows that an open (a hole) should be made in the

Figure 4–4 (a) A digital volt-ohm-milliampere meter, or VOM meter (Courtesy of Triplett Corporation); **(b)** a clamp-on ammeter that is used to read high currents (Courtesy of Triplett Corporation).

(a)

(b)

circuit when a milliampere measurement is made, and each terminal of the ammeter should be placed on the wires where the hole has been made in the circuit. This places the meter in series with the circuit, and so all the electrons flowing in the circuit must also go through the meter because the ammeter has become part of the circuit.

Since the ammeter must become part of the circuit to measure the current, it must have very low internal resistance so that it does not change the total resistance of the circuit. When the VOM meter is set to read milliamps, it is very vulnerable to damage because it has very small internal resistance.

SAFETY NOTICE!
If you mistakenly place the meter leads across a power source (as you would to take a voltage reading) when the meter is set to read current, the meter will be severely damaged. You could also be severely burned and the meter could explode, because it is a very low-resistance component when it is set to read current. Placing the meter leads across a power source would be similar to taking a piece of bare wire and bending it so it could be inserted into the two holes of an electrical outlet.

As a technician you will be expected to use both the milliampere meter and the clamp-on ammeter. Milliampere meters are used to measure small amounts of current, and clamp-on meters are used to measure larger currents, such as the amount of current a motor is using. If a motor is using (pulling) too much current, you will be able to determine that it is overloaded and that it will soon fail. Later chapters in this book explain how to use the clamp-on ammeter to measure motor current.

Digital-type VOM meters and some analog VOM meters have circuits that measure both DC and AC current. These types of meters have several ranges for measuring very small current in the range of milliamps up to 10 A.

4.4
CREATING A MULTIPLIER WITH A CLAMP-ON-TYPE AMMETER

Sometimes the amount of current that is measured with the clamp-on ammeter is very small, and the needle will not deflect sufficiently to create an accurate measurement. Because the claws of the clamp-on ammeter are actually a transformer, the amount of current in the reading can be multiplied by wrapping the wire in which you are measuring the current around the claw several times to cause the reading to be multiplied. For example, if you wrap the wire around the claw two times and the wire has 1.25 A flowing through it, the meter reading will indicate 2.5 A, which is double the actual amount. You will need to divide the meter reading by 2 to get the exact current. If you wrap the wire around the claw four times, the meter reading will be 5 A, and you will need to divide the meter reading by 4 to get the true current flow in the wire. Creating a multiplier will make very small current measurements easier to read more accurately. Figure 4–6 shows an example of a wire twisted around the claws of a clamp-on ammeter to create a multiplier.

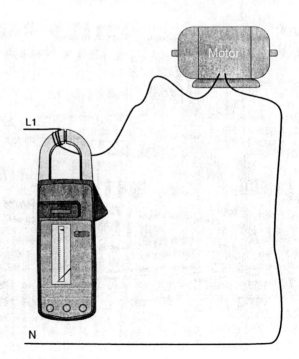

Figure 4–6 A clamp-on ammeter with wire twisted around its claws to create a multiplier.

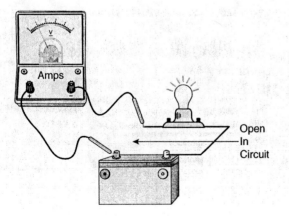

Figure 4–5 The proper way to place the meter probes to measure current.

4.5

MEASURING ELECTRICAL RESISTANCE

The VOM-type meter is also able to measure resistance when the selector switch is moved to one of the ohms positions, such as $R \times 1$, $R \times 10$, or $R \times 1,000$. When the VOM meter is set in the ohms configuration, it uses an *internal battery* to provide voltage to make the meter movement deflect the needle. When this voltage is applied to a resistance, a small amount of current will flow, which the meter detects.

> **SAFETY NOTICE!**
> Because the ohm meter provides an internal power source (1.5–30-V battery) for the current that the meter movement will read when it is set to the ohms ranges, it is important that the components being tested are not connected to any other source of voltage. Otherwise the meter will be severely damaged and could cause a short-circuit current in the circuit, which in turn could cause severe flash burns to the person holding the meter leads. It is also a good practice to remove the component from the circuit when resistance measurements are made. When a component is disconnected from a circuit and all sources of power are removed, it is said to be *isolated*.

The ohmmeter can be used to measure the resistance of a wire or fuse to ensure that the resistance is near zero ohms (0 Ω). A wire is said to have *continuity* and it is considered to be a good wire when it has near 0 Ω. If the wire has extremely high resistance (several million ohms), it is considered to be bad, because it must have an *open* that is causing the high resistance. The same terms apply to fuses. If a fuse is good, it will be described as having continuity, and if it is bad, it will be described as an *open fuse*. When a wire or fuse is open, it is also said to have *infinite resistance*. The word *infinite* indicates the amount of resistance is very high due to the fact the wire or fuse has an open. Because a typical VOM meter cannot measure resistance that is more than 2 million Ω, its scale is marked with the sign for infinity (∞) at the highest point that the reading is made. Figure 4–7a shows an example of the ohmmeter indicating 0 Ω when a good wire is tested, and Fig. 4–7b shows the ohmmeter indicating infinite ohms when the meter is testing a wire with an open in it.

The electrical loads in industrial systems such as motors or heating elements may have some amount of resistance that falls between 0 Ω and infinite ohms. For example, a motor may have 12 Ω of resistance in its run

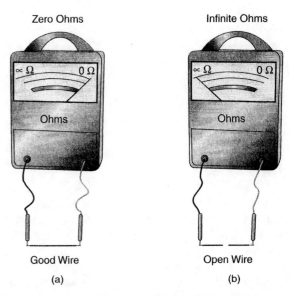

Figure 4–7 (a) An ohmmeter indicates 0 Ω when a good wire is tested; (b) an ohmmeter indicates infinite ohms when a wire with an open is tested.

winding and 30 Ω of resistance in its start winding. The ohmmeter can be used to measure each winding accurately and determine whether it has 12 or 30 Ω. The ohmmeter can be used to test motor leads and determine if they are open; if they are not open, the run and start windings can be distinguished by the amount of resistance each has. You will see in later chapters that the start winding of an AC induction motor will always have more resistance than its run winding, and the ohmmeter allows you easily to tell the run winding from the start winding.

4.6

READING THE SCALES OF A VOM METER

The scale of a typical VOM meter is shown in Fig. 4–8. As you can see in this figure, the face has three separate scales (lines that are shown in the shape of an arc). The bottom-most scale is used to measure AC amps and will not be discussed at this time. The middle scale is identified by the acronyms AC DC, which appear at both ends of the scale. This scale is used to measure AC and DC voltages and DC milliamps. Remember that *milli* is a prefix that represents the value 1/1,000, so the milliamp scale is designed to measure thousandths of an ampere.

The top scale measures resistance, and it has the ohms symbol (Ω) at each end. The needle that is attached to the meter movement should rest at the far left corner of the meter face when voltage, current, or resistance is not being

Figure 4–8 The face of a typical analog-type VOM meter that shows the voltage, current (amperes), and resistance (ohms) scales.

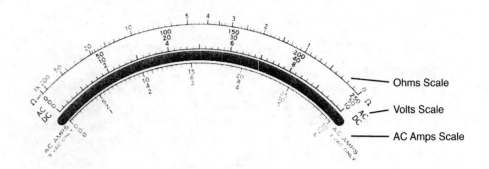

measured. The VOM meter can measure volts, ohms, and milliamperes with the same meter movement because a number of resistors are switched into and out of the circuit that connects the two probes to the meter movement. The switch is moved by a large knob on the front of the meter. Figure 4–9 shows the knob and the range selections on the face of the meter. If we use the hands of a clock to explain the range selections, you can see that the AC voltage circuit is located between 1 and 3 o'clock. The AC voltage circuit has five ranges: 0–2.5, 0–10, 0–50, 0–250, and 0–1,000 V.

The DC milliamp circuits are located between the 3 and 6 o'clock positions. The ranges for milliamps are 0–10, 0–100, and 0–1,000 mA. Additional current ranges include the 0–100-μA (micro amps) range, where *micro* is the prefix for the value 1/1,000,000 (one-millionth), and the 0–10-A range, which allows the meter to measure larger currents up to 10A.

The ohms circuits that are used to measure resistance are located between the 6 and 9 o'clock positions. Notice that "×" precedes each of the ranges, because you will need to multiply the reading on the scale of the meter by the multiplier factor that is identified by the switch. You should get into the habit of saying the word *times* when you encounter the ×. For example, the first range of the ohms scale is identified as ×1. You will call this range the *times 1 range*. The other ohms ranges are ×10 (times 10), ×100 (times 100), ×1,000 or ×1 k (times 1,000), and ×100,000 or ×100 k (times 100,000). When the range is ×1,000 or ×100,000, note that k is substituted for three zeros. The letter k stands for *kilo*, which represents 1,000. The k is often used when there is not enough space to write in all the zeros.

The DC voltage ranges are located between the 9 and 12 o'clock positions. The ranges for DC voltage are 0–2.5, 0–10, 0–50, 0–250, and 0–1,000 V.

4.7

MEASURING DC VOLTS AND READING THE VOM SCALES

When you are making a DC voltage measurement, you place the meter leads (probes) in the sockets identified as (V Ω A) and (COM), and you set the selector

switch on the face of the meter to the highest range. For the meter in Fig. 4–9, the selector switch would be set to 1,000 V. Setting the selector switch to 1,000 V protects the meter when you are not sure how large the voltage is that you are measuring. When you touch the tips of the probes to a voltage source, the needle that is attached to the meter movement begins to move from the position at the far left, where it rests when no voltage is being sensed, toward the middle. If you have the positive meter probe touching the negative voltage source, the meter movement will try to deflect farther to the left. If this occurs, simply switch the two probes so that the positive meter probe is touching the positive voltage source. This is called *changing the polarity* of the meter. If the voltage that is being measured is small, the needle will not move very far, and you can change the switch on the meter face to the next lower scale. If the needle still does not move sufficiently, you can continue to move the switch to a lower scale. Figure 4–10 shows an example of a meter being used to measure the amount of DC voltage in a circuit that has approximately 4 V. Because the needle would not move sufficiently when the switch was set on the 0–1000-V scale, the switch was changed down through the lower ranges until it was switched to the 0–10-V range and the needle moved to the position shown in the example.

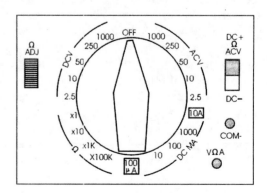

Figure 4–9 The range selector switch on the VOM meter, shows the range the meter is set to, and the knob attached to the switch connects the meter movement to different circuits inside the meter when it is rotated.

When you are ready to read the position of the needle on the face of the meter, you start by looking at the position of the switch. Because the switch in this example is at the 10-VDC setting, you must use the scale marked AC DC for this reading. Figure 4–11 shows an enlarged diagram of only the face of the meter. The AC-DC scales are located directly below the ohms scale, which is the top scale. The voltage scales are highlighted in this figure so they are easier to identify. Notice that the scale marked AC DC has three numbers shown at the far right of the scale: 250, 50, and 10. Because the selector switch is set on 10 V, we will use the scale that has the numbers 0–10

marked on it. These numbers are highlighted in this example.

When you look closely at the meter face, you can see that the needle points to the 4 on this scale. This means that the meter is measuring 4 VDC. You may be confused at first because the numbers 100 and 20 also occur at this same position. You can be sure that the reading is 4 V by starting at the far right side of the scale and locating the number 10. Because the number 10 is the bottom number, you should use the bottom set of numbers for this scale. (*A good habit to get into is to start with the number on the right side of the scale and count backward as you locate the exact position to which the needle points. For this example you would start at 10 and count backward 8, 6, 4 as you move to the left, where the needle is pointing.*)

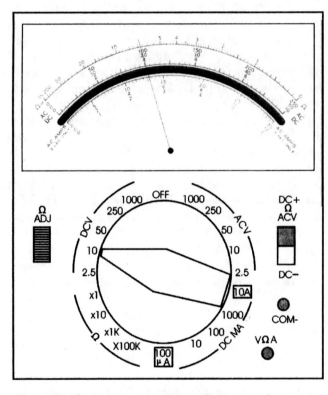

Figure 4–10 Voltmeter with its selector switch set on the 0–10-V range and its needle indicating 4 V is being measured.

4.8

MEASURING AND READING THE MILLIAMP SCALES ON THE VOM

When you are making a current reading with a milliampere meter, you must open the circuit and connect the meter probes so that the meter becomes part of the circuit. You should set the selector switch to the highest milliampere setting (the 1,000 DC mA range). If the needle does not deflect enough so that you can read it easily, you can switch the meter to the next-lower range. When the selector switch is set on this setting, you will need to read the amount of current from the same AC-DC scale that you used for the DC voltage reading.

NOTICE!
The scale marked AC AMPS is used only when measuring AC current; it cannot be used for DC current readings.

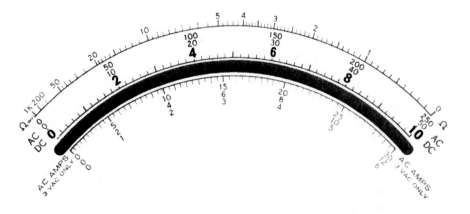

Figure 4–11 Example of the face of a VOM meter with the 0–10-VDC scale highlighted in bold.

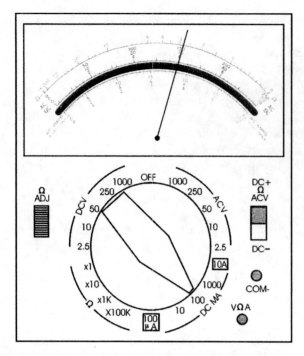

Figure 4–12   Example of a meter reading 60 DC mA.

In the example shown in Fig. 4–12 the selector switch is set on the 100 DC mA setting. Because the scale does not end with 100, you will have to use the 0–10 scale and move the decimal point or multiply the reading by 10.

For example, in the figure you can see that the needle is pointing to the number 6. You can determine this by finding the 0–10 scale and begin counting backward (10, 8, 6, 4, 2) until you reach the place where the needle is resting. Because the selector switch is set to the 100-mA setting, you must treat the 10 as 100. You can do this by multiplying the original reading by 10 or by counting backward from 100 to 60.

If the needle is pointing to the same position and the selector switch is pointing at 10 mA, the value the meter is measuring is 6 mA. The major point to remember when reading the milliampere scales is that all three ranges use the numbers 1 and 0; the difference between each range is in the placement of the decimal point.

4.9

MEASURING AND READING RESISTANCE SCALES ON THE VOM

When you are making a resistance measurement with the VOM meter, you must ensure that the device or component that you are measuring does not have power connected to it and that it is isolated from other components. You should remember that this is a necessary step, because the VOM meter uses an internal battery as the power source when it is used to measure resistance. When the meter is set to read resistance, the voltage from the battery is rather small, 3, 9, or 30 V, depending on the brand of the meter. The meter will be seriously damaged if you touch its probes to a power source, such as 110 V, by mistake when the meter is set for resistance.

Also, when using the VOM meter to read resistance, you must zero the meter. Because the meter uses a battery for the power source, the voltage in the battery must be able to move the needle to the 0-V mark on the scale when the two probes are touched together. This process is called *zeroing the meter*, and it should be done prior to all resistance readings. If the battery is weak, the needle will not be able to move to the zero point, and all resistance readings will be faulty. This step is necessary because the actual voltage in the battery changes constantly as the battery is used and it starts to drain down.

The first step in the process of zeroing the meter is to place the selector knob on the ohms range you want to use. Next you need to touch the two meter probes together, and the meter needle should swing toward the right side of the scale. Because the VOM meter is being used to measure resistance, the very top scale—identified by the symbol Ω for ohms Ω—you can see this scale clearly in Figure 4–11. If the meter is zeroed correctly, the needle will rest directly on the 0-V setting. If the needle does not move to this position, you can adjust the needle position by changing the Ω ADJ knob, located to the left of the selector switch. The Ω ADJ knob is actually a potentiometer, and when you adjust it, you are changing the amount of resistance in the meter movement circuit, which changes the amount of current the battery is sending the meter movement. If you change the Ω ADJ and the meter needle does not reach all the way to the 0-V position, the meter has a weak battery, and it should be replaced.

NOTICE!
You must zero the VOM meter any time you switch resistance scales or any time you change from one resistance scale to another, because a different amount of internal resistance is used for each circuit. You do not need to zero the VOM meter when it is used to make voltage or current measurements, because these types of readings do not use the meter's internal battery.

It is important to understand that you only need to "zero" the VOM when it is used to measure resistance (ohms) because the meter uses a small battery to provide the current in this setting. The reason you must "zero" the meter prior to making a resistance measurement is to compensate for the battery losing its charge over time. You also need to re-zero the meter when it is set to ohms and you change the range from X1 through X1k. The reason you need to re-zero the meter each time you change the ohms range is that each range takes a different amount of current to cause the needle to swing completely to the zero mark. Also some VOM meters use a 1.5 volt battery on the lowest ohms ranges and a 30 volt battery on the highest range and the level of charge in each battery will be different.

Figure 4–13 shows an example of the VOM meter measuring resistance. In this example, the meter selector switch is set to the ×1 (times 1) setting. Because the meter is being used to measure resistance, the top scale is used. The needle is resting on the value 10 on the top scale. The selector knob is on the ×1 (times 1) scale, so you will need to multiply the reading by 1. This means that the VOM meter is measuring 10 Ω. If the selector switch is on the ×10 (times 10) range and the needle is resting in the same position, the meter is measuring 100 Ω, because you need to multiply the scale reading by 10.

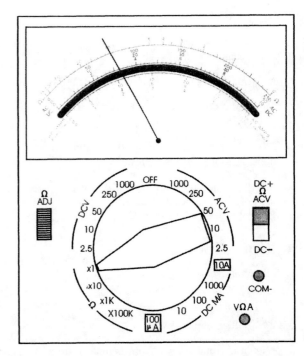

Figure 4–13 VOM meter used to measure resistance. The scale indicates the needle is resting on the 10-Ω value. The meter knob is set to the ×1 (times 1) range, and the needle is pointing to the number 10. (10 times 1 equals 10 Ω)

4.10

Making Measurements with Digital VOM Meters

When you make a measurement with a digital VOM meter, you should follow the same rules as when you use an analog VOM meter. The major difference is that the measurement is presented as a digital number on the meter's display. All you need to do is observe the range-selector switch to determine the units that the meter is measuring and use the value that is displayed as the measurement. Be sure to observe polarity (+/−) if the voltage or current is DC. Some digital meters also provide an audible signal for resistance measurements, which is useful when you are locating wires in multi-conductor cables.

Questions

Short Answer

1. Explain the term *infinite resistance*.

2. Use the diagrams in Fig. 4–3 to explain where you would place the probes of a voltmeter when making voltage measurement.

3. Use the diagram in Fig. 4–5 to explain where you would place the probes of an ammeter when making a current measurement.

4. List two things you should be aware of when making resistance measurements.

5. Explain why you should zero an ohmmeter every time you change scales or make a measurement.

6. Explain why you should set the voltmeter and ammeter to the highest setting when you first make a measurement of an unknown voltage or current.

7. Explain why the VOM only needs to be "zeroed" when it is set to measure resistance.

True or False

1. All voltage should be turned off when making a resistance measurement.

2. Infinity is the highest resistance reading for the ohmmeter.

3. The VOM meter must be zeroed prior to making voltage, current, or resistance measurements.

4. The voltage should be turned off for safety reasons when making a voltage measurement.

5. The ohmmeter uses an internal battery to supply the power for all resistance readings.

Multiple Choice

1. Milliamps are _____.
   a. 1/1,000 of an ampere
   b. 1/1,000,000 of an ampere
   c. 1 million amperes

2. When a current measurement is made with a VOM-type meter, you should _____.
   a. place the meter probes across the load
   b. turn off all power to the circuit during the current reading so you don't get shocked
   c. create an open in the circuit and place the meter probes so that the meter is in series with the load

3. When a voltage measurement is made, you should _____.
   a. place the meter probes across the load
   b. turn off all power to the circuit during the voltage reading so you don't get shocked
   c. create an open in the circuit and place the meter probes so that the meter is in series with the load

4. When a resistance measurement is made, you should _____.
   a. keep all voltage applied to the component you are measuring so you can determine the true resistance
   b. turn off all power to the circuit and isolate the component you want to measure
   c. create an open in the circuit and place the meter probes so that the meter is in series with the load

5. When you are measuring AC current with a clamp-on-type ammeter, you should _____.
   a. open the claws of the meter and place them around the wire where you want to measure the current
   b. create an open in the circuit and place the meter probes so that the meter is in series with the load
   c. turn off all power to the circuit so that you don't get shocked

PROBLEMS

1. Use the diagram in Fig. 4–10 to determine the voltage reading for the meter if the selector switch is set on 50 V.

2. Use the diagram in Fig. 4–12 to determine the current reading for the meter if the selector switch is set on 10 mA.

3. Use the diagram in Fig. 4–13 to determine the resistance reading for the meter if the selector switch is set for 1 K.

4. Determine the amount of current flowing in a wire if a clamp-on ammeter has four turns of the wire twisted around its claw and the reading on the scale is 2 A.

5. List the kind of problems that could be present in the circuit if a milliamp meter reading indicates 0 mA is present in the circuit.

CHAPTER 5

Series Circuits

OBJECTIVES

After reading this chapter you will be able to:

1. Explain the operation of switches that are connected in series.

2. Explain the operation of electrical loads that are connected in series.

3. Explain the operation of a variable resistor.

4. Utilize Ohm's law for series circuit calculations.

5. Calculate the voltage, current, resistance, and power in a series circuit.

6. Troubleshoot a series circuit using the voltage-loss method to locate faults.

5.0

INTRODUCTION

Complex equipment-control systems are made up of series and parallel circuits. It is important to understand that a basic set of rules called Ohm's law can be used to predict the voltage, current, or resistance in a series circuit when you are troubleshooting. This chapter explains those rules as they apply to voltage, current, resistance, and power in a series circuit. In this chapter we will use simple circuits consisting of resistors to explain the relationship between voltage, current, and resistance in series circuits. It should be pointed out at this time that resistors are not common loads in equipment found on the factory floor, but they are frequently used to explain and calculate how voltage, current, resistance, and power function in simple series circuits. You will find resistors or resistance in heating elements in plastic molding machines, and you will also find them in modern electronic circuit boards used to control a

wide variety of equipment found in factories including motor drives. The motors and other loads often found in factory equipment have coils of wire that react in some part as resistance, so it is important for you to understand the relationship between multiple resistors in series circuits. Understanding resistance in series circuits and in parallel circuits will also provide the initial theory that you will require to understand solid-state components that are commonly used in solid-state controls. As a technician, you must understand the effects of switches and electrical loads in series circuits. The first part of this chapter explains the relationships between switches that are connected in series. The second part of this chapter explains the effects of electrical loads that are connected in series.

The relationship between voltage, current, and resistance in series circuits is introduced in the latter part of this chapter. It will be easier to understand these relationships when a known value of resistance is used. A component called a *resistor* will be used to provide the resistance for these simple series circuits. Each resistor has a color code consisting of colored bands that are painted on the resistor to identify the amount of resistance the resistor has. Ohm's law is also introduced to help you understand how voltage and current should react in all types of series circuits. The basic rules given in this chapter will be used to help you troubleshoot larger circuits that are frequently found in electrical control systems.

5.1

EXAMPLES OF SERIES CIRCUITS

All electrical circuits have at least a power source, one load, and conductors to provide a path for current. As circuits become more complex, switches are added for control. These switches can provide a variety of functions,

Figure 5–1   Example of a series circuit that has three switches connected in series. If any of the three switches are open, the current flow to the motor is stopped and the motor is de-energized.

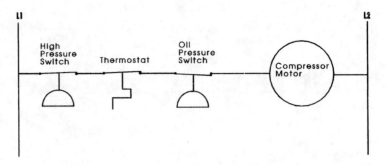

depending on the way they are connected in the circuit. In Fig. 3–1 we saw a diagram of a simple circuit that included a power source, a switch for control, and a motor as the load. This circuit is called a *series* circuit because there is only one path for the current to travel to the load and back to the power source.

When additional switches are added to the circuit to control the load, they can be connected so that all current continues to have only one path to travel around the circuit. Figure 5–1 shows a typical circuit that controls power to a compressor for a small air conditioner used to provide cooling to an electrical cabinet that has a controller inside it. This circuit has a thermostat that controls the temperature where the compressor turns on and off, a high-pressure switch that protects the compressor against excessive pressure if the condenser coils get dirty, and an oil-pressure switch that protects the compressor against loss of oil pressure.

This circuit is called a series circuit because the current has only one path to follow from L1 through the switches to the load and back to L2. If any of the three switches are opened, current is interrupted and the motor is turned off. When all the switches are closed, the compressor motor will run. When the electrical cabinet cools to 80°, the thermostat (temperature switch) will open and turn the compressor motor off. When the cabinet becomes warm again, the thermostat will close and energize the compressor again. This means the thermostat is an *operational control*, because its job is to operate the system at the set-point temperature. If the condenser becomes dirty and causes the high-side pressure (head pressure) to become excessive, the high-pressure switch will open and turn the compressor motor off. The same is true of the oil-pressure switch. If the oil pressure becomes too low, the oil-pressure switch will open and turn the motor off. The high-pressure and oil-pressure switches are in the circuit for safety purposes.

Because all three of these switches are connected in series, each is capable of opening and causing current to stop flowing to the compressor motor. If additional operational or safety switches need to be added to this circuit, they would be connected in series with

the original switches. It is important to remember that when switches are connected in series, any one of them can open and stop current flow in the entire circuit.

Another example of switches connected in series is the door safety switches that are used in many of the machines found on the factory floor. If any of the doors are opened while the machine is running, the door switch will open and cause current flow to the machine to stop, and the machine will turn off. The important point to remember is that any switch in a series circuit can stop current flow through the entire circuit.

5.2
SERIES SWITCHES

When you are working on equipment in a factory, you will find that the switches that control the equipment are connected in the circuit in series or parallel. In this section, you will learn what happens when switches are connected in series in a circuit. In the next section, you will learn how loads that are connected in series will react.

Figure 5–2a shows a single switch in series with a lamp that is used as a load. The reason the lamp is used for this load is so that you can build this circuit in a lab and easily see when the load is energized because the lamp will be illuminated. Since the circuit has only one switch and it provides the only path for electrons to reach the load, any time it is open, the flow of electrons (current) is stopped and the lamp will not be illuminated. Any time the switch is in the closed position, current is allowed to flow to the lamp, and it is illuminated. Likewise, if any part of the wiring in the circuit gets a hole in it (an open), the flow of electrons will stop, and the lamp will not be illuminated. Figure 5–2b shows a wiring diagram of the same circuit that is in Fig. 5–2a. The wiring diagram is designed to show the position or location of all the components and wires in the circuit. In this diagram you can see that the wire starts from the positive terminal of the power supply and is connected to the switch. The next wire connects the switch to the

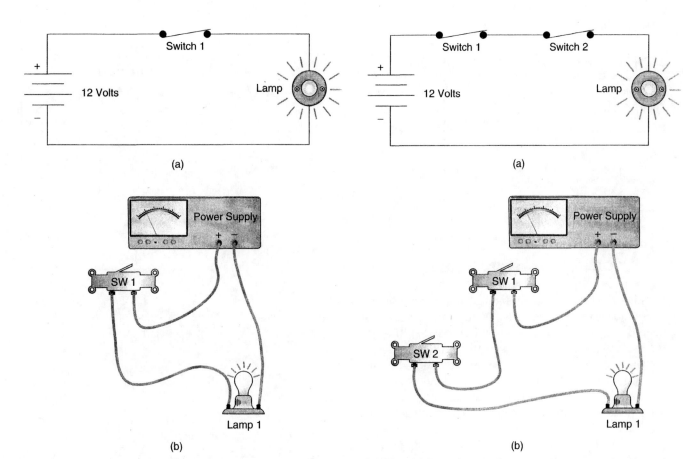

Figure 5–2 (a) A switch connected in series with a lamp; (b) wiring diagram of one switch connected in series with one lamp. The wiring diagram shows the position of each component.

Figure 5–3 (a) Two switches connected in series with a lamp; (b) wiring diagram of two switches wired in series to a lamp.

lamp, and the final wire connects the lamp to the negative terminal on the power supply. You will need to be able to use the diagram shown in Fig. 5–2a to connect the components in the circuit as shown in Fig. 5–2b. You will also need to use the wiring diagram when you troubleshoot a circuit.

It is also important to understand that the rules that involve series switches will be constant regardless of the number of switches or the types of switches used in a circuit. Any time a switch is connected in series with a load or with other switches and it is open, it will stop the flow of current to the lamp and everywhere else in the circuit.

5.2.1
Connecting Two Switches in Series with a Lamp

Figure 5–3a shows two switches connected in series with the lamp. In this circuit you can see a second switch has been added, and it is connected in series

with the first switch and the load. This means that all current flow in this circuit must go through both switches. If either of the switches is opened, the flow of current to the lamp will be stopped, and the lamp will no longer be illuminated. Figure 5–3b shows a wiring diagram of this same circuit. The wiring diagram shows the exact location of each component in the circuit. It also shows the wires as you would find them if you went to this circuit to troubleshoot it. It is important that you be able to use the wiring diagram when you are connecting these components in a lab exercise.

5.2.2
Troubleshooting to Find the Open in a Series Circuit

Another time that you will be working with series circuits is when a machine is broken down and you are asked to troubleshoot it to determine which switch is keeping the machine from running. This section shows you a very simple way to use your voltmeter to determine which switches are open or closed in any circuit.

Figure 5–4 Electrical diagram of two 120-V heaters connected in series so that a 240-V power supply is divided equally between them.

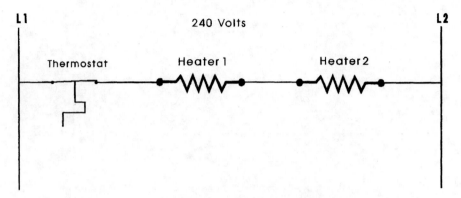

5.3
ADDING LOADS IN SERIES

In industrial systems, loads such as motors are not connected in series with other loads; instead they are connected in parallel with each other so that they all receive the same voltage. Even though loads such as motors are not connected in series, you should understand the problems that exist if motor loads were connected in series. This section explains the problems associated with connecting motor-type loads in series, and Section 5.10 explains how resistors are connected in series to create useful voltage-drop circuits for electronic circuit boards and components used to control voltage and current in circuits.

There are some exceptions to connecting loads in series, such as electric heaters used on plastic molding machines, where two heating elements are connected in series on purpose. Figure 5–4 shows an example of two heating elements connected in series. If the amount of resistance for each heater is the same, the supply voltage will be split equally between them. In this example the supply voltage to the heaters is 240 V, and each heater will receive 120 V. This means that each heater's rating is 120 V. This type of circuit allows smaller heaters to be used on larger-voltage sources. In some applications, two heater elements, each rated for 240 V, can be connected in series and will use a supply voltage of 480 V. In theory you could connect four 120-V heaters together in a series circuit and supply 480 V, and each heater would only see 120 V. Another problem with connecting loads in series is that if one of the loads has a defect and develops an open circuit, the current to all the other loads in the circuit will be interrupted. An example of this type of circuit is the strings of lights that are used to decorate a Christmas tree. If all the lights are connected in series, they all go out if one of the lights burns out. The advantage of connecting 50 lights in series is that the lightbulbs will each receive approximately 2 V when the circuit is plugged into a 110-V power source.

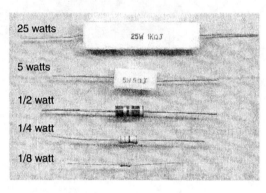

Figure 5–5 Examples of resistors commonly found in electronic circuit boards used in industrial systems.

5.3.1
Using Resistors as Loads

A resistor is a component that is specifically manufactured with a precise amount of resistance. Some circuits that you encounter have resistors that are used to change the amount of voltage and current in a circuit. These types of circuits are commonly found in the solid-state control boards that are now used in many industrial systems. These solid-state boards are used to control motor speeds, to time-delay functions, and to convert small-signal voltages from sensors to larger voltages that can energize motor controls. Figure 5–5 shows several resistors often found in electronic circuit boards. The relative size for each resistor is shown in this figure.

5.4
RESISTOR COLOR CODES

Since most resistors are very small, a color code system has been designed to show the amount of resistance each resistor has. The amount of resistance is identified on each resistor by color stripes or bands.

The resistor color bands are shown in their respective positions in Fig. 5–6. From this figure you can see

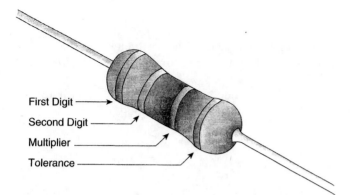

Figure 5–6 The location of color bands on a resistor.

Table 5–1

Color	Digit Value	Multiplier	Tolerance
Black[1]	0	1	
Brown	1	10	
Red	2	100	
Orange	3	1,000	
Yellow	4	10,000	
Green	5	100,000	
Blue	6	1,000,000	
Violet[2]	7		
Gray[2]	8		
White[2]	9		
Silver		0.01	10%
Gold		0.1	5%

[1]Cannot be used in the first band (first digit).
[2]Maximum resistance is 22 M. Therefore these colors cannot be used as multipliers.

that the resistor has four basic color bands that are grouped at one end of the resistor. (*Note:* A few resistors have fifth and sixth bands that represent other manufacturing information.) The color bands are used to represent numbers that indicate the amount of resistance each resistor has. The first color band represents the first digit of the number, and the second color band represents the second digit of the number. Each digit can be a value from 0 to 9. The third band is a multiplier band. This is basically the number of zeros that you place with two-digit numbers to show numbers greater than 99. The fourth band is called the *tolerance band*. The actual amount of resistance in the resistor is close to the amount identified by the color bands, but it may be slightly more or less depending on the amount of tolerance that is used. Typical tolerances are 5% and 10%. This means that the actual amount of resistance can be plus or minus the percentage identified by the tolerance band.

The colors that are assigned to the color code are listed in Table 5–1, and you can see each number from 0 to 9 is represented by a color. The tolerances are also represented by the colors silver and gold. A silver band means the resistor has a tolerance of 10%, and a gold band means the resistor has a tolerance of 5%.

5.5
DECODING A RESISTOR

A resistor with a brown first band, blue second band, and red third band is decoded as follows: The first two colors, brown and blue, mean that the first two numbers are 1 and 6. The third band, which is the multiplier band, is red, so you need to add two zeros to the first two digits. Thus, the total resistance is 1,600 Ω. To use the multiplier for red, simply multiply the first two digits by 100 (16 × 100 = 1,600 Ω).

A second resistor has a violet first band, a white second band, and a green third band. This resistor is decoded as follows: The first two colors, violet and white, mean the first two numbers are 7 and 9. The third band, which is the multiplier band, is green, so you need to add five zeros to the first two digits. This makes the total resistance 7,900,000 Ω. Again, to use the multiplier for green, multiply 79 by 100,000 (79 × 100,000 = 7,900,000 Ω).

▶ *Example 5–1*
Determine the total resistance of a resistor that has a color code of yellow, brown, and orange.

Solution:
The first and second colors are yellow and brown, so the first two digits are 4 and 1. The multiplier is orange, so three zeros must be added to 41, giving 41,000 Ω. To use the multiplier for orange, multiply 41 × 1,000 = 41,000 Ω.   ◀

5.6
WHY RESISTOR TOLERANCE IS IMPORTANT

It is very important to understand how resistors are manufactured and why their tolerance is important. When resistors are mass-produced, the conducting and insulating material is mixed much like you would mix flour and sugar in a food recipe. The actual amounts of each ingredient that is inside each resistor may vary slightly, which will give it a slightly different amount of resistance from its color code. The industry has developed standards that indicate what percentage above or below the color code value the actual resistance can be and still be considered a good resistor that is within tolerance. The important part about calculating the resistor tolerance is to accurately determine what is the highest amount and lowest amount of resistance. After you determine the highest and lowest value the resistor can have, you should measure the resistance with an ohmmeter to determine if it is within the tolerance values. If the measured resistor value is higher or lower than the calculated tolerance, the resistor is considered to be out of tolerance and should not be used. In the next section you will learn how to calculate the resistor tolerance.

5.6.1
Calculating the Resistor Tolerance

The first resistor that was decoded in the previous section had color bands of brown, blue, and red and a total resistance of 1,600 Ω. If this resistor has a fourth band of silver, the resistor tolerance is calculated as 10%. The first step in calculating this tolerance is to multiply the total resistance by 10%.

$$1,600 \ \Omega \times 0.10 = 160 \ \Omega$$

The second step in calculating the tolerance is to add 160 Ω to and subtract 160 Ω from 1,600 Ω.

$$
\begin{array}{cc}
1,600 \ \Omega & 1,600 \ \Omega \\
+160 \ \Omega & -160 \ \Omega \\
\hline
1,760 \ \Omega & 1,440 \ \Omega
\end{array}
$$

These calculations show that the total resistance of the resistor with color bands of brown, blue, red, and silver can be from 1,760 to 1,440 Ω. If you were measuring this resistor with an ohmmeter and its total resistance fell between 1,440 and 1,760 Ω, it would be considered a good resistor. If the total resistance were less than 1,440 Ω or greater than 1,760 Ω, the resistor would be considered out of tolerance, and it would not be usable. If the fourth band was gold instead of silver, the multiplier would be 5% instead of 10%. The calculation would be completed the same way as with a tolerance of 10%, except you would add 5% to and subtract 5% from the total to get the upper and lower limits.

▶ *Example 5–2*

Determine the total resistance, including tolerance, of a resistor whose color code is yellow, red, orange, and gold.

Solution:

Because the first two colors are yellow and red, the first two digits are 4 and 2. The multiplier is orange, so you need to add three zeros to make the total resistance 42,000 Ω.

The tolerance band is gold, and so the tolerance is 5%. You find 5% of 42,000 Ω by multiplying

$$42,000 \ \Omega \times 0.05:$$
$$42,000 \times 0.05 = 2,100$$

The next step requires that you add 2,100 Ω to 42,000 Ω to get the upper value and subtract 2,100 Ω from 42,000 Ω to get the lower value:

$$
\begin{array}{cc}
42,000 \ \Omega & 42,000 \ \Omega \\
+2,100 \ \Omega & -2,100 \ \Omega \\
\hline
44,100 \ \Omega & 39,900 \ \Omega
\end{array}
$$

The upper value of this resistor is 44,100 Ω, and the lower value is 39,900 Ω.   ◀

5.7
VARIABLE RESISTORS

Variable resistors are used on radios and TVs as volume controls. As an electrical technician, you will also find variable resistors in a variety of electrical circuits. The symbols for the variable resistor are shown in Fig. 5–7. The symbol shown in Fig. 5–7a uses the same symbol as a fixed resistor, but with an arrow going through it. The symbol in Fig. 5–7b is shown with an arrow touching the resistor. The arrow is used in electrical symbols to indicate the device is variable.

Figure 5–8 shows a variable resistor with its adjustment knob protruding out the front of the component. The three terminals of the device are also shown on the

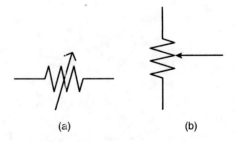

(a)   (b)

Figure 5–7   Electrical symbols of a variable resistor. (a) The symbol is a fixed resistor symbol with an arrow through it. (b) Here the symbol is a fixed resistor symbol with an arrow touching it.

back of the resistor in this figure and also in the variable resistor shown in Fig. 5–9. The picture and the symbols show that the variable resistor is a three-terminal device. Typically, when you turn the knob, you can adjust the amount of resistance between either end terminal and the middle terminal.

A variable resistor is identified by the total resistance that is found between its two end leads. For example, if a variable resistor is called a 1-kΩ pot, it will have 1 kΩ of resistance between the two ends. When the adjustment knob is rotated, the amount of resistance between either end terminal and the middle terminal will be 0–1 kΩ. An additional rating for the variable resistor is the amount of wattage it can safely control.

The variable resistor may be called a "pot" because it is often used as a potentiometer, or it may be called a rheostat when it is capable of controlling larger currents. In general, a rheostat is designed to handle larger currents than the potentiometer. Variable resistors are used in everyday circuits to control the volume on your radio or television. Another use of variable resistors is to control the level of light in the interior lights in your automobile. These resistors are also used to make fine adjustments during the calibration of electrical instruments. You will find rheostats used to control the speed of some DC motors by adjusting the field current in the motor.

5.7.1
Operation of a Variable Resistor

As a technician, you will need to identify the terminals of a variable resistor and be able to test it to ensure that it functions correctly. Figure 5–10a shows a 1,000-Ω variable resistor, also called a 1-k Ω pot, with its cover removed. You can see that the inside of a variable resistor (which is made of resistance composite material) is shaped in a semicircle with the two ends of the semicircle connected to the two outside terminals. The center terminal is connected to a metal ring located at the center of the device. Figure 5–10b shows a rotating contact that is connected to the center terminal and rides on the resistance composite material. In this figure the adjustment knob is set at midpoint so that each of the outside terminals to the center terminal has 500 Ω of resistance, which is exactly half of the total end-to-end resistance of 1,000 Ω.

Figure 5–8 A variable resistor with the adjustment knob in the front and three terminals on the back.

Figure 5–9 A close-up view of the three terminals on the back of a variable resistor.

Figure 5–10 (a) 1000-Ω variable resistor with its cover off. 1,000 Ω would be measured between the two outside terminals; (b) 1,000-Ω variable resistor with the variable arm set halfway so that 500 Ω is measured between each end to the center terminal; (c) 1,000-Ω variable resistor set at 70%. 700 Ω would be measured from the left outside terminal to the middle terminal. 300 Ω would be measured from the right outside terminal to center terminal.

(a)

(b)

(c)

As the center contact is rotated, it will come into contact with more or less resistance material, which allows the amount of resistance between the center terminal and outside terminal to become variable. Figure 5–10c shows the adjustment knob set so that the center terminal to the left outside terminal has 700 Ω and the center and right outside terminals has 300 Ω. You can see that the adjustment knob allows the center contact to move and vary the amount of resistance from zero Ω to the maximum amount of resistance between the two outside terminals.

When the adjustment knob is moved very close to one end of its rotation, the amount of resistance between the center terminal and one of the outside termi-

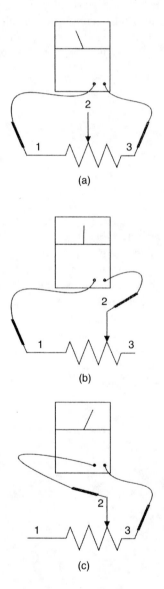

(a)

(b)

(c)

Figure 5–11   (a) An ohmmeter measuring 1,000 Ω from terminal 1 to 3; (b) an ohmmeter measuring 700 Ω from terminal 1 to 2; (c) an ohmmeter measuring 300 Ω from terminal 2 to 3.

nals will be at its minimum value. When this occurs and the variable resistance is placed in a circuit under power, it is important to understand that the current in the circuit at this time will be at its maximum. This may cause the variable resistor to burn up since you are drawing maximum current through a minimum amount of resistance material, which will cause the material to heat up rapidly and break down in a failed condition. Sometimes a small fixed resistor is connected in series with the variable resistor to ensure the total resistance in the circuit does not drop below a value that will allow the current in the circuit to exceed the power (wattage) rating of the variable resistor and cause it to become permanently damaged.

5.7.2
Troubleshooting a Variable Resistor Using an Ohmmeter

At times when you are in the field, you will suspect that a variable resistor is not operating correctly. You can check the variable resistor with an ohmmeter as shown in Figure 5–11a–c. In this figure you can see that the two end terminals of the 1-kΩ variable resistor are arbitrarily identified as 1 and 3. The variable terminal is identified as terminal 2. In Figure 5–11a you can see that the variable resistor is adjusted so that it is at midpoint. In this position the ohmmeter will measure 1,000 Ω across terminals 1 and 3, and it will measure approximately 500 Ω between terminals 1 and 2 or 2 and 3.

Figure 5–11b shows the resistor adjusted to the 70% position. In this position you would measure 700 Ω between terminals 1 and 2 and the remaining 300 Ω between terminals 2 and 3, as shown in Figure 5–11c. You will see that the resistance between 1 and 2 and the resistance between 2 and 3 will always equal the resistance rating of the variable resistance, which you can measure between terminals 1 and 3.

You should make one additional check to see that the variable resistance changes smoothly as the adjustment knob is moved through its complete range. If you encounter any abrupt changes in resistance as the variable resistor is moved through its entire range, it more than likely has a burnt spot on it and it is not functioning properly and should be changed.

5.7.3
Connecting a Variable Resistor into a Circuit as a Potentiometer

You may need to connect a variable resistor into a circuit to create a variable voltage source. This can be done by connecting the variable resistor in parallel

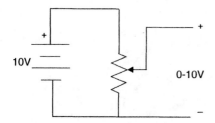

Figure 5–12 A variable resistor (potentiometer) connected in parallel with a power supply. This circuit will provide 0–10 V at the + and − terminals.

with the power supply as a potentiometer or connecting it in series with the power supply as a rheostat. Figure 5–12 shows a diagram of two end terminals of the variable resistor connected in parallel with the power supply. The center terminal of the variable resistor becomes the positive terminal of the variable power supply, and the negative terminal is connected also to the negative terminal of the battery. In this diagram you would connect the load to the center terminal of the variable resistor, which is the positive terminal of the variable power supply and the negative terminal. The output voltage would be adjustable between zero and 10 V of the original power supply. This circuit allows very accurate control of voltage for circuits where a variable voltage is needed. For example, this circuit is very useful in testing analog control where a motor drive needs a 0–10 V signal. You could connect a variable resistor across a 9-V battery and make a variable voltage power supply that is useful as a test voltage.

5.7.4
Connecting a Variable Resistor into a Circuit as a Rheostat

One problem with connecting a variable resistor to a power supply to create a variable voltage is that the variable resistor must have a large enough current rating to supply the circuit current requirement. In some cases the current requirement is very large, so the variable resistor is designed for larger currents. This type of variable resistor is called a rheostat. Figure 5–13 shows a picture of a wire-wound rheostat, and Fig. 5–14 shows a diagram of a variable resistor connected in series with the positive terminal of a power supply as a rheostat. In this circuit, all the current going to the load must go directly through the rheostat. One example of a rheostat in a circuit is the field rheostat for a DC motor or DC generator that controls the field current. As the field current changes in the DC motor, the speed of the motor will change.

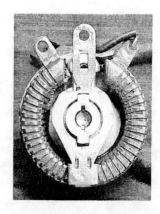

Figure 5–13 A wire-wound variable resistor called a rheostat.

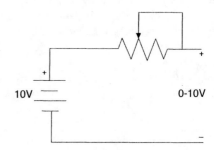

Figure 5–14 Electrical diagram of a variable resistor connected in series with the power supply as a rheostat.

At this point you may become a little confused about whether the variable resistor changes voltage or current in a circuit. A better way to think about the variable resistor is to think of two resistors in a series circuit. If one of the resistors (the variable resistor) increases its resistance, the total resistance of the circuit will increase, causing the current in the circuit to decrease, and the voltage drop across the fixed resistor will drop proportionally. Sometimes this concept is used to increase or decrease the current in a circuit, and other times the same components are used to increase or decrease the voltage across one of the resistors, which could be the load.

5.8
USING OHM'S LAW TO CALCULATE OHMS, VOLTS, AND AMPS FOR RESISTORS IN SERIES

In some electronic circuits, all the loads are resistors. These resistors are sized so that the supply voltage will be dropped into smaller increments. These types of circuits are widely used in electronic circuits for industrial equipment, and they are also useful for learning the concepts of how voltage and current are affected by changes in resistance. Several circuits that have resistors connected in series will be used to explain these concepts. Figure 5–15 shows an example of three resistors connected in series with a voltage source (battery). Because the resistors are connected in series, there is only one path for the current. The resistors are numbered R_1, R_2, and R_3, and the supply voltage is identified as E_T. The formula for calculating total resistance in a series circuit is $R_T = R_1 + R_2 + R_3$. If more than three resistors are used in the circuit, the additional resistors are added to the first three to get a total. The arrows show conventional current flow. [Electron flow would show the flow of electrons (current) moving from the negative battery terminal to the positive terminal. This text uses conventional current flow in its explanations and diagrams.]

Figure 5–16 shows a series circuit with the size of each resistor identified. Resistor R_1 is 30 Ω, R_2 is 20 Ω,

and R_3 is 50 Ω. The total resistance for this circuit can be calculated as follows:

$$R_T = R_1 + R_2 + R_3$$
$$R_T = 30\ \Omega + 20\ \Omega + 50\ \Omega$$
$$R_T = 100\ \Omega$$

5.9
SOLVING FOR CURRENT IN A SERIES CIRCUIT

The total current can be calculated by the Ohm's law formula:

$$I = \frac{E}{R}$$
$$I = \frac{300\ V}{100\ \Omega}$$
$$I = 3\ A$$

The total current in a series circuit is the same everywhere in the circuit, because there is only one path for the current. This means that once you find the total resistance of a series circuit, you also have determined the current that flows through each resistor. The current (3 A) has been listed with each resistor on the diagram in Fig. 5–16. Because the current is the same everywhere in the series circuit, it may be identified as I_T or I_1 where it is shown at resistor R_1.

5.10
CALCULATING THE VOLTAGE DROP ACROSS EACH RESISTOR

The voltage drop across each resistor can be calculated by the Ohm's law formula ($E = IR$). It is important to remember that the current in the series circuit is the same in all places, so I_T is equal to 3 A, and I_1, I_2, and I_3 also equal 3 A. In this case the voltage drop across resistor R_1 is calculated by

$$E_1 = I_1 R_1$$
$$E_1 = 3\ A \times 30\ \Omega$$
$$E_1 = 90\ V$$

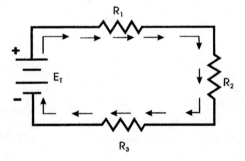

Figure 5–15   A series circuit that has three resistors connected so current can flow in only one path.

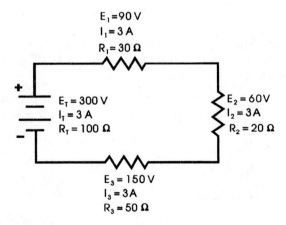

Figure 5–16   A series circuit with both the total resistance calculated and the voltage and current for each resistance calculated.

If you placed the probes from a voltmeter on each side of resistor R_1, you would measure 90 V. This means that the resistor is causing 90 V to *drop* when current is flowing through it. The voltage drop across resistor R_2 can be calculated by

$$E_2 = I_2 \times R_2$$
$$E_2 = 3\,A \times 20\,\Omega$$
$$E_2 = 60\,V$$

The voltage drop across resistor R_3 can be calculated by

$$E_3 = I_3 \times R_3$$
$$E_3 = 3\,A \times 50\,\Omega$$
$$E_3 = 150\,V$$

From these calculations you can see that the voltage dropped across R_1 is 90 V, that across R_2 is 60 V, and that across R_3 is 150 V. If you add all these voltage drops, you will find that the sum equals the supply voltage. This means that $E_1 + E_2 + E_3 = E_T$ (90 V + 60 V + 150 V = 300 V).

▶ **Example 5–3**

Figure 5–17 shows three resistors connected in series. Resistor R_1 is 40 Ω, R_2 is 30 Ω, and R_3 is 20 Ω. Calculate the total resistance for this circuit. After you have calculated the total resistance, calculate the total current. When you have determined the total current for this circuit, calculate the voltage drop that would be measured across each resistor.

Solution:
R_T can be calculated as follows:

$$R_T = R_1 + R_2 + R_3$$
$$R_T = 40\,\Omega + 30\,\Omega + 20\,\Omega$$
$$R_T = 90\,\Omega$$

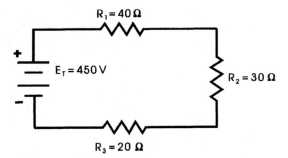

Figure 5–17   Series circuit for Example 5–3.

I_T is calculated by

$$I_T = \frac{E_T}{R_T}$$
$$I_T = \frac{450\,V}{90\,\Omega}$$
$$I_T = 5A$$

Because current is the same in all parts of the circuit,

$$I_T\,5 = A \qquad I_1 = 5\,A \qquad I_2 = 5\,A \qquad I_3 = 5\,A$$

The voltage that is dropped across each resistor from the current flowing through it is calculated as follows:

$$E_1 = R_1 \times I_1 \qquad E_2 = R_2 \times I_2 \qquad E_3 = R_3 \times I_3$$
$$E_1 = 40\,\Omega \times 5\,A \quad E_2 = 30\,\Omega \times 5\,A \quad E_3 = 20\,\Omega \times 5\,A$$
$$E_1 = 200\,V \qquad E_2 = 150\,V \qquad E_3 = 100\,V$$

You can check your answers by adding the drops to see that they equal the total supply voltage:

$$E_T = E_1 + E_2 + E_3$$
$$E_T = 200\,V + 150\,V + 100\,V$$
$$E_T = 450\,V \qquad\qquad ◀$$

5.11

CALCULATING THE POWER CONSUMPTION OF EACH RESISTOR

The power consumed by each resistor or the power consumed by the total circuit can be calculated by any one of three formulas for power (wattage). The formula that you elect to use will be determined by which two of the three variables (volts, ohms, or amps) that you know for sure. For example, you would use this formula if you know the voltage and current at a resistor: $P = EI$. You would use this formula if you know the current and resistance: $P = I^2R$. Or you could use this formula if you know the voltage and resistance: $P = E^2/R$.

We will calculate the power at resistor R_1 in the circuit shown in Example 5–3 using each of these three formulas just to prove they all work and will provide the same answer.

In the circuit for Example 5–3, we already calculated the voltage, current, and resistance for each resistor. We will use the values for R_1 where the voltage drop across R_1 is 200 V, the current through R_1 is 5 A, and the resistance of R_1 is 40 Ω.

$$P = EI \qquad\qquad P = I^2R \qquad\qquad P = \frac{E^2}{R}$$

$$P = 200\text{ V} \times 5\text{ A} \quad P = (5\text{ A})^2 \times 40\ \Omega \quad P = \frac{(200\text{ V})^2}{40\ \Omega}$$

$$P = 1{,}000\text{ W} \qquad P = 1{,}000\text{ W} \qquad P = 1{,}000\text{ W}$$

Because all three formulas give the same results, you can see that you would choose any of the three, depending on which variables you have calculated or measured.

5.12
CALCULATING THE POWER CONSUMPTION OF AN ELECTRICAL HEATING ELEMENT

The power formula can also be used to calculate the power consumption of any other type of resistance used in a circuit. The electrical heating elements used in an electric furnace are resistances, and so if you know the amount of resistance in the element and the amount of voltage applied to the circuit, you can calculate the current the element uses and the amount of power it consumes. For example, if the heating element has 10 V of resistance and it is connected to 220 V, it would draw 22 A. This element would consume 4,840 W of power.

▶ Example 5–4

A heating element in an electric furnace has resistance of 2.5 Ω. Calculate the current and the power consumption of this heating element if the furnace is connected to 240 V AC.

Solution:
The amount of current is 240 V/2.5 Ω = 96 A. The power can be calculated by $P = I^2R$ or $P = IE$:
  Using $P = I^2R$,

$$P = (96\text{ A})^2 \times 2.5\ \Omega$$
$$P = 23{,}040\text{ W}$$

Using $P = IE$,

$$P = 96\text{ A} \times 240\text{ V}$$
$$P = 23{,}040\text{ W}$$

Note that this answer can also be expressed as 23.040 kW (kilowatts). ◀

5.13
TROUBLESHOOTING SERIES SWITCHES

As a technician, you must be able to use a voltmeter to test any malfunctioning circuit and determine where the problem exists. It is important to understand that the fastest and most accurate test is with voltage applied. You may wonder why it is important to test a circuit with voltage applied to the circuit. You may even think that this is a dangerous practice—that it would be safer and easier to test the entire circuit for continuity with power off. However, you will find that it is a very safe way to test a circuit, and it is also the most reliable.

The problem with testing a circuit for continuity is that it is a very slow process because all the wires must be disconnected before they can be reliably tested and because it is very easy to have faulty readings from parallel circuits that you are not aware of. If you expect to become a qualified electrical technician, you must be able to repeat the voltage-loss test explained in the previous section without making mistakes. You must also be able to make the test quickly and accurately.

5.13.1
Troubleshooting the Circuit for Loss of Voltage

We will use the circuit shown in Fig. 5–18, which has two switches in series to explain this troubleshooting procedure. You should understand that the control circuit in an industrial machine may contain several switches

Figure 5–18   A circuit with two series switches. Each point that will be tested in the circuit is identified with letters A–H.

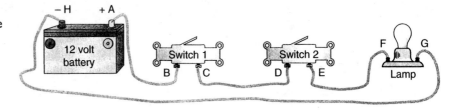

that may be faulty or not operate correctly. The problem could be any one of several things, such as the loss of the power supply, a loose or broken wire, a faulty switch, or a faulty load (lamp). This means that you must troubleshoot the complete circuit in a structured procedure instead of picking on individual components and testing them at random. The test that is outlined in the following procedure is called the *voltage-loss test*, and you should learn to use this test whenever you are trying to locate a faulty component or wire in a circuit that is not operating.

Start the procedure by testing for voltage across the load terminals (the two terminals of lamp F-G), as shown in Fig. 5–19. If voltage is present at the terminals of the lamp, you can be sure that the power supply and all the pilot switches and interconnecting wires are working correctly. This being the case, the remainder of the troubleshooting test should focus on the lamp. Since voltage is available at the lamp but the lamp is not illuminated,

you should suspect the lamp is bad, and you can remove and replace the lamp to fix the problem quickly.

If no voltage is present at the terminals of the lamp (F-G), the test should focus on the loss of voltage somewhere between the source and where it travels through the switches to the lamp. Your next test should be at the power source, as shown in Fig. 5–20. If voltage is not present at the power supply, be sure to check for a blown fuse, open disconnect, or other power supply problems. At this point in the test, all that you are concerned with is that the supply voltage (12 V) is the same voltage that is specified on the diagram.

After you have determined that the proper amount of supply voltage is available, the remainder of the tests should focus on the switches and interconnecting wires. The fastest and most accurate way to locate an open wire, loose terminal, or faulty switch is to test for a voltage drop (loss). This test requires that power remain applied to the circuits while you touch the meter

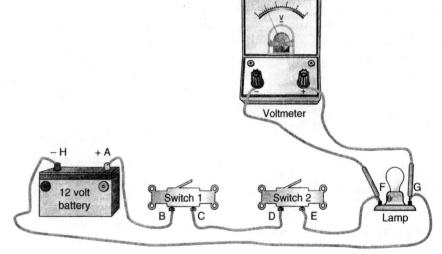

Figure 5–19   The series circuit with the voltmeter probes touching the two terminals of the load (terminals F–G). If voltage is present at this point and the lamp is not illuminated, you should suspect the lamp is burned out. You can also presume the switches are closed and the wires are working correctly to get voltage to that point.

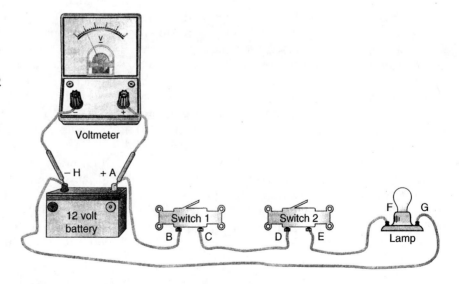

Figure 5–20   The voltmeter measuring voltage at points A–H, which is across the positive and negative terminals of the power supply. Your meter should measure 12 V at this point.

Figure 5–21   This test shows the location of voltmeter probes to test the wire between points A and B.

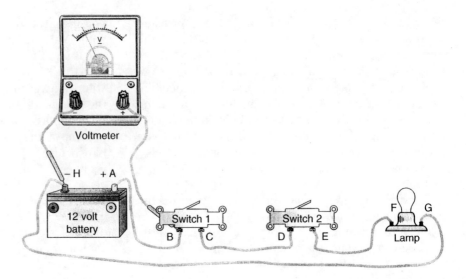

probes to each test point. Remember that this test can be used to locate problems in any circuit, regardless of the number and types of switches or similar devices. **You must be aware of where you are placing your hands, tools, and meter probes at all times because severe electrical shock could result if you come in contact with the electrical circuit.** Even though leaving power applied to a circuit while you test it presents an electrical safety condition, it is necessary because control circuits tend to be complex and you may not be able to find the problem with the power off.

5.13.2
Leaving One Meter Terminal on the Negative Power Terminal as a Reference

Figure 5–21 shows the test points where you should place the voltmeter probes to execute the test of the wire between the positive battery terminal and the first switch (between points A and B). It is important to understand that one of the meter leads must stay on the negative terminal of the power supply (point H) as a reference. If you do not leave one meter lead as a reference, you will not be able to make an accurate test.

The first test should be made with the voltmeter probes touching points B and H, as shown in Fig. 5–21. If no voltage is available at point B and you had voltage at point A, this test indicates the wire between the switch and the power supply is open and not passing power. Knowing this, you can turn off the power and test the wire for continuity, and you should be able to verify that the wire has an open. You should remember that a continuity test uses the resistance range on the VOM

meter, and if the wire has an open, the meter will indicate infinite resistance. If the wire is good, it is said to have continuity and the meter will indicate low resistance. It is also important to remember that the wire must be isolated (one end removed from its terminals) so that you do not measure resistance from a parallel circuit that may be connected to a wire you are testing. If your continuity test indicates the wire has an open, you should replace the wire to put the circuit back in service.

If you measured voltage when the meter probes were on points B and H, the test indicates that the circuit is good to point B. You can move the meter probes to points C and H, to test the next part of the circuit, as shown in Fig. 5–22. If you do not have voltage at points C and H and you had voltage at points B and H, the test indicates that switch 1 is faulty or open. Recheck the switch position to ensure it is in the closed position. If the switch is in its closed position and no power is present at point C, the switch is faulty and must be replaced. You can turn power off and remove the switch from the circuit and test it for continuity with your ohmmeter. Replace the switch, and the circuit should work correctly and cause the lamp to illuminate.

The next test for a loss of voltage in this circuit is to move the voltmeter probes to points D and H, as shown in Fig. 5–23. If the voltmeter indicates that voltage is present at points C and H, and not at points D and H, your test indicates that there is an open in the wire that connects the first switch to the second switch. Since the voltage test indicates a loss of voltage at points D and H, you can turn off power to the circuit and test this wire for continuity.

If you have voltage at points D and H, you have determined that the circuit is operating correctly to this point, and you can move to points E and H. You need to

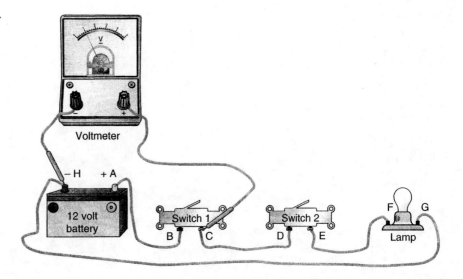

Figure 5–22 Location of voltmeter terminals at points C-H to test switch 1.

ensure that the second switch is closed before making the test at points E and H. If you do not have voltage at points E and H when the switch is depressed, you should turn off power and test the switch for continuity. If the switch is faulty during the continuity tests, you should replace it.

If you have voltage between points E and H, your next test should be at points F and H, as shown in Fig. 5–24. If you do not have voltage at points F and H, you can predict that the wire between the second switch and the lamp has an open in it. You can turn off power and test this wire for continuity.

Figure 5–25 shows the location of the voltmeter probes to make the final test of the wires leading from the positive power supply terminal to the lamp. This test also tests the wire between E and F. If you have voltage at points F and H, you have proved that all the wires and switches between the positive terminal of the battery and the lamp are good. Since the lamp is not

illuminated and the very first reading across terminals F and G indicate that you do not have a voltage reading directly across the lamp, you should now suspect that the wire from point G to the negative side of the battery is open. You can turn off power to the circuit and remove the meter lead from point H to point G. When the circuit is working correctly, you can move the voltmeter probes so they are touching both sides of the lamp, and you should get a reading of 12 V, the voltage provided by the battery. If you have full voltage applied to terminals F and G, you will know that all the wires and switches are working correctly. This is perhaps the most important point in troubleshooting. The main function of the wires and switches in every circuit is to provide the proper voltage to the load. In this circuit the load is the lamp, and if you measure voltage at its terminals, it should be illuminated, and you can predict the circuit wires and switches are operating correctly.

Figure 5–23 Location of voltmeter probes at points D and H to test the wire between C and D.

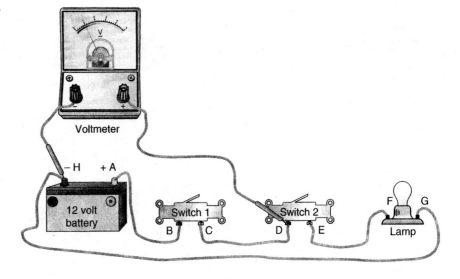

Figure 5–24 The location of voltmeter probes at points E and H to test switch 2.

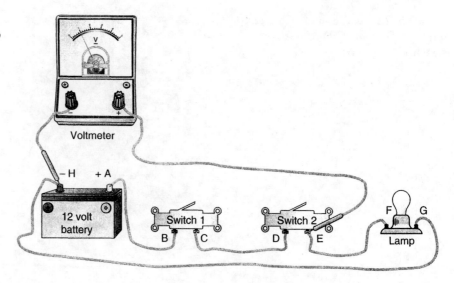

The final two tests to locate the problem in the circuit are perhaps the most difficult to understand, because it appears as though the circuit is operating correctly when your test determined that voltage is present at point F to H. In reality, what has happened is that your tests have proved that the wire and switches on the left side of the lamp are operating correctly. Since the lamp will not illuminate, you can then predict that the problem is in the wire between points G and H.

The best way to locate the fault in this part of the circuit is to move the meter lead that started on the positive terminal to point F and leave it there for the remainder of the these two tests. Then resume the test, leaving the other meter lead at point H and then moving it to point G. If you had voltage between points F and H and do not have it between points F and G, it indicates the wire between G and H has an open. Turn off the power and test the wire for continuity. Replace the wire if it is faulty.

5.14

REVIEW OF SERIES CIRCUITS

In a series circuit all the loads are connected in such a way that there is only one path for current to follow. If one or more switches are used to control current flow to the loads in the circuit, they must also be connected so that only one path exists. This also means that the current will be the same in every part of the circuit. The total resistance in a circuit can be calculated by adding all the resistor values together ($R_T = R_1 + R_2 + R_3 + \ldots$). (Notice that the three dots mean that any number of resistors can be used in this formula because they are all added.)

The total current of a series circuit can be calculated using the formula $I = E/R$, and again it is important to remember that the current will be the same in all parts of the circuit. This means that once you measure or calculate current at any point or through any com-

Figure 5–25 Location of the voltmeter terminals at points F and H that test the wire between E and F.

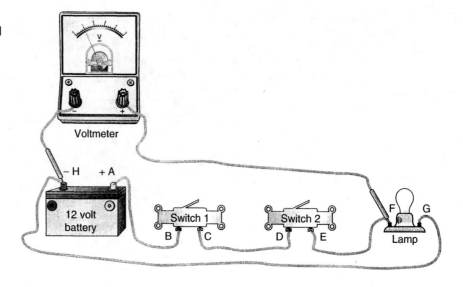

ponent, it will be the same at all other components in the circuit. It also means that if there is an open any place in the circuit, no current will flow anywhere in the series circuit.

The voltage that is measured across each component in the series circuit can be different. If the resistance values of any two resistors in a circuit are the same, the amount of voltage measured across them will be the same. The amount of voltage that is measured across each resistor is caused by the value of the resistor and the amount of current flowing through it. The formula for calculating voltage is $E = IR$. The amount of voltage measured across each resistor in a series circuit is also called the *voltage drop*

for that resistor. The total of all the voltage drops in a series circuit will always be equal to the supply voltage for that circuit. The power consumed by each component in a series circuit can be calculated by the formula $P = I \times E$ *or* $P = I^2R$. The total power consumed by the components in a series circuit can be calculated by the same formula or by adding the amount of power consumed by each component. For example, $P_T = P_1 + P_2 + P_3$.

When current flows through a resistor, voltage is dropped across the resistor. This voltage drop can be measured with a voltmeter. The total of the voltage drop in a circuit equals the applied voltage in a circuit.

QUESTIONS

Short Answer

1. Discuss the advantage of connecting switches in series with a load.

2. Explain why loads are typically not connected in series with each other.

3. In some cases loads may be connected in series with each other. Explain when this would be advantageous and what type of loads would be used.

4. Explain what a resistor is and where you would find it.

5. Explain what each of the four color bands on a resistor is used to indicate.

True or False

1. When a series circuit has an open, current flow stops in all parts of the circuit.

2. The main reason the switches in Fig. 5–1 are connected in series is so that if any one of the switches opens current to the load will be interrupted.

3. It is a good practice to connect two electrical loads such as motors in series so that when one quits, the other will stop.

4. Two heating elements should be connected in series if you want them to split the amount of voltage applied to them.

5. All resistors have the same amount of resistance if they are the same physical size.

Multiple Choice

1. In a series circuit the current in each part of the circuit is _____.
   a. always zero
   b. the same
   c. equal to the voltage

2. The amount of resistance a resistor has _____.
   a. is the same if the resistors are the same size
   b. can be determined by color codes
   c. changes as the amount of current that flows through it changes

3. When three resistors are connected in series, the amount of voltage that is measured across each one is _____.
   a. determined by the amount of current flowing through it and the amount of resistance it has
   b. determined by the wattage rating of each resistor
   c. impossible to determine by calculations

4. If a thermostat, high-pressure switch, and oil-pressure switch are connected in series with a compressor motor and the oil-pressure switch is opened because of low oil pressure, the compressor motor will _____.
   a. still run because two of the other switches are still closed
   b. not be affected because no other loads are connected to it in series
   c. stop running because current flow will be zero

5. If the resistance of a heating element increases from 2 Ω to 5 Ω, the current it draws will _____.
   a. increase if the voltage stays the same
   b. decrease if the voltage stays the same
   c. not change if the voltage does not change

PROBLEMS

1. Decode the following resistors color codes and place the value for the color of each band in the table. Show the tolerance value in the fourth band as a percentage.

First Band	Second Band	Third Band	Fourth Band	First Band Value	Second Band Value	Third Band Value	Fourth Band Value
a. Red	Blue	Brown	Gold				
b. Green	Gray	Red	Silver				
c. Orange	Black	Orange	Gold				

2. Calculate the value of each resistor and show its upper tolerance and lower tolerance value.

First Band	Second Band	Third Band	Fourth Band	Resistor Value	Upper Tolerance Value	Lower Tolerance Value
a. Red	Blue	Brown	Gold			
b. Green	Gray	Red	Silver			
c. Orange	Black	Orange	Gold			

CHAPTER 6

Parallel and Series-Parallel Circuits

OBJECTIVES

After reading this chapter you will be able to:

1. Explain the operation of components connected in parallel.

2. Calculate the voltage, current, resistance, and power in a parallel circuit.

3. Explain Ohm's law as it applies to parallel circuits.

4. Calculate the voltage, current, resistance, and power in a series-parallel circuit.

6.0

INTRODUCTION

Parallel circuits are used frequently in factory electrical control. Figure 6–1 shows a sample circuit, which shows the difference between the series circuit and the parallel circuit: All the loads in a parallel circuit have the same voltage supplied, whereas in a series circuit, each load has a different amount of voltage. The load components in systems such as conveyor motors and fan motors must all be provided with the same voltage, so they must be connected in parallel when they are in the same circuit. Each point where a resistor is connected in parallel in this circuit is called a *branch circuit*. The parallel circuit can have any number of branch circuits. The current in a parallel circuit allows current along more than one path to return to the power supply. These paths are identified by the arrows in the diagram.

The current in a parallel circuit is additive; the formula for calculating total current is $I_T = I_1 + I_2 + I_3 +$ ... (remember the dots mean that additional currents are added to the total). Another point to understand is that the current in a parallel circuit gets larger as more loads are added to the circuit. The parallel circuit also provides a means by which any load can be disconnected from the power source while still allowing voltage to remain supplied to the remaining loads. This is accomplished by placing a switch in each branch circuit just ahead of each resistor.

Series-parallel circuits are used in electrical systems where some parts of the circuit are series in nature and other parts of the circuit are parallel. For example, the fuse and disconnect for the circuit must be able to interrupt all power that is supplied to the loads in a circuit, so the fuse and disconnect must be connected in series with the power supply. If the system has two conveyor motors that are both rated for 220 V, they must have the same amount of voltage supplied to each of them, so they must be connected in parallel with each other and the power supply. The important point to remember when working with series-parallel circuits is that the part of the circuit that is series uses the formulas for series circuits and the part

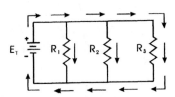

Figure 6–1 A parallel circuit. Notice in this circuit that all the resistors have the same amount of voltage supplied to each.

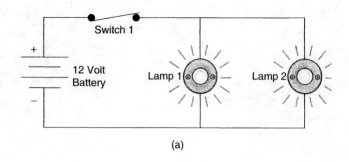

(a)

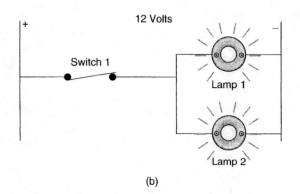

(b)

Figure 6–2 Two lamps connected in parallel that are controlled by a switch connected in series. (a) wire diagram; (b) ladder diagram.

of the circuit that is parallel uses the formulas for parallel circuits. It is also possible to combine parts of the circuit to make it a simplified series or simplified parallel circuit.

6.0.1
Parallel Switches and Loads

In Chapter 5 we explained how switches and loads can be connected in series. In this section we will focus on switches and loads connected in parallel. You will learn how voltage, current, resistance, and power react in parallel circuits. You will also see how to calculate voltage, current, resistance, and power in parallel circuits using Ohm's law. This information will be very useful when you are required to troubleshoot circuits that have parallel switches or loads.

Figure 6–2 shows two lamps connected in parallel that are controlled by a single switch. Figure 6–2a shows a wiring diagram of the circuit, and Figure 6–2b shows a ladder diagram of the same circuit. When the switch is closed, both lamps will be illuminated, and when the switch is open, both lamps will be turned off. When the switch is closed, both lamps will have the same amount of voltage supplied to each of them. If the

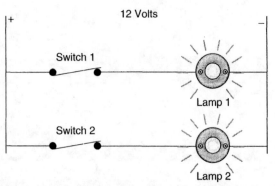

Figure 6–3 A ladder diagram of two series circuits connected in parallel. Each switch controls only the lamp that is in series with it.

lamps are identical, they will both have the same amount of current flowing through each of them. If one lamp is a 100-W bulb and the other is a 50-W bulb, the amount of resistance the 50-W bulb will show to the circuit will be smaller, and the 50-W bulb will have less current flowing through it than the 100-W bulb.

6.0.2
Two Parallel Circuits

Another type of parallel circuit that you will encounter has two branches, each of which is basically a series circuit. Figure 6–3 shows a ladder diagram of this type of circuit. In the figure, switch 1 connected in series with lamp 1 is the first branch circuit, and switch 2 connected in series with lamp 2 is the second branch circuit. In this type of parallel circuit, each branch circuit can operate independently of the other. When switch 1 is opened, only lamp 1 will be turned off. When switch 2 is opened and closed, only lamp 2 is affected. Since the two branch circuits are in parallel with each other and the branch circuit is supplied with 12 V, each lamp will receive 12 V. The amount of current each branch circuit has will depend on the amount of resistance in the load, or you could also calculate the current if you knew the wattage of each lamp. When you are troubleshooting this type of circuit, you need to know which branch (lamp) is affected. Then you can troubleshoot that individual circuit with the voltage-loss method you learned in Chapter 5.

Figure 6–4 shows a wiring diagram of the two series circuits connected in parallel. You can see that the wiring diagram looks more complex than the ladder diagram, but it may be necessary to locate the components when you are troubleshooting. The ladder diagram shows the sequence of the operation of the

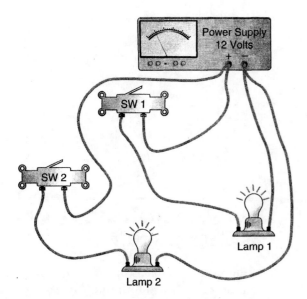

Figure 6–4   A wiring diagram of the circuit shown in Figure 6–3. Notice the wiring diagram shows the location of each component.

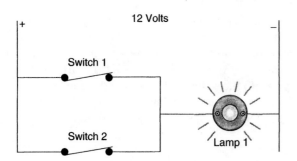

Figure 6–5   A ladder diagram of two switches connected in parallel with a single lamp.

circuit, while the wiring diagram shows the location of each component.

6.0.3
Two Parallel Switches and a Single Lamp

Still another type of parallel circuit you will encounter in electrical controls has two switches connected in parallel with one lamp. Figure 6–5 shows a ladder diagram of this type of circuit. You can see that if switch 1 is opened and switch 2 is closed, the lamp will remain illuminated. If either switch remains closed, the lamp will remain illuminated. Both switches must be opened to cause the lamp to turn off.

6.1
CALCULATING VOLTAGE, CURRENT, AND RESISTANCE IN A PARALLEL CIRCUIT

Voltage, current, and resistance can be calculated in a parallel circuit just as in a series circuit by using Ohm's law. Figure 6–6 shows a sample circuit with the voltage, current, and resistance calculated at each point in the circuit. From this diagram you can see that the supply voltage is 300 V. Because the supply voltage is 300 V, you can determine that the voltage of each branch circuit (across each resistor) is also 300 V, because the voltage at each branch circuit in a parallel circuit is the same as the supply voltage.

The current that is flowing through each resistor can be calculated by using the Ohm's law formula for current, $I = E/R$. The current in each branch circuit is calculated from this formula, and it is placed in the diagram beside each resistor. Total current could also be calculated by the formula E_T/R_T if total resistance is known.

$$I_1 = \frac{E}{R_1} \qquad I_2 = \frac{E}{R_2} \qquad I_3 = \frac{E}{R_3}$$

$$I_1 = \frac{300\text{ V}}{30\ \Omega} \qquad I_2 = \frac{300\text{ V}}{20\ \Omega} \qquad I_3 = \frac{300\text{ V}}{60\ \Omega}$$

$$I_1 = 10\text{ A} \qquad I_2 = 15\text{ A} \qquad I_3 = 5\text{ A}$$

$$I_T = \frac{E_T}{R_T} \quad \text{or} \quad I_T = I_1 + I_2 + I_3$$

$$I_T = \frac{300\text{ V}}{10\ \Omega} \quad \text{or} \quad I_T = 10\text{ A} + 15\text{ A} + 5\text{ A}$$

$$I_T = 30\text{ A} \qquad I_T = 30\text{ A}$$

If the voltage and current were given and the resistance needed to be calculated, the Ohm's law formula for resistance, $R = E/I$, can be used:

$$R_1 = \frac{E}{I_1} \quad R_2 = \frac{E}{I_2} \quad R_3 = \frac{E}{I_3} \quad R_T = \frac{E}{I_T}$$

$$R_1 = \frac{300\text{ V}}{10\text{ A}} \quad R_2 = \frac{300\text{ V}}{15\text{ A}} \quad R_3 = \frac{300\text{ V}}{5\text{ A}} \quad R_T = \frac{300\text{ V}}{30\text{ A}}$$

$$R_1 = 30\ \Omega \quad R_2 = 20\ \Omega \quad R_3 = 30\ \Omega \quad R_T = 10\ \Omega$$

If the total resistance and the total current in a circuit are known, you can calculate the voltage for the circuit. If you know the current and resistance at any branch circuit, you can also calculate the voltage using the Ohm's law formula $E = IR$. The nice part about calculating voltage is that once you determine the voltage at any branch circuit, the same amount of voltage occurs at every other branch circuit and at the supply. The same is true if you calculate the voltage at the supply; you do not need to calculate the voltage at any other branch, because it is the same.

$$E_T = I_T R_T$$
$$E_T = 30 \text{ A} \times 10 \text{ }\Omega$$
$$E_T = 300 \text{ V}$$

$$E_1 = I_1 R_1$$
$$E_1 = 10 \text{ A} \times 30 \text{ }\Omega$$
$$E_1 = 300 \text{ V}$$

$$E_2 = I_2 R_2$$
$$E_2 = 15 \text{ A} \times 20 \text{ }\Omega$$
$$E_2 = 300 \text{ V}$$

$$E_3 = I_3 R_3$$
$$E_3 = 5 \text{ A} \times 60 \text{ }\Omega$$
$$E_3 = 300 \text{ V}$$

▶ *Example 6–1*

Use the circuit in Fig. 6–7 to calculate the individual branch current, total current, and total resistance.

Solution:
Use the following formulas to calculate the current for each branch.

$$I_1 = \frac{E}{R_1}$$
$$I_1 = \frac{240 \text{ V}}{60 \text{ }\Omega}$$
$$I_1 = 4 \text{ A}$$

$$I_2 = \frac{E}{R_2}$$
$$I_2 = \frac{240 \text{ V}}{30 \text{ }\Omega}$$
$$I_2 = 8 \text{ A}$$

$$I_3 = \frac{E}{R_3}$$
$$I_3 = \frac{240 \text{ V}}{15 \text{ }\Omega}$$
$$I_3 = 16 \text{ A}$$

Use the following formula to calculate the total current, I_T.

$$I_T = I_1 + I_2 + I_3$$
$$I_T = 4 \text{ A} + 8 \text{ A} + 16 \text{ A}$$
$$I_T = 28 \text{ A}$$

Use the following formula to calculate total resistance R_T.

$$R_T = \frac{E_T}{I_T}$$
$$R_T = \frac{240 \text{ V}}{28 \text{ A}}$$
$$R_T = 8.57 \text{ }\Omega \qquad \blacktriangleleft$$

6.2

CALCULATING RESISTANCE IN A PARALLEL CIRCUIT

At times you will need to calculate the total resistance of a parallel circuit when only the branch resistance and supply voltage are provided. You can calculate the individual currents at each branch circuit and then calculate the total with the formula $I_T = I_1 + I_2 + I_3$. After you have determined the total current, you can use the formula $R_T = E_T/I_T$.

Another method, called the *product over the sum method*, requires you to use a formula for calculating total resistance in a parallel circuit. This method is called the product over the sum method because you multiply the two resistors to get the product, and then you add the two resistors to get the sum. The third step in the calculation includes dividing the product by the sum (product over sum). The formula is

$$R_T = \frac{R_1 \times R_2}{R_1 + R_2}$$

With this formula you can only calculate the total resistance of two resistors at a time. Because this circuit has three resistors, you need to find the total of the

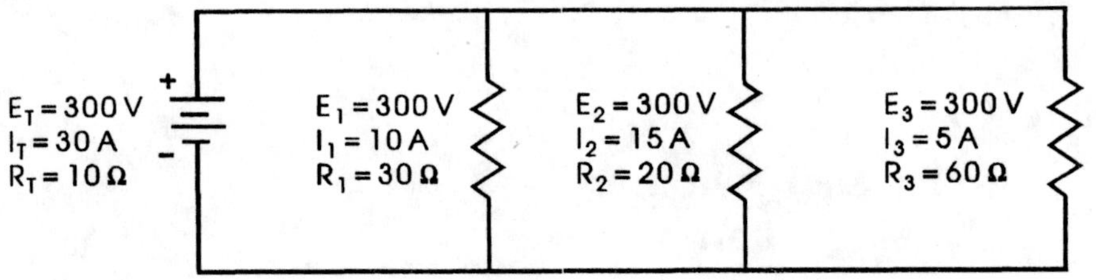

Figure 6–6 Example of a parallel circuit with the voltage, resistance, and current shown at each load.

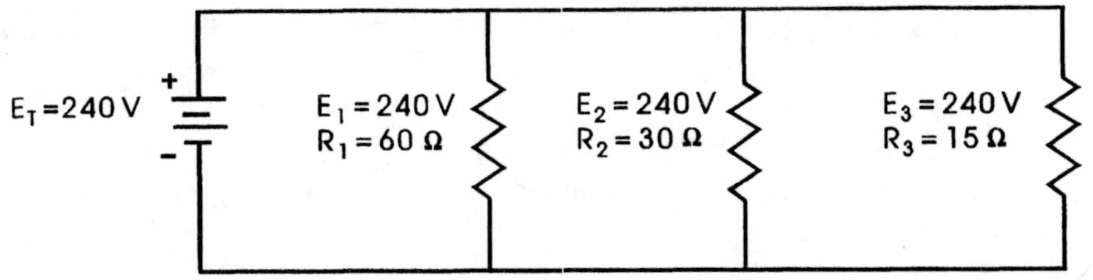

$E_T = 300 \text{ V}$
$I_T = 30 \text{ A}$
$R_T = 10 \text{ }\Omega$

$E_1 = 300 \text{ V}$
$I_1 = 10 \text{ A}$
$R_1 = 30 \text{ }\Omega$

$E_2 = 300 \text{ V}$
$I_2 = 15 \text{ A}$
$R_2 = 20 \text{ }\Omega$

$E_3 = 300 \text{ V}$
$I_3 = 5 \text{ A}$
$R_3 = 60 \text{ }\Omega$

Figure 6–7 Parallel circuit for Example 6–1.

$E_T = 240 \text{ V}$

$E_1 = 240 \text{ V}$
$R_1 = 60 \text{ }\Omega$

$E_2 = 240 \text{ V}$
$R_2 = 30 \text{ }\Omega$

$E_3 = 240 \text{ V}$
$R_3 = 15 \text{ }\Omega$

first two resistors in the branch and then use the formula again with that total and the third resistance to find the *grand total* resistance. We will use this method to calculate the total resistance for the resistors shown in the circuit in Fig. 6–8.

$$R_{T_{R1R2}} = \frac{R_1 \times R_2}{R_1 + R_2}$$

$$R_{T_{R1R2}} = \frac{30\ \Omega \times 20\ \Omega}{30\ \Omega + 20\ \Omega}$$

$$R_{T_{R1R2}} = \frac{600\ \Omega}{50\ \Omega}$$

$$R_{T_{R1R2}} = 12\ \Omega$$

$$R_{T_{R1R2}} = \frac{R_{T_{R1R2}} \times R_3}{R_{T_{R1R2}} + R_3}$$

$$R_{T_{R1R2}} = \frac{12\ \Omega \times 60\ \Omega}{12\ \Omega + 60\ \Omega}$$

$$R_{T_{R1R2}} = \frac{720\ \Omega}{72\ \Omega}$$

$$R_T = 10\ \Omega$$

From these calculations you can see that the total parallel resistance is 10 Ω. *It is important to understand that in all parallel circuits, the total resistance will always be smaller than the smallest branch circuit resistance.* You can see that in this circuit, the smallest resistance in the branch circuits is 20 Ω, and the total resistance is 10 Ω.

The next calculations show the total resistance, determined from the formula

$$R_T = \frac{1}{\dfrac{1}{R_1} + \dfrac{1}{R_2} + \dfrac{1}{R_3}}$$

This formula is designed to be used with a calculator. If you do not have a calculator, it is recommended that

you use the previous method. If you use a calculator, you should use the specified keystrokes to get an answer from this formula.

$$R_T = \frac{1}{\dfrac{1}{R_1} + \dfrac{1}{R_2} + \dfrac{1}{R_3}}$$

$$R_T = \frac{1}{\dfrac{1}{30\ \Omega} + \dfrac{1}{20\ \Omega} + \dfrac{1}{60\ \Omega}}$$

Keystrokes
(The boxes indicate keys on the calculator to use, and the numbers indicate the number keys to use.)

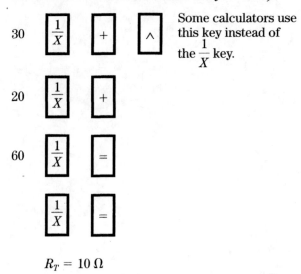

$$R_T = 10\ \Omega$$

▶ Example 6–2
Use the parallel circuit in Fig. 6–9 to calculate the current through each resistor, the total current, and the total resistance for this circuit. The supply voltage is 100 V.

Figure 6–8 Parallel circuit for calculating total resistance.

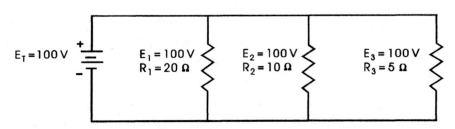

Figure 6–9 Parallel circuit for Example 6–2.

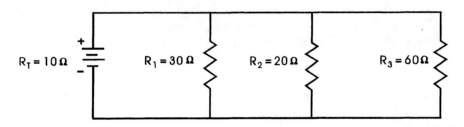

Solution:

$$I_1 = \frac{E}{R_1} \qquad I_2 = \frac{E}{R_2} \qquad I_3 = \frac{E}{R_3}$$

$$I_1 = \frac{100 \text{ V}}{20 \text{ }\Omega} \qquad I_2 = \frac{100 \text{ V}}{10 \text{ }\Omega} \qquad I_3 = \frac{100 \text{ V}}{5 \text{ }\Omega}$$

$$I_1 = 5 \text{ A} \qquad I_2 = 10 \text{ A} \qquad I_3 = 20 \text{ A}$$

$$I_T = I_1 + I_2 + I_3$$

$$I_T = 5 \text{ A} + 10 \text{ A} + 20 \text{ A}$$

$$I_T = 35 \text{ A}$$

$$R_T = \frac{E_T}{I_T}$$

$$R_T = \frac{100 \text{ V}}{35 \text{ A}}$$

$$R_T = 2.86 \text{ }\Omega$$

$$R_T = \frac{1}{\dfrac{1}{R_1} + \dfrac{1}{R_2} + \dfrac{1}{R_3}}$$

$$R_T = \frac{1}{\dfrac{1}{20 \text{ }\Omega} + \dfrac{1}{10 \text{ }\Omega} + \dfrac{1}{5 \text{ }\Omega}}$$

$$R_T = 2.86 \text{ }\Omega \qquad \blacktriangleleft$$

6.3

CALCULATING POWER IN A PARALLEL CIRCUIT

The formula for calculating power in a parallel circuit is the same as the formula for a series circuit: $P_T = P_1 + P_2 + P_3$. The formula for power at each resistor is found from the original Watt's law formula, $P = IE$. This means that you must calculate the power consumed by each branch resistor and then add them all together to get the total power consumed. For example, in the parallel circuit that was shown in Fig. 6–6, the voltage at R_1 is 300 V and the current is 10 A, the voltage at R_2 is 300 V and the current is 15 A, and the voltage at R_3 is 300 V and the current is 5 A. The following calculations determine the power consumed by each resistor and the total power used by the whole circuit:

$$P_1 = I_1 E_1 \qquad P_2 = I_2 E_2 \qquad P_3 = I_3 E_3$$
$$P_1 = 10 \text{ A} \times 300 \text{ V} \quad P_2 = 15 \text{ A} \times 300 \text{ V} \quad P_3 = 5 \text{ A} \times 300 \text{ V}$$
$$P_1 = 3{,}000 \text{ W} \qquad P_2 = 4{,}500 \text{ W} \qquad P_3 = 1{,}500 \text{ W}$$

$$P_T = P_1 + P_2 + P_3$$
$$P_T = 3{,}000 \text{ W} + 4{,}500 \text{ W} + 1{,}500 \text{ W}$$
$$P_T = 9{,}000 \text{ W}$$

You can also calculate the total power consumed by this circuit as follows:

$$P_T = I_T E_T$$
$$P_T = 30 \text{ A} \times 300 \text{ V}$$
$$P_T = 9{,}000 \text{ W}$$

6.4

REVIEWING THE PRINCIPLES OF PARALLEL CIRCUITS

Parallel circuits are used in electrical systems because these types of circuits provide the same amount of voltage to each motor in the circuit. For example, if a system has a conveyor motor that requires 240 V and a hydraulic pump motor that requires 240 V, a parallel circuit will provide both motors this voltage if they are connected in parallel. A parallel circuit provides more than one path for current to return to the power source. Each path is called a branch circuit. The amount of current in each branch of the parallel circuit can be determined by dividing the amount of voltage supplied to that branch by the amount of resistance for the load in that branch. The amount of total resistance in a parallel circuit will always be smaller than the smallest branch resistance. This also means that as additional loads (resistances) are added to a parallel circuit, the total amount of resistance will become smaller. Since the amount of resistance is reduced as loads are added, the total amount of current will increase.

- Current in a parallel circuit is added: $I_T = I_1 + I_2 + I_3 \ldots$.
- Voltage in a parallel circuit is the same for each branch circuit.
- The total resistance in a parallel circuit gets smaller as resistance is added.
- Power in a parallel circuit is added: $P_T = P_1 + P_2 + P_3$.

6.5

REVIEW OF USING PREFIXES WITH UNITS OF MEASURE

At times in previous examples you have seen the amount of resistance exceed 1,000 Ω or the amount of power exceed 1,000 W. In such cases it is useful to substitute a letter for the large number of zeros. For example, we learned that the word for 1,000 is *kilo*, so the letter k can be substituted for the value 1,000. This means that if the answer to a problem is 1,000 Ω, we can use the term 1 k Ω to indicate the same amount. The letter k in this case replaces three zeros.

When resistor values are calculated, the amount of resistance can be as large as 300,000 Ω. This amount can be represented by the term 300 kΩ, because the letter k represents 1,000. If the value of a resistor is determined to be 5,000,000 (5 million), the six zeros can be replaced by the letter M. The letter M is the abbreviation for the word *mega*, which is the word for million. Thus 5,000,000 Ω can be represented by the value 5 MΩ. When you see the value 5 MΩ, you say that you have measured 5 megohms.

If the amount of current is one-thousandth ampere, you can show this value as 0.001 A or 1/1,000 A. The prefix *milli* is used to represent one-thousandth. This prefix is represented by the lowercase letter m. This means that the value 0.001 A can be written as 1 mA. (Notice that a lowercase m is used to specify *milli* and a capital A is used to represent *amperes*.)

If the amount of current is measured as one-millionth of an ampere, it can be shown as 0.000001 A or 1/1,000,000 A. The prefix *micro* and the Greek letter μ are used to represent this amount. Thus the value 0.000001 A can be represented as 1μ A.

Figure 6–10 lists the common prefixes you will use most often in electricity. The exponential value of each prefix is also shown. The exponent is another way of expressing the number without showing a large number of zeros. For example, the prefix *mega* represents the value 1 million. To show the value 6 kilohms, you can use 6 kΩ. Because k represents 10^3, 6 kΩ = 6×10^3 Ω = $6 \times 10 \times 10 \times 10$ Ω = 6,000 Ω. Thus, to multiply 6 kΩ times 20 A, you can simply multiply 6,000 × 20 to get 120,000 V.

The exponential value for a million is 10^6, which means that 1,000,000 is 10 times itself 6 times ($10 \times 10 \times 10 \times 10 \times 10 \times 10$). The representation of the value, 1,000,000 as 1×10^6, makes it easier to see that the value has 6 zeros. The other advantage of exponents is that when you multiply numbers that have exponents in them, you can simply add the exponents to get the correct answer. For example, to multiply 5,000,000 times 4,000, you can do it longhand as shown:

$$\begin{array}{r} 5,000,000 \\ \times\ 4,000 \\ \hline 20,000,000,000 \end{array}$$

You can also do it with exponents: Simply multiply 5 × 4 to get 20, and then add the exponents in 10^6 and 10^3, giving 10^9. Thus, the answer is 20 followed by 9 zeros (20,000,000,000).

$$\begin{array}{r} 5 \times 10^6 \\ \times\ 4 \times 10^3 \\ \hline 20 \times 10^9 \end{array}$$

Most calculators have exponent buttons. This button is identified as

EE

Prefix	Symbol	Number	Exponent
Mega-	M	1,000,000	10^6
Kilo-	k	1,000	10^3
Milli-	m	0.001	10^{-3}
Micro-	μ	0.000001	10^{-6}

Figure 6–10 Table of common prefixes and their numerical and exponential values.

The EE on this button stands for *enter exponent*. If you used a calculator to solve the previous problem, you would use the following keystrokes:

5 EE 6 ×

4 EE 3 =

and the answer would show 20^9.

Note: For some calculators the exponent key is identified as

EXP

(It is important to remember that the exponent will be shown in the upper right corner of the calculator's display as a number only; there is not room to display the × 10.)

If the value 2 mA is shown as a number of amperes, using exponents, the exponent will be displayed as a negative number (2×10^{-3}). The negative number (-3) that is used to indicate the exponent means that the decimal point should be moved three places to the left. Because the original number is 2, you can write 2.0 to show where the decimal point is. Then, the exponent -3 means the decimal point should be moved three places to the left, as shown: 0.002. (It is also important to understand that the first zero that is shown to the left of the decimal point does not add any value to the number. It is used so that you do not confuse the decimal point with a period.) The value 7 μA (microamps) can be shown in terms of amps using an exponent of -6 (7×10^{-6}A). To multiply 15 mA times 4 MΩ, you can use a calculator and use exponents instead of trying to remember how many zeros to use and where to place the decimal point. The keystrokes are as follows:

15 EE -3 ×

4 EE 6 ×

and the answer would show 60^3. The answer indicates 60,000 V. Because the exponent is 3, you can also give the answer as 60 kV.

When you are working with electrical circuits, you will find the prefixes commonly used to save space. On most meters, the kilohms scale is shown using $k\Omega$ and the megohm scale is shown with $M\Omega$. The milliamp scale is shown with mA, and the microamp scale is shown with A. Some companies that manufacture meters use the capital letter K to represent kilo instead of lowercase k because it is easier to read. The correct symbol for the prefix kilo is lowercase k.

It is also important to understand that when you perform calculations with your calculator using units involving micro, milli, kilo, and mega, your answer may contain either positive or negative exponential values from 0 to 6. The problem with an answer that has an exponential value of 2 or 5 is that the VOM meter has scales that recognize only the exponents of 10^{-6}, 10^{-3}, 10^{+3}, and 10^{+6}. Thus you must write all the answers to your calculations in terms of one of these exponents if you plan to make the measurement with the VOM.

6.6
SERIES-PARALLEL CIRCUITS

As a technician or maintenance person, you will work on circuits for a wide variety of factory systems. These systems will have a number of switches and loads that are connected in series and parallel. A typical circuit is shown in Fig. 6–11. In this diagram you can see that a conveyor motor and a hydraulic pump motor are connected in parallel. These two motors are connected in parallel so that they will each receive the same amount of voltage.

These loads are also connected in series with three control switches: a high-pressure switch, a thermostat, and an oil-pressure switch. These switches are designed to control problems with the hydraulic part of

the system. Any time the hydraulic part of the system is de-energized, it is also important to stop the conveyor. If any of these three switches open, the current to both loads will be stopped. The switches are connected in series for this reason. Because this circuit has some switches connected in series and two loads connected in parallel, it is called a *series-parallel circuit*.

The parts of the circuit that are connected in series will follow all the rules for series circuits, and the parts of the circuit that are connected in parallel will follow all the rules for parallel circuits. This series-parallel circuit will provide the best of series circuits and the best of parallel circuits to supply voltage to the loads and control their operation. This type of series-parallel circuit is very common in field equipment that you will encounter.

It will be easier to see the operation of series-parallel circuits by following the voltage, current, and resistance of a circuit that has resistors connected in series-parallel. Figure 6–12 shows a typical series-parallel circuit that uses six resistors. You should notice that the overall shape of this circuit is similar to a large parallel circuit. You can better envision the outline of the parallel circuit if you combined R_1 and R_2 into one equivalent resistance, combine R_3 and R_4 into one equivalent resistance, and combine R_5 and R_6 into one equivalent resistance.

The first branch of this circuit consists of a 20-Ω resistor connected in series with a 30-Ω resistor. The second branch consists of a 10-Ω resistor with a 30-Ω resistor, and the third branch consists of a 40-Ω resistor with a 30-Ω resistor. The supply voltage for this circuit is 200 V. This means that each branch has 200 V applied, but each resistor will have a different voltage applied, because each is connected in series with one other resistor.

6.6.1
Calculating the Amount of Resistance for Each Branch

The amount of voltage that is measured across each resistor and the amount of current flowing through each

Figure 6–11 An electrical diagram of a hydraulic pump motor and conveyor motor connected in parallel and a high-pressure switch, thermostat, and oil-pressure switch in series.

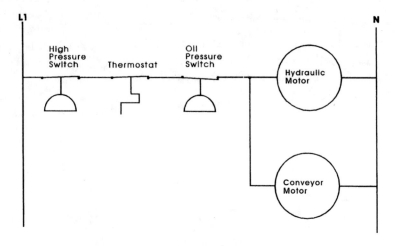

Figure 6–12   A series-parallel
circuit consisting of six resistors.

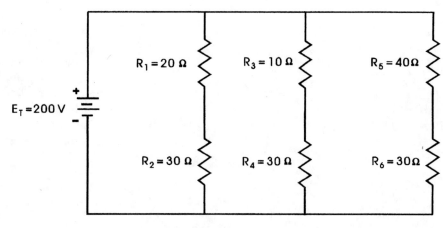

Figure 6–13   The series-parallel
circuit that shows the series resistors
in each branch combined. The value of
the equivalent resistance is identified
as Req_1, Req_2, and Req_3.

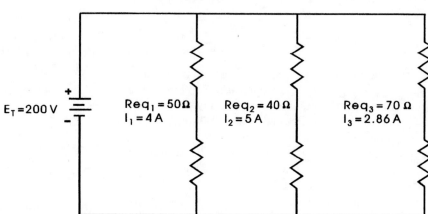

resistor can be calculated by using the formulas that you
have previously learned for series and parallel circuits.
Figure 6–13 shows the next step in calculating the voltage and current for each resistor. This step includes calculating the total resistance for each branch circuit and
the total resistance for the circuit. In this diagram you
can see that the equivalent resistance of each branch is
calculated and shown at each branch. Equivalent resistance is very similar to making change for a dollar bill.
For example, if you have two 50¢ pieces, you can replace
them with a dollar bill if you are solving a problem. The
same is true if you have four quarters; you can show an
equivalent amount of money by using two 50¢ pieces, or
you could use a dollar bill. When you are working with
change and dollar bills, you basically use the form that
is the simplest to handle for each different operation.

The same is true of combining resistor values to
make a circuit easier to understand. For example, in
Fig. 6–13 when you combine the 20-Ω and 30-Ω resistors, you can substitute the equivalent value of 50 Ω; the
two resistors (R_1 and R_2) can be replaced with one resistance called Req_1 (*resistance equivalent* 1) that has
the same value as the two resistors in series (50 Ω). Resistors R_3 (10 Ω) and R_4 (30 Ω) are combined to form the
equivalent resistance Req_2 equal to 40 Ω. Resistors R_5
(40 Ω) and R_6 (30 Ω) are combined to form the equivalent resistance Req_3 equal to 70 Ω.

6.6.2
Calculating the Current
for Each Branch

The current in each branch circuit of the circuit in
Fig. 6–13 can be calculated using Ohm's law ($I = E/R$).
The following calculations show how the branch circuit
current was determined. You should notice that because the overall design of the circuit is a parallel circuit, the supply voltage of 200 V is also supplied to each
of the branch circuits.

$$I_1 = \frac{E_1}{Req_1} \quad I_2 = \frac{E_2}{Req_2} \quad I_3 = \frac{E_3}{Req_3} \quad I_T = I_1 + I_2 + I_3$$

$$I_1 = \frac{200}{50\ \Omega} \quad I_2 = \frac{200}{40\ \Omega} \quad I_3 = \frac{200}{70\ \Omega} \quad I_T = 4\,\text{A} + 5\,\text{A} + 2.86\,\text{A}$$

$$I_1 = 4\,\text{A} \quad I_2 = 5\,\text{A} \quad I_3 = 2.86\,\text{A} \quad I_T = 11.86\,\text{A}$$

6.6.3
Calculating the Total Resistance
for the Circuit

The total resistance for this circuit can be calculated using the formula

$$R_T = \frac{E_T}{I_T}$$

$$R_T = \frac{200\text{ V}}{11.86\text{ A}}$$

$$R_T = 16.86\ \Omega$$

or the formula

$$R_T = \frac{1}{\dfrac{1}{Req_1} + \dfrac{1}{Req_2} + \dfrac{1}{Req_3}}$$

$$R_T = \frac{1}{\dfrac{1}{50\ \Omega} + \dfrac{1}{40\ \Omega} + \dfrac{1}{70\ \Omega}}$$

$$R_T = 16.87\ \Omega$$

You should notice a slight difference in the total resistance when you use different formulas, but this is due to rounding the numbers and it really does not affect the circuit. Regardless of which method you use, the total resistance will determine the total current the circuit requires. The total resistance is determined by the connections of each load (resistor). Because the circuit looks overall like a parallel circuit, the more branch circuits that are added, the more current is required for the loads. It should also be obvious that as more loads are added to this circuit in parallel, the total resistance will continue to drop.

6.6.4
Calculating the Voltage Drop across Each Resistor

The voltage drop across each resistor can be calculated by determining the current through each branch circuit and multiplying it by the resistance in each branch circuit. Because the two resistors in each branch circuit are connected in series, the Ohm's law formulas are valid. Figure 6–14 shows the current and resistance for

each resistor in each of the three branch circuits. The following formula will be used to calculate the voltage that is measured across each resistor. As you know, the voltage across each resistor is also called the *voltage drop* for that resistor. The total current and total resistance are also shown in this circuit.

In the first branch circuit the current (I_1) is 4 A. When this current flows through the 20 Ω of resistor R_1, the voltage drop across R_1 is 80 V.

$$E_{R1} = I_1 \times R_1$$
$$E_{R1} = 4\text{ A} \times 20\ \Omega$$
$$E_{R1} = 80\text{ V}$$

Because the second resistor (R_2) is connected in series with R_1, it will also see 4 A of current. The voltage drop across R_2 is 120 V. It is found by multiplying 4 A times 30 Ω.

$$E_{R2} = I_1 \times R_2$$
$$E_{R2} = 4\text{ A} \times 30\ \Omega$$
$$E_{R2} = 120\text{ V}$$

In the second branch circuit the current is 5 A. When the 5-A current flows through the 10-Ω resistor R_3, the voltage drop across the resistor is 50 V.

$$E_{R3} = 12 \times R_3$$
$$E_{R3} = 5\text{ A} \times 10\ \Omega$$
$$E_{R3} = 50\text{ V}$$

Because the fourth resistor (R_4, 30 Ω) is connected in series with R_3, the current flowing through R_4 will also be 5 A. The voltage drop across R_4 is 150 V, and it is found by multiplying 5 A $\times$ 30 Ω.

$$E_{R4} = I_2 \times R_4$$
$$E_{R4} = 5\text{ A} \times 30\ \Omega$$
$$E_{R4} = 150\text{ V}$$

Figure 6–14 Series-parallel circuit with voltage, current, and resistance calculated for each resistor and for the total circuit.

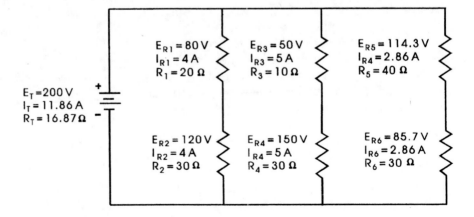

In the third branch circuit the current is 2.86 A. When this current flows through the 40-Ω fifth resistor (R_5), the voltage drop across the resistor is 114.4 V.

$$E_{R5} = I_3 \times R_5$$
$$E_{R5} = 2.86\,\text{A} \times 40\,\Omega$$
$$E_{R5} = 114.4\,\text{V}$$

Since the sixth resistor (R_6, 30 Ω) is connected in series with R_5, the current flowing through R_6 will also be 2.86 A. The voltage drop across R_6 is 85.8 V.

$$E_{R6} = I_3 \times R_6$$
$$E_{R6} = 2.86\,\text{A} \times 30\,\Omega$$
$$E_{R6} = 85.8\,\text{V}$$

After all the voltage current and resistance are calculated for each part of the circuit, you can begin to see patterns that will help you when you troubleshoot a series-parallel circuit that has motors and other types of loads instead of simple resistors. For example, in the previous circuit, a 30-Ω resistor was placed in each of the branches to show that in a series-parallel circuit the voltage drop across each of these resistors will not necessarily be the same. The only time the voltage drop across two resistors that have the same resistance value will be the same is when they have an equal amount of current flowing through them. This could occur in a series-parallel circuit only if the total resistance of two branch circuits were the same. The main point to remember is that it is generally possible to measure the voltage, current, and resistance of all components in a circuit. If you know some of these values, it is possible to use the Ohm's law formulas to calculate other values in the circuit to predict its behavior. These concepts are built upon in later chapters when specific troubleshooting examples are provided.

6.7

SERIES-PARALLEL CIRCUITS IN FACTORY ELECTRICAL SYSTEMS

Series-parallel circuits exist in a variety of factory electrical equipment. These types of circuits are also used in heating equipment. You can use the basic information that you learned about series-parallel circuits to make some simple observations about the size of wires and switches in regard to the amount of voltage and current for which they will need to be rated. Figure 6–15 shows an example of a series-parallel circuit for a plastic injection molding machine with two

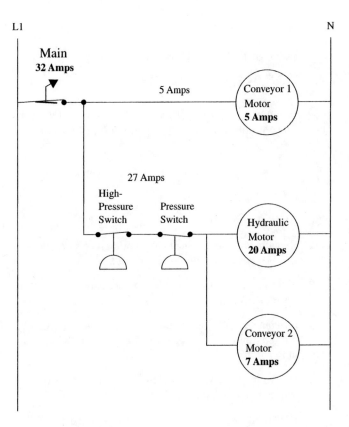

Figure 6–15 Electrical diagram of a typical series-parallel circuit for a system.

conveyors that move parts from the machine. From this diagram you can see that the circuit has some switches in series with motors, and the motors are connected in parallel with each other. For example, the disconnect switch at the top of the circuit in both L_1 and N is connected in series with the entire circuit. This means that if you open the disconnect switch, all power in the entire circuit is disconnected.

The main control switch is connected in series with the hydraulic pump motor and the parallel circuit that has the conveyor 1 and conveyor 2 motors connected in parallel in this circuit. The conveyor 1 motor draws 5 A, the hydraulic pump motor draws 20 A, and the second conveyor motor draws 7 A; therefore, the main control switch must be rated for at least 32 A, because all this current will flow through it. The high-pressure switch and oil-pressure switch are connected in series with the hydraulic pump motor and the conveyor 2 motor, so they must be rated to carry the 27 A used by these two motors. This also means the current to these motors will be interrupted any time either of these switches is opened. All three motors are connected in parallel with each other, so they will all receive the same amount of voltage, but the amount of current each will draw will be different unless their load characteristics are identical. The amount of total current needed to supply the three motors can be calculated by the formula $I_T = I_1 + I_2 + I_3$.

6.8

REVIEW OF SERIES-PARALLEL CIRCUITS

When you are working with a series-parallel circuit, you should remember that any part of the circuit that is a se- ries circuit will operate as a series circuit and any part of the circuit that is parallel will operate as a parallel circuit. You will need to be able to see or think about the sections of the circuit as either series or parallel, and then you can apply the rules of series circuits or parallel circuits.

QUESTIONS

Short Answer

1. Explain why the voltage at each branch of a parallel circuit is the same as the supply voltage.

2. What is the advantage of using exponents when you do Ohm's law calculations?

3. Explain what happens to the voltage at conveyor motor 2 if the hydraulic pump motor has an open in the diagram in Fig. 6–15.

4. Determine the total current that flows through the main switch in the diagram in Fig. 6–15.

5. Explain the symbols M, k, m, and μ and provide an example of how each is used.

True or False

1. Current in each part of a parallel circuit is always the same.

2. Voltage in the branches of parallel circuits is always the same.

3. Total resistance in a parallel circuit becomes smaller as more resistors are added in parallel.

4. Total current in a parallel circuit becomes smaller as more resistors are added in parallel.

5. The prefix *milli* (m) means one-millionth.

Multiple Choice

1. Current in a parallel circuit _____.
   a. increases as additional resistors are added in parallel
   b. decreases as additional resistors are added in parallel
   c. may increase or decrease when resistors are added in parallel depending on their size

2. Voltage in a parallel circuit _____.
   a. increases as additional resistors are added in parallel
   b. decreases as additional resistors are added in parallel
   c. stays the same across parallel branches as additional resistors are added in parallel

3. Resistance in a parallel circuit _____.
   a. increases as additional resistors are added in parallel
   b. decreases as additional resistors are added in parallel
   c. may increase or decrease when resistors are added in parallel, depending on their size

4. When one branch circuit of a multiple-branch parallel circuit develops an open, voltage in other branch circuits _____.
   a. decreases to zero
   b. increases because fewer resistors are using up the voltage
   c. stays the same as the supply voltage

5. Electrical meters and circuits use exponents 10^6, 10^3, 10^{-3}, and 10^{-6} because _____.
   a. these numbers are easier to use than other exponents
   b. these exponents are the values for the symbols M, k, m, and μ
   c. exponents must be multiples of 3 or 6

PROBLEMS

Note: Please show the formula and all work for each problem.

1. Draw a circuit that has three resistors (20 Ω, 40 Ω, and 80 Ω) connected in parallel, and calculate the total resistance for this circuit. Voltage for the circuit is 160V.

2. Calculate the current for each branch of the circuit that you made for Problem 1. Voltage for the circuit is 160V.

3. Calculate the total current for the circuit that you made for Problem 1. The voltage for the circuit is 160V.

4. Calculate the power consumed by each resistor in the circuit you made for Problem 1. Voltage for the circuit is 160 V.

5. Determine the current capacity for each switch in the circuit for Fig. 6–15 if conveyor motor 1 uses 8 A, the hydraulic pump motor uses 15 A, and conveyor motor 2 uses 5 A.

CHAPTER 7

Magnetic Theory

OBJECTIVES

After reading this chapter you will be able to:

1. Describe a magnet and explain how it works.

2. Explain the difference between a permanent magnet and an electromagnet.

3. Identify ways to increase the strength of an electromagnet.

4. Explain what flux lines are and where you would find them.

5. Explain electromagnetic induction.

6. Explain how electromagnetic theory is used to operate a basic generator to create AC voltage.

7. Explain how electromagnetic theory is used to create a simple motor.

8. Explain how electromagnetic theory is used to create a simple relay.

7.0

INTRODUCTION TO MAGNETIC THEORY

The operation of all types of transformers, motors, and relay coils can be explained with several simple magnetic theories. As a technician or maintenance person that works on equipment on the factory floor, you will need to fully comprehend all magnetic theories so that you will understand how these components operate. You must understand how a component is supposed to operate before you can troubleshoot it and perform tests to determine if it has failed. Understanding magnetic theory will make this job easier. It is also impor-

tant to understand that some of the magnetic theories rely on AC voltage. These theories are introduced in this chapter, and more detail about AC voltage and magnetic theories that use AC voltage follows in the next chapter. In this chapter you learn about permanent magnets first, and then you learn about electromagnets.

Magnet is the name given to material that has an attraction to iron or steel. This material was first found naturally about 4,000 years ago as a rock in a city called Magnesia. The rock was called magnetite and was not usable at the time it was discovered. Later it was discovered that pieces of this material could be suspended from a wire and it would always orient itself so that the same ends always pointed the same direction, which is toward the earth's North Pole. Scientists soon learned from this phenomenon that the earth itself is magnetic. At first the only use for magnetic material was in compasses. It was many years before the forces caused by two magnets attracting or repelling could be utilized as part of a control device or motor.

As scientists gained more knowledge and as equipment became available to study magnets more closely, a set of principles and laws evolved. The first of these showed that every magnet has two poles, called the north pole and south pole. When two magnets are placed end to end so that similar poles are near each other, the magnets repel each other. It does not matter if the poles are both north or both south, the result is the same. When the two magnets are placed end to end so that the south pole of one magnet is near the north pole of the other magnet, the two magnets attract each other. These concepts are called the *first and second laws of magnets.*

When sophisticated laboratory equipment became available, it was found that this phenomenon is due to the basic atomic structure and electron alignments. By studying the atomic structure of a magnet, scientists determined that the atoms in the magnet were grouped in regions called *domains,* or *dipoles.* In material that is not magnetic or that cannot be magnetized, the alignment of the electrons in the dipoles is random and

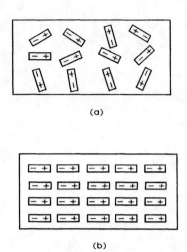

(a)

(b)

Figure 7–1 (a) Example of metal in which dipoles are randomly placed, which makes it a very weak magnet or not magnetic at all; (b) example of metal in which the dipoles are aligned to make a very strong magnet.

usually follows the crystalline structure of the material. In material that is magnetic, the alignments in each dipole are along the lines of the magnetic field. Because each dipole is aligned exactly like the ones next to it, the magnetic forces are additive and are much stronger. In material where the magnetic forces are weak, it was found that the alignment of the dipole was random and not along the magnetic field lines. The more closely this alignment is to the magnetic field lines, the stronger the magnet is. Today we refer to a piece of soft iron in which all dipoles are aligned as a *permanent magnet*. The term *permanent magnet* is used because the dipoles remain aligned for very long periods of time, which means the magnet will retain its magnetic properties for long periods of time. Figure 7–1a shows a diagram of nonmagnetic metal, in which the dipoles are randomly placed, and Figure 7–1b shows a piece of metal that is magnetic, in which the dipoles align to make a strong magnet.

7.1

A TYPICAL BAR MAGNET AND FLUX LINES

Figure 7–2 shows a bar magnet that is made of soft iron that has been magnetized. The magnet is in the shape of a bar, and its north and south poles are identified. Because the bar remains magnetized for a long period of time, it is a permanent magnet. The magnet produces a strong magnetic field because all its dipoles are aligned.

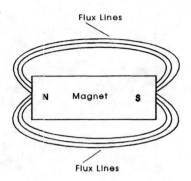

Figure 7–2 Example of a bar magnet. Notice that the poles are identified as north (N) and south (S). Flux lines are shown emanating from the north pole to the south pole.

The magnetic field produces invisible *flux lines* that move from the north pole to the south pole along the outside of the bar magnet. The diagram in Fig. 7–2 shows these flux lines are lines of force that form a slight arc as they move from pole to pole.

Because the flux lines are invisible, you will need to perform a simple experiment to allow you to see that the flux lines do exist and what they look like as they surround the bar magnet. For this experiment you will need a piece of clear plastic film, such as the plastic sheets used for overhead transparencies, and some iron filings. Place the plastic sheet over a bar magnet, making the plastic as flat as possible, and sprinkle iron filings over it. The filings will be attracted by the invisible flux lines as they extend in an arc from the north pole to the south pole along the outside of the magnet. Because the flux lines begin at one pole and stretch to the other, the highest concentration of flux lines will be near the poles. The iron filings will also concentrate around the poles, but a definite pattern of flux lines can be seen along each side of the bar magnet. If an overhead projector is available, the image of the flux lines can be projected onto a projector screen or blackboard so that they can be seen more easily. The pattern of these filings will look similar to the diagram in Fig. 7–2. The number of flux lines around a magnet is directly related to the strength of the magnet. A stronger magnet will have more flux lines than a weaker magnet. The strength of a magnet's field can be measured by the number of flux lines per unit area. Because the strength of a magnet's field is based on the alignment of the magnetic dipoles, the number of flux lines will increase as the alignment of the magnetic dipoles increases.

Some materials, such as alnico and Permalloy, make better permanent magnets than iron, because the alignment of their magnetic domains (dipoles) remains

Figure 7–3 (a) Example of flux lines around a straight wire that is carrying current; (b) flux lines around a coil of wire that is carrying current. Notice that the number of flux lines increases when the wire is coiled.

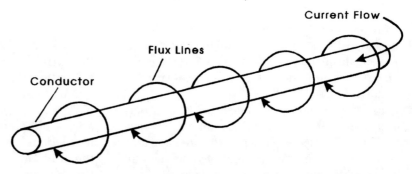

(a)   Few flux lines around conductor that is not coiled

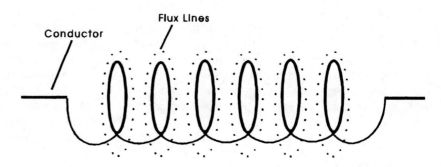

(b)   Flux lines become more concentrated when wire is coiled

consistent even after repeated use. You may find these materials used in some expensive controls and motors, but normally the permanent magnet will be made of soft iron. The reason permanent magnets are useful in many types of controls, especially in motors and generators, is because the soft iron produces residual magnetism for long periods of time over many years. Permanent magnets have several drawbacks, however. One of these problems is that the magnetic force of a permanent magnet is constant and cannot be turned off, if it is not needed. This means that if something is attracted to a magnet, it will remain attracted until it is physically removed from the force of the flux lines. Another problem with a permanent magnet's flux field being constant is that it cannot easily be made stronger or weaker if circumstances so require.

7.2

ELECTROMAGNETS

An electromagnet is produced by passing current through a coil of wire. One type of electromagnet is made by connecting a coil of wire to an electric cell (battery). The electromagnet has properties that are similar to those of a permanent magnet. When a wire conductor is connected to the terminals of the battery, current will begin to flow, and magnetic flux lines will form around the wire like concentric circles. If the wire is placed near a pile of iron filings while current is flowing through it, the filings will be attracted to the wire just as if the coil were a permanent magnet. Figure 7–3 shows several diagrams indicating the location of magnetic flux lines around conductors. Figure 7–3a shows flux lines will occur around any wire when current is flowing through it. You can set up several simple experiments to demonstrate these principles. In one experiment you can insert a current-carrying conductor through a piece of cardboard and place iron filings around the conductor on the cardboard. When current is flowing in the wire, the filings will settle around the conductor in concentric circles, showing where the flux lines are located. As the amount of current is increased, the number of flux lines will also increase. The flux lines will also concentrate closer and closer to the wire until the current reaches *saturation*. When the flux lines reach the saturation point, additional increase of current in the wire will not produce any more flux lines.

When a straight wire is coiled up, the flux lines will concentrate and become stronger. Figure 7–3b shows an example of flux lines around a coil of wire that has current flowing through it. Because the flux lines are much stronger in a coil of wire, most of the electromagnets that you will encounter will be in the form of coils. For example, coils are used in transformers, relays, solenoids, and motors.

One advantage an electromagnet has over a permanent magnet is that the magnetic field can be energized and de-energized by interrupting the current flow through the wire. The strength of the magnetic field can also be varied by varying the strength of the current flow through the conductor that is used for the electromagnet. This theory is perhaps the most important theory of magnetism, because it is used to change the strength of magnetic fields in motors, which causes the motor shaft to produce more torque so it can turn larger loads. Because the flux lines are invisible, you may need to perform an experiment to prove that the magnetic field becomes stronger as current flow through the coil is increased. Figure 7–4 shows how to set up this experiment. Wrap a wire into a coil, connect the ends to a dry cell battery, and place it near a pile of iron filings. Add a variable resistor to the circuit to increase or decrease the current. When the current is set to a minimum, as in Fig. 7–4a, the magnetic field around the coil of wire will attract only a few filings. As the current is increased, as in Fig. 7–4b, the number of filings the magnetic field will attract also increases, until the current causes the magnetic field in the wire to reach *saturation*. When the saturation point is reached, any additional current flowing in the wire will not produce additional flux lines. When a switch is added to this circuit, the magnetic field can be turned on and off by turning the switch on and off to interrupt the flow of current in the coil. When the switch is opened, current is interrupted and no flux lines are produced, so the magnetic field will not exist.

Components such as electromechanical relays and solenoids use the principle of switching the magnetic field on and off. When current flows in the coil, the magnetic field causes the contacts in the relay or the valve in the solenoid to close. When the current to the coil in these devices is interrupted and the magnetic field is turned off, springs cause the relay contacts or valves to open. More information about relays and solenoids is provided in later chapters.

This principle is also used to turn motors on and off. When current flows through the coils of a motor, its shaft will turn. When current is interrupted, the magnetic field will diminish and the motor will stop rotating. This point is also important to remember if the wire that is used in the coil develops an open. When current stops flowing in the coil due to the open circuit, the shaft of the motor stops turning.

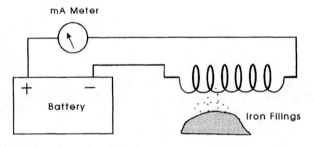

a) Few filings are attracted when small current passes through the conductor.

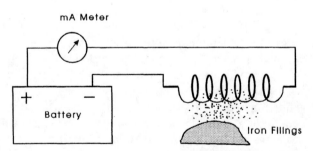

(b)   Large number of filings are attracted as current is increased.

Figure 7–4   (a) A small amount of current flowing in a coil of wire creates a small number of flux lines that attract iron filings; (b) a large amount of current flowing in a coil of wire creates a large number of flux lines that attract iron filings.

7.3

ADDING COILS OF WIRE TO INCREASE THE STRENGTH OF AN ELECTROMAGNET

Another advantage of the electromagnet is that its magnetic strength can be increased by adding coils of wire to the original single coil of wire. The increase of the magnetic field occurs because the additional coils of wire require a longer length of wire, which provides additional flux lines. The magnetic field will be stronger when the coil is more tightly wound because the flux lines are more concentrated. Thus very fine wire is used in some electromagnets to maximize the number of coils. As smaller wire is used, however, the amount of current flowing through it must be reduced so that the wire is not burned open.

You will learn that some motors use this principle to increase their horsepower and torque ratings. These motors have more than one coil that can be connected in various ways to affect the torque and speed of the motor's shaft. *Torque* is defined as the amount of rotating force available at the shaft of a motor. You will also learn that coils can be connected in series or in parallel to affect the torque and speed of a motor.

7.4
USING A CORE TO INCREASE THE STRENGTH OF THE MAGNETIC FIELD OF A COIL

The strength of a magnetic field can also be increased by placing material inside the helix of the coil to act as a core. The farther the core is inserted into the coil, the stronger the flux field becomes. When the core is removed completely from the coil, it is considered to be an *air coil magnet*, and the magnetic field is at its weakest point. If a soft iron is used as the core, it will strengthen the magnetic field, but it also creates a problem because it has excessive residual magnetism, which is unwanted. Residual magnetism means the core will retain magnetic properties when current is interrupted in the coil, which will make it like a permanent magnet. This problem can be corrected by using laminated steel for the core. The laminated-steel core is made by pressing sheets of steel together to form solid core. Figure 7–5 shows an example of layers of laminated-steel pressed together to form a core. When current flows through the conductors in the coil, the laminated-steel core enhances the magnetic field in much the same way as the soft iron, and when current flow is interrupted,

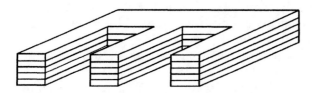

Figure 7–5 Example of pieces of steel pressed together to form a laminated steel core for use in electromagnets.

the magnetic field collapses rapidly because each piece of the laminated steel does not retain sufficient magnetic field.

7.5
REVERSING THE POLARITY OF A MAGNETIC FIELD IN AN ELECTROMAGNET

When current flows through a coil of wire, the direction of the current flow through the coil will determine polarity of the magnetic field around the wire. The polarity of the magnetic field around the coil of wire is important because it determines the direction a motor shaft turns in an AC or DC motor. If the direction of current flow is reversed, the polarity of the magnetic field is reversed, and the direction a motor shaft is turning is reversed. In some motors used in factory systems, such as fan motors and pump motors, the direction of rotation is very important. In these applications you will be requested to change the connections for the windings in the motor or the supply voltage for a three-phase motor to make the motor rotate in the opposite direction. The changes you are making take advantage of changing the direction of current flow through a coil or changing the polarity of the supply voltage with respect to the other phases so that the motor will reverse its rotation.

As you are learning advanced theories about motors and other electromagnetic components, diagrams will be presented to explain more complex concepts. When these diagrams are used to explain the direction of current flow through a wire, they will use a method of identifying the direction of current flow that has been developed and universally accepted. Figure 7–6a shows a diagram that indicates the location of the flux lines in a coil of wire and shows the direction the flux lines travel around a straight, current-carrying conductor. A dot or a cross (X) is used to mark the conductor to indicate the direction current is flowing. In Fig. 7–6a you should notice that a dot is used to indicate that current is flowing toward the observer. The flux lines in this diagram show their flow is in a clockwise motion. In Fig. 7–6b, a cross is used to indicate that current is flowing away from the observer and the flux lines are shown moving in the counterclockwise direction. If you know the direction of current flow, the advanced theories will allow you to determine the polarity of the magnetic field and predict the direction a motor's shaft will rotate.

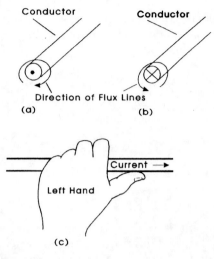

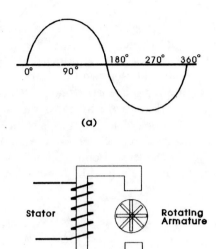

Figure 7–6   (a) The direction of flux lines around a wire when current is flowing in the wire toward you. The dot indicates the direction of current flow is toward the observer. (b) The direction of flux lines around a wire when current is flowing in the wire away from you. The "X" indicates current is flowing in the wire away from the observer. (c) An example of the left-hand rule. The thumb is pointing in the direction of the current flow in the wire, and the fingers are pointing in the direction of the flux lines as they move around the wire when current is flowing.

Figure 7–7   (a) A sine wave for AC voltage; (b) a simple alternator that has a moving coil and a stationary coil. The alternator produces the AC sine wave.

Figure 7–6c shows another way to determine the directions of the flux lines, the direction of current flow, and which pole is the north pole. This method is called the *left-hand rule*. From this diagram you can see that you need to know either the direction of the current flow or the direction of the flux lines to determine the polarity of the coil. Normally the direction of the current flow is easy to determine with a voltmeter by determining which end of the wire is positive and which is negative. Current (electron flow) is from negative to positive.

7.6

ELECTROMAGNETIC THEORY USED TO OPERATE A BASIC GENERATOR TO CREATE AC VOLTAGE

One of the reasons you are learning electromagnetic theory is so you can understand how electromagets are used in a basic generator to create AC voltage. This

section provides an overview of how AC voltage is generated in a basic generator, and the next chapter provides an in-depth explanation of AC voltage. AC stands for *alternating current*. This term is derived from the fact that AC voltage alternates, being positive for half of a cycle and then negative for the other half of the cycle. Figure 7–7a shows the characteristic AC waveform. The AC waveform shown in this figure is called a *sine wave*.

Since this kind of generator creates an AC waveform, we will refer to it as an alternator for the remainder of this section. Basically we use the term *generator* to mean a machine that creates DC voltage and *alternator* to mean a machine that creates AC voltage. Here we will explain the operation of a very basic alternator, pictured in Figure 7–7b, and in later chapters you can learn more details about the operation of more complex alternators. The simple alternator has two basic parts: the *rotor*, which is a rotating coil of wire mounted on a shaft, and the *stator*, which is a stationary coil that has current flowing through it so that it creates a magnetic field. Sometimes the stationary coil is just called the field. A voltage is applied to the stationary coil (field) to ensure a current is flowing through it to create a very strong electromagnetic field.

The coils of wire in the rotor are mounted to the shaft of the alternator. This shaft is rotated by an energy source such as a steam turbine. When the rotating coil of wire passes through the magnetic field that is created by current flowing through the stator, an electron flow is created in the coil of wire. Because the coil of wire rotates, it will pass the positive magnetic field and then the negative magnetic field during each complete rotation of 360°. This action causes the voltage sine wave to be produced, and a complete sine wave occurs in 360°. In the United States, the speed of rotation of the alternator is maintained at a constant rate so that the sine wave will have a frequency of 60 cycles per second, which is also called 60 hertz (H_3).

Since the rotor shaft is constantly turning, a set of carbon brushes that will conduct electricity is used to make contact between the stationary part of the alternator and the rotor. The brushes are held in a mechanism that uses springs. The springs ensure that sufficient tension is placed on the brushes to make contact with the rotor shaft so that the generated voltage can be moved from the rotating part of the alternator to the stationary frame, where it can be insulated and connected for use by the loads that are connected to the alternator. It is also important to understand that the amount of voltage that is applied to the stator can be adjusted, to adjust the amount of voltage that is produced by the alternator. Later chapters provide a very detailed explanation of the operation of both DC generators and AC alternators. In those chapters you can see how additional windings and other components are added to the machines to make them more efficient and more reliable.

It may seem strange to use a voltage force that is applied to the stator coils just to get voltage out of the armature. Why, you might wonder, would you not just use the small amount of voltage applied to the field coil as the output voltage instead of putting this voltage into the alternator? The reason is that the alternator is a very good energy converter, and so only a very small amount of voltage is needed to cause the stator to create the current to make the magnetic field, whereas the actual amount of voltage produced by the alternator may be several hundred volts. The actual power (volts amps) that the alternator produces will be nearly equal to the amount of power that is put into its shaft. For example, if a large amount of steam is used to turn the shaft of the alternator, it can produce several hundred times the amount of electrical voltage that is used to excite the alternator to create the magnetic field. It is also important to understand that we can convert very large amounts of coal, natural gas, or nuclear power to create the heat to make the steam to turn the alternator. In some locations, water that is stored behind dams is used to turn the shaft. These sources allow large amounts of energy to be converted to electricity and be transmitted over long distances to provide power to industries where it is used to turn

motors or be converted to heat energy for industrial applications.

You may be familiar with another power converter of electicity—the generator or alternator used on your automobile. The generator or alternator supplies the voltage that your vehicle uses for lights, windshield wipers, power windows, and other electrical components. Most of these components use DC voltage, and so a generator is used to produce DC voltage, or diodes are used with an alternator to change AC voltage to DC voltage.

7.7
MAGNETICS USED IN THE BASIC THEORY OF A SIMPLE DC MOTOR

Another application of magnetism is the operation of a simple DC motor. Later chapters in this text provided a very detailed explaination about the operation of both DC and AC motors. As you can see from Fig. 7–8, this

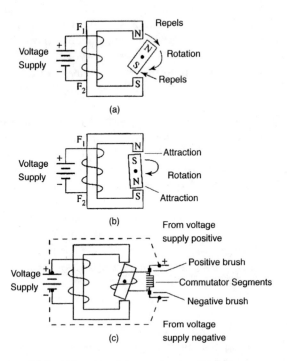

Figure 7–8 Magnetic diagram that explains the operation of a simple DC motor. (a) The rotating magnet moves clockwise because like poles repel. (b) The rotating magnet is being attracted because the poles are unlike. (c) The rotating magnet is now shown as the armature coil, and its polarity is determined by the brushes and commutator segments.

motor consists of two basic parts. One part is the stationary field, which is a coil of wire that has voltage applied to it to create the north and south pole of a magnet. The second part is the moving shaft, called the rotor. The rotor also is a magnet that is identified by its north and south pole. The magnet in the figure is a simple bar magnet, but in a more complex motor the rotor may have multiple magnetic fields built up in it. Just as in the example of the alternator in the previous section, a set of carbon brushes is used to make the connection between the rotating section of the motor and the stationary part. The carbon brushes can conduct electricity very easily and do not wear out too quickly.

Figure 7–8a shows the motor when voltage is first applied to it and the stationary field that looks like the letter C. When voltage is applied to the stationary field, current flows through it, and it becomes magnetized so that the north pole of the field appears at the top of the C and the south pole appears at the bottom of the C. The north pole of the rotor, or bar magnet in this simple example, is near the north pole of the field magnet. This will cause the rotating bar magnet to begin to rotate because the laws of magnets cause like poles to repel, and unlike poles to attract.

The bar magnet represents the *armature*, and the coil of wire represents the *field*. The arrow shows the direction of the armature's rotation. Notice that the arrow shows the armature starting to rotate in the clockwise direction. The north pole of the field is repelling the north pole of the armature, and the south pole of the field coil is repelling the south pole of the armature.

As the armature begins to move, the north pole of the armature comes closer to the south pole of the field, and the south pole of the armature comes closer to the north pole of the field. As the two unlike poles near each other, they begin to attract each other. This attraction becomes stronger until the armature's north pole moves directly in line with the field's south pole and its south pole moves directly in line with the field's north pole (Fig. 7–8b).

When the opposite poles are at their strongest attraction, the armature will be "locked up" and will resist further attempts to continue spinning. For the armature to continue its rotation, the armature's polarity must be switched. Since the armature in this diagram is a permanent magnet, it would lock up during the first rotation and would stop working. If the armature is an electromagnet, its polarity can be changed by changing the direction of current flow through it. For this reason, the armature must be changed to a coil (electromagnet), and a set of commutator segments must be added to provide a means of making contact between the rotating member and the stationary member. This is shown in Fig. 7–8c. One commutator segment is provided for each terminal of the magnetic coil. Because this armature has only one coil, it will have only two terminals, and the commutator has only two segments.

Since the armature is now a coil of wire, it will need direct current flowing through it to become magnetized. This presents another problem: The armature will be rotating, and so the DC voltage wires cannot be connected directly to the armature coil. A stationary set of carbon brushes is used to make contact with the rotating armature. The brushes ride on the commutator segments to make contact so that current will flow through the armature coil.

In Fig. 7–8c the DC voltage is applied to the field and to the brushes. Since negative DC voltage is connected to one of the brushes, the commutator segment that the negative brush rides on will also be negative. The armature's magnetic field causes the armature to begin to rotate. This time when the armature gets to the point where it becomes locked up with the magnetic field, the negative brush begins to touch the end of the armature coil that was previously positive, and the positive brush begins to touch the end of the armature coil that was negative. This action switches the direction of current flow through the armature, which also switches the polarity of the armature coil's magnetic field at just the right time so that the repelling and attracting continues. The armature continues to switch its magnetic polarity twice during each rotation, which causes it to continually be attracted to and repelled by the field poles.

This is a simple two-pole motor that is used primarily for instructional purposes. Since the motor has only two poles, the motor will operate rather roughly and not provide too much torque. Additional field poles and armature poles must be added to the motor for it to become useful for industry. Chapters 13, 14, and 15 show detailed information about AC single-phase motors, AC three-phase motors, and DC motors.

7.8

ELECTROMAGNETIC INDUCTION USED IN TRANFORMERS

Another use of magnetics in industrial components is called *electromagnetic induction*. This magnetic principle is used in transformer and motor theory. This theory states that if two coils of wire are placed in close proximity to each other and a magnetic field is created in one of the coils so that flux lines are created around it, the flux lines will collapse across the second coil when the current is interrupted in the first coil. When these flux lines collapse across the coils of the second coil, they will cause a current to begin to flow in the second coil. This current flow will be 180° out of phase with the current flow that created the magnetic field in the first coil.

Because AC voltage periodically builds voltage to a positive peak and interrupts it when the sine wave reaches 180°, it is ideal for creating the magnetic field that builds and collapses. When the current flows, it causes the flux lines to build in the coil; and when the current flow is interrupted, it causes the flux lines to collapse. Because AC voltage in America is generated at 60 Hz, the magnetic field in the first coil will build and collapse 60 times a second. The flux lines will also build and collapse across the windings of the second coil 60 times a second.

One component that uses two coils is a *transformer.* The first coil in the transformer is called the *primary winding,* and the second coil in the transformer is called the *secondary winding.* The ratio of the number of turns of wire in the primary winding to the number in the secondary winding will determine the ratio of the voltage in the primary winding and secondary winding. Another important point to remember is that because the two coils are electrically isolated, the voltage produced in the second coil by induction is totally isolated from the voltage in the first coil. The isolation allows the number of turns of wire in the two coils to be different so that a different amount of voltage can be created in the second coil. If the number of turns of wire in the first coil and second coil are identical, the amount of voltage induced in the second coil will be approximately the same as the voltage in the first coil. Chapter 9 provides a detailed explanation of all types of transformers.

7.9
MAGNETICS USED IN RELAYS AND CONTACTORS

Magnetics are also used in relays, a use that is an important industrial application. A *relay* consists of one or more sets of contacts that are closed or opened with a magnetic coil. Figure 7–9a shows the simple operation of a relay. In this diagram you can see that the relay has one set of normally open contacts and a coil. The contacts consist of a movable contact mounted on an arm with a pivot at the left end and a stationary contact mounted below it. The movable

contact arm has a piece of metal attached to it that becomes attracted to the magnetic field created by the coil. When the coil is energized with current and becomes magnetic, the movable contact is pulled downward as the piece of metal is attracted to the magnetic coil. This action stretches the spring and causes the upper movable contact to pull down against the stationary contact, which causes the contacts to go to the closed position. When voltage to the coil is removed, the coil is no longer magnetic, and the spring pulls the upper arm upward so that the contacts become open. Figure 7–9b shows the electrical ladder diagram of the relay with a switch controlling voltage to the coil and the contacts controlling voltage to the motor. The contacts act as a switch to a 2-hp motor that draws 12 A at 230 V. It is not safe to run this type of voltage and current through a small switch that you open and close by hand, because you might get a shock. The contacts in a relay are closed by the magnetic field that is created when current flows through a coil placed directly under the contacts. You can see in the figure that a spring pushes the contacts to their open position any time the coil is not energized. When voltage is applied to the coil, current flows through it, and it becomes a strong magnet and pulls the contacts to their closed position. Now 230 V and 12 A flow through the contacts to the motor, while less than 24 V and 1 A flow through the switch your hand touches. The 24 V and 1 A are sufficient to create the strong magnetic field in the coil that causes the contacts to close. The large voltage and current from a separate power source flows through the contacts to cause the motor to run. When the small switch is moved to the open position, then the coil is no longer magnetized, the spring forces the contacts to their open position, the voltage and current are interrupted to the motor, and the motor stops.

A small switch is used to control the small amount of voltage and current that flows to the coil to cause it to become magnetized. The small switch in the coil circuit allows you to safely control larger loads without exposing your hands to the large voltage and current the motor or load needs to run. The relay also allows the switch that controls the coil to be mounted several hundred feet from the relay and motor because small wire can safely handle the small voltage and current that flows between the switch and the coil, and larger wire is used to carry the large voltage and current between the contacts and the motor. A contactor and motor starter are larger versions of relays. These larger devices are used to control larger motors, heating loads, and lighting loads. The magnetic coil is also used to open and close solenoid valves to control fluid flow through the device. You

Figure 7–9 (a) A relay with normally open contacts and coil shown. The coil becomes a magnet when current flows through it and pulls the metal plate mounted on the contact arm downward. Since one end of the contact arm has a pivot, the contacts on the other end of the arm close. A spring pulls the contacts open when the coil is not energized. (b) A relay coil controlled by 24 V. The contacts control a motor that is powered by 230 V.

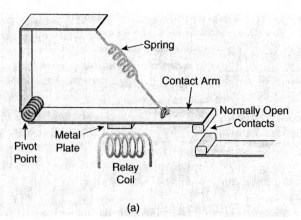

(a)

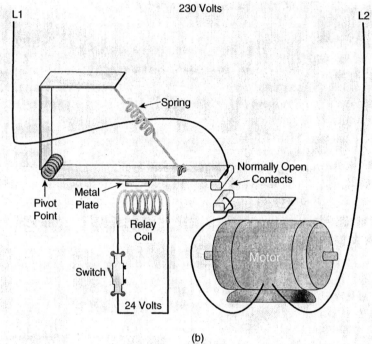

(b)

will learn more about relays, contactors, and solenoids in Chapter 11 and more about motor starters in Chapter 12.

7.10
REVIEW OF MAGNETIC PRINCIPLES

When you are learning about components that utilize electromagnetic principles, you can return to this section to review the basic concepts about them. The important electromagnetic principles are as follows:

- The dipoles of a magnet are aligned.
- Every magnet has a north pole and a south pole.

- Like magnetic poles repel, and unlike magnetic poles attract.
- A permanent magnet is useful because its magnetic field is residual and can remain strong over a number of years.
- When current passes through a conductor, magnetic flux lines form around the conductor in concentric circles.
- The strength of the magnetic field around a conductor is proportional to the amount of current flowing through it until the amount of current reaches the saturation point of the electromagnet.
- The magnetic field of an electromagnet can be turned on and off by interrupting the current flow through the coil of wire.
- The polarity of an electromagnetic coil is determined by the direction of current flow through the conductor.

- If the conductor is wrapped in coils, the strength of the magnetic field is increased.
- When more coils of wire are added to an electromagnet, the magnetic field strength is increased.

- A core will cause the field of an electromagnet to be strengthened.
- The core must be made from laminated steel when AC voltage is used to power the coil, so that the magnetic field can quickly build and collapse in the core as the AC voltage changes polarity every 1/60 s.

QUESTIONS

Short Answer

1. Explain the difference between a permanent magnet and an electromagnet.

2. Define the term *saturation* as it applies to an electromagnet.

3. Explain the principle of electromagnetic induction.

4. Identify two ways to increase the strength of an electromagnet.

5. Explain why it is important that the magnetic field of an electromagnet can be turned on and off.

True or False

1. When dipoles are aligned in a material, it is a good magnet.

2. Any time current flows through a conductor, magnetic flux lines are created around the conductor.

3. The polarity of the magnetic field of an electromagnet is determined by the direction of current flow through the conductor.

4. Like poles of magnets attract.

5. The strength of the magnetic field for a permanent magnet can be changed easily.

Multiple Choice

1. _____ can have the strength of its field changed easily.
   a. A permanent magnet
   b. An electromagnet
   c. A dipole

2. The polarity of an electromagnet can be changed by _____.
   a. changing the direction of current flow through the coil of wire
   b. changing the amount of current flow through the coil of wire
   c. changing the frequency of the current flow through a coil of wire

3. A laminated-steel core is used in electromagnets that have AC voltage applied to them because _____.
   a. laminated steel is more economical to use than soft iron
   b. laminated steel is easily formed to any shape, which makes it more usable in complex components
   c. laminated steel allows the magnetic field to build and collapse quickly

4. A magnetic field is created in a coil of wire when _____.
   a. current flows through the wire
   b. current stops flowing through a wire
   c. the coil has a core

5. The left-hand rule is used to _____.
   a. determine the amount of magnetic flux in a coil
   b. determine the number of coils in an electromagnet
   c. determine the polarity of the magnetic field in an electromagnet

PROBLEMS

1. Draw a sketch of a permanent magnet and show where flux lines occur.

2. Draw a sketch of an electromagnetic coil and show where flux lines occur.

3. Draw a sketch of an electromagnetic coil with an open somewhere in the coil of wire. Explain what effect the open will have on the magnetic field.

4. Draw two permanent magnets with their poles placed so the two magnets will attract each other.

5. Perform the experiment presented in Section 7.1 and explain why the flux lines are concentrated around the ends of the permanent magnet.

CHAPTER 8

Fundamentals of AC Electricity

OBJECTIVES

After reading this chapter you will be able to:

1. Explain the term *alternating current.*

2. Explain the terms *peak-to-peak* and *RMS voltage.*

3. Calculate the frequency and period of an AC sine wave.

4. Explain the operation of an oscilloscope.

5. Explain the effects of a capacitor and inductor in an AC circuit.

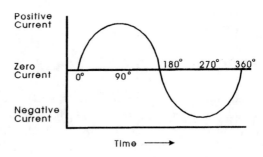

Figure 8–1   AC sine wave shown moving through 360°.

8.0
WHAT IS ALTERNATING CURRENT?

This section reviews some of the principles of AC voltage that you learned in the previous chapter. *Alternating current* (AC) is identified by its characteristic waveform. The AC waveform is shown in Fig. 8–1. The sine wave has a positive half-cycle and a negative half-cycle. The device that creates AC electricity is called an *alternator.* The alternator has coils of wire mounted on its rotating part, which is called the *rotor.* The alternator also has coils of wire that are used to create strong magnetic fields. These coils of wire are mounted in the *stationary part* of the alternator, which is called the *stator.* The magnetic field in the stator, like all other magnets, has flux lines that move between its north and south poles. When the shaft of the alternator is rotated, the coils of wire in the rotor move past the flux lines from the strong magnetic field in the stator. Because the coils pass the north pole of the magnetic field and then the south pole, they create a sine wave in

the rotor, which becomes the AC voltage. This voltage is used to power a wide variety of electrical components in the factory. AC voltage is the primary voltage used in factory equipment. AC electricity is generated by utility companies and transmitted to commercial and residential users.

8.1
WHERE DOES AC VOLTAGE COME FROM?

AC voltage is generated at a number of power stations around the country. The voltage that is used in your city is generated within a range of 600 mi. In some areas the energy source that turns the rotor of the alternator to produce this voltage comes from burning coal, burning fuel oil, or burning other fossil fuels to make steam. In some areas of the country the steam is generated by nuclear power stations. The steam is used to turn a turbine wheel that turns an alternator shaft. In other parts of the country the turbine wheel is turned by water that is stored behind hydroelectric dams. When the alternator shaft turns, AC voltage is produced.

The voltage that is produced at large utility power stations is generated from three equal fields in the alternator. Thus the generated voltage is *three-phase voltage*. The voltage is sent to transformers at the generating station, where it is stepped up to several hundred thousand volts so it can be transmitted over long distances. When the voltage arrives at a city, it is transformed down (stepped down) to a value of approximately 40,000 V. When the voltage reaches an industrial or commercial area, the voltage is stepped down again to 480 or 240 V. The voltage that is used in residential areas is stepped down to 240 V.

8.2

FREQUENCY OF AC VOLTAGE

The most important feature of AC voltage that is different from DC voltage is that AC voltage has a frequency. The typical frequency for AC voltage in the United States and much of the rest of North America is 60 Hz. The frequency of 60 Hz is determined by the rotating

Frequency = Number of cycles in 1 second

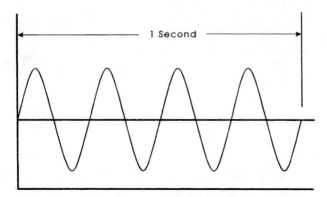

Figure 8–2   The frequency of AC voltage is calculated from the number of cycles that occur in 1 s.

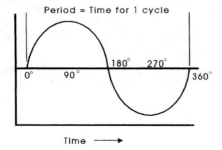

Figure 8–3   The period of AC voltage is calculated as the time it takes for one cycle to occur.

speed of the alternator when the voltage is generated. The frequency of AC voltage that is used to supply voltage for heating and air-conditioning equipment is constant. *Frequency* is defined as the number of cycles (sine waves) that occur in 1 s. Figure 8–2 shows a number of sine waves occurring in 1 second(s).

The *period* of a sine wave is the time it takes one sine wave to start from the zero point and pass through 360°, as seen in Fig. 8–3. A period represents one complete cycle, which is one complete revolution of the alternator. Because the shaft of the alternator rotates one full circle to produce the sine wave, we equate one complete cycle to 360°. Thus the sine wave can be described in terms of 360°. In Fig. 8–3 you can see the sine wave starts at 0°, it reaches a positive peak at 90°, returns to zero at 180°, reaches a negative peak at 270°, and finally returns to 0 at 360°. The number of degrees will be used to identify points of the sine wave in the future discussions. The period of a sine wave can be described as $P = \frac{1}{frequency}$, and frequency can be defined as $F = \frac{1}{period}$.

▶ *Example 8–1*

Determine the period of a 60-Hz sine wave.

Solution:
Use the formula $P = \frac{1}{F}$ (where F = frequency):

$$P = \frac{1}{F} = \frac{1}{60} = 0.016 \text{ s} \quad \blacktriangleleft$$

8.3

PEAK VOLTAGE AND RMS VOLTAGE

Figure 8–4 shows a sine wave, including the point where peak voltage occurs. *Peak voltage* is the highest point of the sine wave. You can see that the peak voltage is 60 V for this example. The negative peak voltage is also identified as -60 V in this example. The total voltage peak to peak (PP) is 120 V. The peak voltage can be measured only with a peak-reading voltmeter or an oscilloscope.

The voltage that you read with a VOM-type meter is called *RMS voltage*. The term *RMS* stands for *root-mean-square*. Root, mean, and square are the mathematical functions used to calculate the voltage that a typical RMS meter reads. The major difference between root-mean-square voltage and peak voltage is that the meter compensates for the RMS voltage being less than peak at various times in any given cycle. The formula for calculating RMS voltage from peak voltage (V_p) is

$$\text{RMS} = 0.707 \times \text{peak voltage}$$

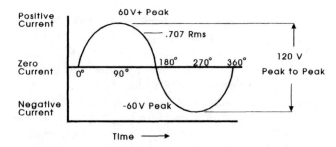

Figure 8–4 Sine wave with positive and negative peaks identified. The peak-to-peak (PP) voltage is also identified.

As you can see, the RMS voltage is approximately 70% of the peak voltage. This means that if the peak voltage is 100 V, the RMS voltage is 70 V. In the example in Fig. 8–4, the peak voltage is 120, so the RMS voltage is 84 V.

▶ *Example 8–2*

Calculate the RMS voltage if the peak voltage is 156 V.

Solution:

$$RMS = 0.707 \times peak$$
$$RMS = 0.707 \times 156$$
$$RMS = 110 \text{ V} \qquad ◀$$

▶ *Example 8–3*

Find the peak voltage if RMS voltage is 120 V.

Solution:

Because RMS = 0.707 × peak voltage, peak = 1/0.707 × RMS = 1.414 × RMS.

$$Peak = 1.414 \times RMS$$
$$Peak = 1.414 \times 120$$
$$Peak = 169.68 \text{ V} \qquad ◀$$

8.4

OSCILLOSCOPE

The oscilloscope provides a way to see the waveform of the voltage you are measuring. A VOM or DVM (digital voltmeter) only provides limited information about an AC voltage. It cannot show the peak voltage or frequency of the sine wave. An oscilloscope can show the waveform and indicate its period (the time for one cycle), and some scopes can show the frequency. The oscilloscope can also show if the voltage has any DC voltage embedded in it or if there is any DC offset voltage. In this new age where you will find digital signals embedded on top of analog voltages, it is important to

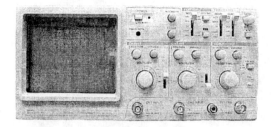

Figure 8–5 The front panel of a typical oscilloscope showing the CRT and the adjustment knobs.

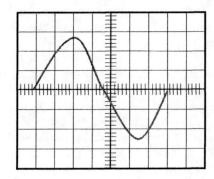

Figure 8–6 A sine wave shown on a CRT screen.

be able to use an oscilloscope to view voltages when problems occur. It will also become necessary to use a scope from time to time when you are troubleshooting variable-frequency drives or electronic circuits. You will also use the oscilloscope when a circuit has two voltages or two waveforms that must be compared. The scope allows you to put both signals on the screen at the same time so you can compare them.

Figure 8–5 shows the front of a basic oscilloscope. On the left side of the scope, a CRT (cathod-ray tube) displays the waveforms just as you would see them on a computer screen or on a television. To the right of the CRT there are adjustment knobs. The knobs across the top control the brilliance and the focus of the signal on the CRT, and under these knobs are the voltage and time controls.

Figure 8–6 shows an example of a sine wave displayed on a CRT. You can see that the CRT has a grid on it. The grid lines are called *gradicules*. The lines that go completely across the screen or completely top to bottom on the screen are the major divisions. Between each major division are four smaller marks called minor divisions. Since there are four minor divisions between each major division, each minor division mark is 0.2 of a division. The horizontal lines allows the measurement of the height of the sine wave, which indicates the peak-to-peak voltage for the sine wave. Figure 8–7 shows that the sine wave from Fig. 8–6 has been moved with the *vertical position knob* so that the bottom of the waveform touches exactly on a full line (major division). The

5.2 Divisions

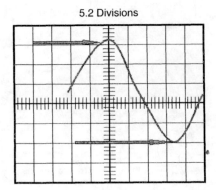

Figure 8–7 Measuring peak-to-peak voltage of a sine wave on the CRT. This sine wave measures 5.2 divisions from top to bottom.

horizontal position knob is used to ensure that the top of the waveform intersects the centerline. Now it is easy to see that the waveform takes up five full (or major) divisions and one minor division. Since each minor division is 0.2 of a major division, the sine wave is exactly 5.2 divisions high. It is important to understand that you can move the sine wave up and down with the *position knob* to make the bottom of the waveform start exactly on a major division, and then you can count whole divisions from the bottom to the top. The remainder of the partial or minor divisions will all be on the top part of the sine wave. Moving the waveform up and down and left and right makes it easier to make an exact measurement, and it does not cause an error in the measurement.

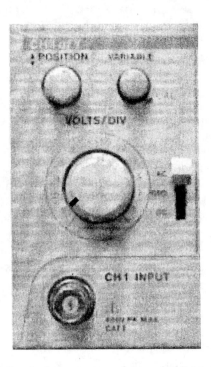

Figure 8–8 A close-up view of the volts/division knob. You can see the Volts/Division knob is set at 5 V.

After you have determined the number of major and minor divisions, you can check where the Volts/Division knob is set. Figure 8–8 shows a close-up view of the Volts/Division knob for the scope which is located on the bottom left set of knobs. You can see the knob is set at 5 Volts/Division. To determine the exact voltage the scope is measuring, we need to multiply the 5.2 divisions by 5 Volts/Division. The total peak-to-peak voltage is 26 V. It is also important to ensure that the *variable knob* is rotated clockwise to the *cal* (calibrate) position, which will lock the knob so the view of the sine wave is accurate. The variable knob allows you to adjust the size of the sine wave if you want to observe one specific part of the sine wave and are not trying to make an accurate measurement.

8.4.1
Converting Peak-to-Peak Voltage to RMS Voltage

Now that you have measured the peak-to-peak voltage, you can calculate the RMS voltage. You should remember that your VOM measures RMS voltage. We will use the formula RMS = $0.707 \times V_p$ to determine the RMS voltage. Since the peak-to-peak voltage is 26 V, we will divide this value by 2 to determine the peak voltage. The peak voltage is 13 V. When we use the calculation $0.707 \times V_p$, the RMS voltage is calculated as 9.19 V.

8.4.2
Measuring Period and Calculating Frequency

Figure 8–9 shows the same sine wave displayed on the CRT. This time we are going to measure the size of the sine wave from left to right, which will show us the time it takes for one sine wave to occur. You will remember from earlier in this chapter that the time for one sine wave to occur is also called the period, and we can convert the period to frequency by inverting the value. This time we will use the horizontal position knob to move the waveform so that the left side of the wave starts exactly on a major division. From the figure you can see that the sine wave takes up eight full (or major) divisions and three minor divisions (left to right). Since each minor division is 0.2 of a major division, the sine wave is exactly 8.6 divisions. After you have determined the number of major and minor divisions, you can check where the time/division knob is set. Figure 8–10 shows the horizontal controls, and they are used to measure the time. Figure 8–11 shows a close-up view of the time/

8.6 Divisions

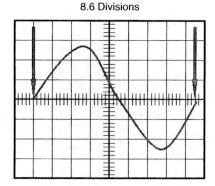

Figure 8–9 The waveform is moved so that its left side starts exactly on a major division. You can see the waveform is exactly 8.6 divisions wide.

Figure 8–10 The horizontal controls for the oscilloscope.

Figure 8–11 A close-up view of the time/division knob on the scope. You can see that the knob is set at 0.2 ms per division.

division knob, which, as you can see, is set on 0.2 ms, or *F*- 0.2 millisecond. The period for this sine wave is 17.2 ms. We will use the formula 1/*P* to find the frequency. In our example, the frequency for this sine wave is 58.1 Hz.

8.5

THE SOURCE OF AC VOLTAGE FOR FACTORY ELECTRICAL SYSTEMS

When you are working on a job in a factory or in a commercial building, you will need to locate the source of incoming voltage. The source of incoming voltage at these locations is typically a *circuit breaker panel*. The circuit breaker panel is also called a *load center*, and it is shown in Fig. 8–12. The load center in a larger commercial building is typically three-phase, and the load center in a smaller commercial building is typically single-phase.

Voltage from the load center is sent to a *disconnect box*, which is mounted near the equipment on which you are working. Figure 8–13 shows a picture of a disconnect

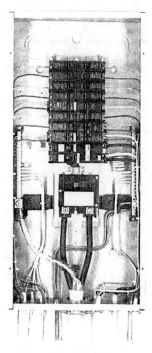

Figure 8–12 A typical load center. (Courtesy of Eaton Electrical)

Figure 8–13 A fused disconnect that is mounted near the equipment. (Courtesy of Eaton Electrical)

box. If the disconnect box has fuses, it is called a *fused disconnect*. The disconnect is used to turn power off to the unit when you are working on the equipment. The fuses in the disconnect are generally cartridge-type fuses that provide protection against too much current for the equipment that is connected to them.

SAFETY NOTICE!
You should always use a lockout, tag-out procedure when you are working on a machine. This procedure includes placing your personal padlock on the disconnect after it is turned to the off position. The padlock will also have a card with it that has your name and picture on it so that others will know you are working on this system. The other part of the lockout, tag-out procedure includes dropping all hydraulic and pneumatic pressure to zero and using the proper blocks on vertical and horizontal dies to ensure that you are not in danger of being pinched in a machine.

8.6
MEASURING AC VOLTAGE IN A DISCONNECT

When you are installing or troubleshooting equipment in the field, you will need to measure the amount of AC voltage available at the load center or at the disconnect. You must be extremely careful when you take these readings because full voltage will be applied to the system at the time you are taking the measurement. Figure 8–14 shows the points in the disconnect where you should place the meter probes to measure high voltage and low voltage. Figure 8–14a shows the points to place the meter leads when you are reading voltage line to line (L1 to L2). From this figure you can see that the meter probes are placed on the line 1 (L1) and line 2 (L2) incoming terminals, and the applied voltage will be 240 V if you are in a residence. If the fused disconnect is located in a commercial location, you may measure 208 V when you place the meter on L1 and L2 or you may measure 480 V. These terminals are called the *line-side terminals*.

Figure 8–14 shows the meter probes on line 1 (L1) and neutral. The meter should measure 120 V if the disconnect is in a residence. This is the lower voltage, which is used on smaller equipment. You may measure a higher voltage, such as 208 V or 240 V, in the disconnect in some cases.

The previous voltage measurements are made at the line-side terminals of the disconnect. You will need to make a second set of measurements at the *load-side terminals*, located at the bottom of the disconnect. If you have voltage at the line-side but not at the load-side ter-

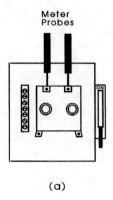

(a)

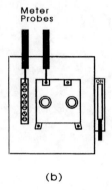

(b)

Figure 8–14 (a) Reading 240 V on the line side of a fusible disconnect; (b) reading 120 V on the line side of a fusible disconnect.

minals, the disconnect may have a blown fuse. In some cases you can determine that the fuse is blown if you can measure full voltage across the top and bottom of a fuse. The reason you can read full voltage across the fuse is that it has an open, and you are reading voltage that backfeeds from the bottom of the fuse through a load such as a transformer winding to the other supply voltage line. If you suspect one or both of the fuses are faulty, you can remove them and check them for continuity.

8.7
VOLTAGE AND CURRENT IN AC CIRCUITS

Voltage waveforms and current waveforms are in phase in an AC circuit that has only resistance in it. This means that both the waveforms start at the same point in time. If the load in the circuit is a resistance heating element, the voltage waveform and current waveform are in phase with each other. If a capacitor or inductor is used in an AC circuit such as a motor circuit, the voltage waveform and current waveform are not in phase.

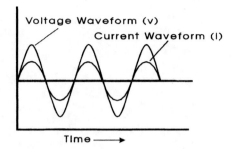

Figure 8–15   Voltage waveform (*v*) and current waveform (*i*) are shown in this example.

Figure 8-15 shows the voltage, identified by the letter *v*, and current waveform, identified by the letter *i*. You can see that these waveforms are in phase, because they start at the same point in time and end each of their cycles at the same time. The voltage in this circuit is larger than the current, so the voltage waveform is larger than the current waveform.

When the voltage and current waveforms are in phase, there are no losses in the circuit due to inductive reactance or capacitive reactance. If an AC circuit has an inductor in it, such as a motor coil, it has *inductive reactance*, and if the circuit has a capacitor in it, such as a start capacitor for a motor, it has *capacitive reactance. Reactance* is an opposition to the AC circuit that is similar to the opposition in a DC circuit caused by resistance. *Inductive reactance* is the opposition caused by the inductor, and *capacitive reactance* is the opposition caused by capacitors. If a circuit has inductors (transformer windings or motor windings) or capacitors, there will be a phase shift between the voltage and current waveforms for that circuit. The amount of phase shift can be defined as a number of degrees, or it can be defined as the total amount of reactance (opposition) in the circuit. As a technician you need to know that reactance exists in these types of circuits and that the effects of reactance can be calculated, predicted, and used to an advantage. For example, the more phase shift that can be created between the start winding and the run winding of an induction motor, the higher starting torque the motor will have. This is useful for helping larger compressor motors to start. The next sections will help you understand the theory about capacitive reactance and inductive reactance. Even though you may never complete reactance calculations when you are troubleshooting a compressor motor, you must comprehend the theory behind capacitive reactance and inductive reactance to understand why additional capacitance may be required to make a single-phase motor start more easily. You will also need a basic understanding of reactance to learn about electronic control circuits that are designed to make pump motors and fan motors run more efficiently.

Figure 8–16   Electrical symbol of a capacitor

8.8
CAPACITORS

A capacitor is an electrical device that has two terminals and has the capability of storing a charge. The capacitor is made of two conducting plates that are separated by an insulator called a dielectric. Figure 8–16 shows the electrical symbol of a typical capacitor, and Figure 8–17a–c shows, respectively, a ceramic, electrolytic, and paper capacitor. If the capacitor is an electrolytic capacitor, one of the terminals is negative and the other is positive. In all other capacitors, the two terminals do not have polarity. The negative terminal is identified by a minus sign, and its terminal looks like a half circle. The positive terminal is identified by a plus sign.

The operation of the capacitor can best be described by applying a DC voltage to it and allowing it to charge and then to discharge. Figure 8–18a shows a capacitor connected in a circuit to a battery power source that supplies DC voltage. The current flow in the circuit is controlled by a single-polo, double-throw switch. Notice that the negative terminal of the battery (DC power source) is connected to the negative terminal of the capacitor when the switch is moved to position A. At first, since the capacitor is not charged, electrons (current) begin to flow from the negative terminal of the battery to the negative terminal of the capacitor when the circuit is complete. It is important to understand that current does not flow through the capacitor, since its two plates are separated by a dielectric, which is a good insulator. Electrons continue to flow until the voltage (charge) on the capacitor plate is equal to the battery voltage. At this point, there is no longer a difference of potential between the capacitor and the battery, so current flow is stopped. Since the dielectric between the positive and negative plate of the capacitor is a good insulator, theoretically it will remain charged indefinitely when the switch is opened and the capacitor is disconnected from the battery.

Figure 8–18b shows the switch moved to position B so that the two terminals of the capacitor are connected to each other. Since there is a large potential difference between the positive plate and negative plate of the capacitor, the electrons will flow from the negative plate to the positive plate until the charge on each plate is equal, and the capacitor is considered discharged. At

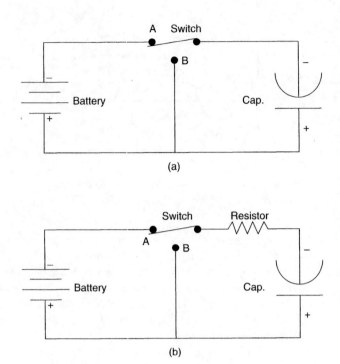

(a)

(b)

Figure 8–18 (a) An electrical diagram of a capacitor charging circuit; (b) capacitor charging circuit with a resistor in series with the capacitor.

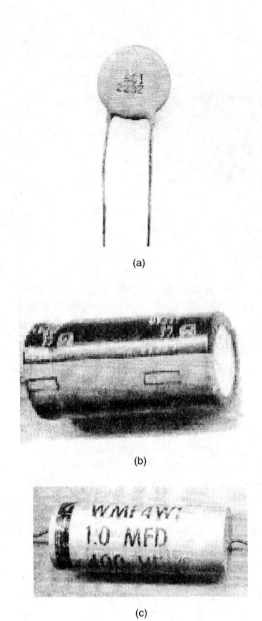

(a)

(b)

(c)

Figure 8–17 (a) Typical round ceramic capacitor; (b) typical electrolytic capacitor; (c) typical paper capacitor.

energy (charge) the capacitor can store on its plates. A farad is equal to 1 ampere of current flowing with a voltage of 1 volt in 1 second. Technically, since we can measure both the positive and negative charge, the term *coulomb* is generally used instead of current since coulomb refers to either the positive or negative charge:

$$Q = C \times V$$

where Q = charge in coulombs
          C = size of capacitance
          V = voltage

It is also important to note that the farad is a relatively large unit, and it would require an electrolytic capacitor to have plates that are approximately the size of a football field. For that reason, the typical size of most capacitors is in the range of a millionth of a farad, called a microfarad. Smaller capacitors may be measured in picofarads. One picofarad would be equal to 0.000,000,000,001 F. This can also be expressed as 1×10^{-12} F. It is important to understand that a microfarad is abbreviated as μF, where the Greek letter μ stands for *micro*. You should also understand that some motor starting capacitors use the abbreviation MF to indicate microfarad. This is actu-

this point, if the switch is moved to position A again, the capacitor will charge again. When the capacitor is placed in a circuit with DC voltage, the capacitor will become charged and remain charged until the circuit is changed to allow the capacitor to discharge. If the capacitor is placed in a circuit with AC voltage, it will charge and discharge at the frequency of the voltage.

8.8.1
Units of Measure for the Capacitor

The main unit of measure for the capacitor is called the farad (F). This measure indicates how much electrical ally an incorrect abbreviation because, in standard SI units of measurement, 'M' is the abbreviation for *mega* and represents a million. So in this case, even though "MF" is used, the units of the capacitor are still microfarads.

The amount of capacitance in a capacitor depends on one of these three factors: the area of the plates, the distance between the plates, or the dielectric constant of the material between the plates. The amount of capacitance increases as the size of the plates increases, or as the distance between the plates increases, or if the dielectric strength increases. It should be noted that newer materials have recently allowed capacitors up to 4 farads to be constructed. These very large capacitors are used in electric vehicles as storage devices.

8.8.2
Types of Capacitors

Capacitors are identified by the type of material that is used as the dielectric. Most capacitors are sealed to keep out air and moisture and prevent the dielectric from breaking down from contamination. Types of capacitors include electrolytic, film, ceramic, paper, air, mica, and surface mount, which are manufactured into printed circuit boards. The electrolytic capacitor can be used in motor starting circuits.

8.8.3
Ratings for Capacitors

All capacitors have a rating for their working-voltage direct current (or WVDC), which is the voltage the capacitor can handle without destroying itself. There is also a WVAC rating for capacitors if they are used in alternating current (AC) circuits. The capacitance rating is indicated on the side of the capacitor

8.8.4
Time Constant

The amount of time a capacitor uses to charge or discharge can be controlled by placing a resistor in the charging or discharge circuit. The charging/discharging time is a constant, which means it will be the same each time the capacitor is charged or discharged with the same resistor in its circuit. If the size of the resistor or capacitor changes, the amount of time it takes to charge it will change. This also means that the time constant can easily be calculated, and so the time it takes the capacitor to fully charge can be used as a solid-state timer.

For a capacitor and resistor circuit, the time constant is the amount of time in seconds that it takes a capacitor to charge to 63.2% of its full charge. The formula

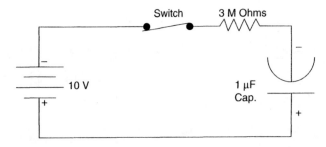

Figure 8–19   Resistor-capacitor charging circuit.

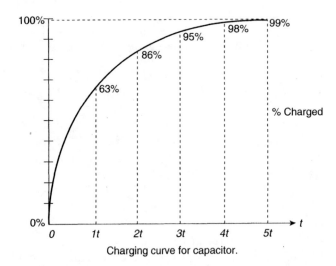

Charging curve for capacitor.

Figure 8–20   The charging curve for a capacitor showing the five time constants needed for a full charge. The capacitor charges 63% of the remaining voltage during each time constant.

for time constant in a resistive capacitive circuit is $T = R \times C$, where time (T) is in seconds, resistor values (R) are in ohms, and capacitance values (C) are in microfarads. Figure 8–19 shows a resistor-capacitor (RC) time-constant circuit where the resistor in the circuit is a 3-MΩ resistor, the capacitor is 1 μF, and the supply voltage is 10 V DC. When the switch is moved to the closed position, current will begin to flow through the resistor into the capacitor. Figure 8–20 shows the curve for the resistor-capacitor charging circuit. The time it will take the capacitor to charge to 6.32 V is 3 s. If the switch remains closed, the capacitor will continue to charge, and during the next 3 s it will charge to 63.2% of the difference between 6.32 V and 10 V. The capacitor will continue to increase its charge by 63.2% of the remaining voltage during each time constant until the capacitor is fully charged. An easier way to think of this is that the capacitor will charge 63.2% during the first time constant, 86.4% for the second time constant, 94.9% for the third time constant, 98% for the fourth time constant, and 99.9% for the fifth time constant. For all practical purposes, the capacitor is considered fully charged when it reaches

RC Time Constant Calculation for 1 Microfarad Capacitor and 3 Megohm Resistor					
	1st TC	**2nd TC**	**3rd TC**	**4th TC**	**5th TC**
Accumulated Time	3 seconds	6 seconds	9 seconds	12 seconds	15 seconds
Applied Voltage	10 Volts	10 Volts	10 Volts	10 Volts	10 Volts
Percentage	63.2%	86.4%	94.9%	98%	99.9%
Calculation	63.2% X 10 V = 6.23 Volts	86.4% X 10 V = 8.64 Volts	94.9% X 10 V = 9.49V	98% X 10 V = 9.8V	99.9% X 10 V = 9.99V
Charged voltage	6.32 Volts	8.64V	9.49V	9.8V	9.9V
Voltage remaining to charge	3.68 Volts	1.35 Volts	.5 Volts	.18 Volts	.07 Volts

Figure 8–21   Table of data for a charging capacitor time constant where the resistor is 3 MΩ and the capacitor is 1 µF.

99.9% charge after the fifth time constant. Figure 8–21 shows a table of the amount of charge during each time constant.

8.8.4.1 Discharging Time Constant

When the capacitor is discharged through a resistor, the same time-constant theory applies. Figure 8–22a shows the resistor-capacitor discharge circuit with the 3-MΩ resistor and the 1-µF capacitor, and Fig. 8–22b shows the discharge curve for this circuit. The rate the capacitor will discharge is 63% during the first time constant, which will leave 37% of the charge remaining on the capacitor. During the second time constant, the capacitor discharges another 63% of the remaining voltage, which leaves 14%; and during the third time constant, the capacitor loses 63% of the remaining voltage, which leaves 5%. During the fourth time constant the capacitor loses 63% of the remaining voltage, which leaves 2%; and during the fifth time constant the capacitor loses 63% of the remaining voltage, which leaves 1%. At this time the capacitor is considered discharged. You should notice that the numbers 37, 14, 5, 2, and 1% are the same ratio as the numbers for the charging time-constant percentages, 63, 86, 95, 98, and 99%.

8.8.5
Capacitors in Series

Capacitors can be connected in series in a circuit, as shown in Fig. 8–23. As you can see in the figure, you are essentially moving the plates of the capacitor farther apart, which decreases the total amount of capacitance. Figure 8–24 shows the formula for solving for the total

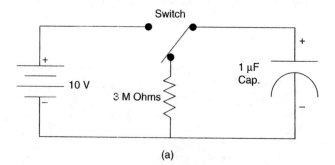

(a)

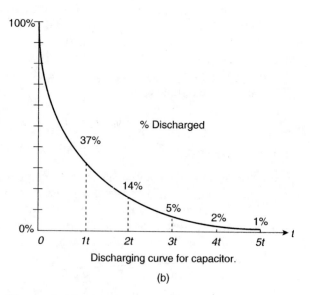

Discharging curve for capacitor.

(b)

Figure 8–22   (a) The discharging circuit for the resistor-capacitor circuit; (b) the discharge curve for a capacitor time-constant circuit.

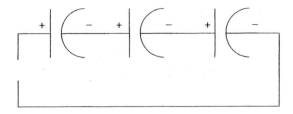

Figure 8–23   Capacitors that are connected in series.

$$C_T = \frac{1}{\frac{1}{C_1} + \frac{1}{C_2} + \frac{1}{C_3} \ldots}$$

Figure 8–24   The formula for calculating capacitors in series.

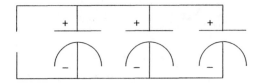

Figure 8–25   Capacitors connected in parallel.

capacitance (C_T) in series. This formula is similar to the formula for solving for the total resistance (R_T) of resistors connected in parallel. Notice in the formula that the three dots (. . .) after the last capacitor means you would continue to add additional capacitors to the formula in the same way as the first three capacitors are shown, so the formula would work for any number of capacitors connected in series.

8.8.6
Capacitors in Parallel

Capacitors can also be connected in parallel. Figure 8–25 shows three capacitors connected in parallel, and you can see that one plate of each capacitor is connected directly to the similar plate of the capacitor next to it. The opposite plates of each capacitor are also connected together. This means that all the negative plates of the capacitors are connected together and all the positive plates are connected together. The total capacitance (C_T) of capacitors connected in parallel is equal to the sum of all the capacitors. Figure 8–26 shows the formula for capacitors in parallel, and you can see it is similar to the formula for resistors in series. Notice in the formula that the three dots (. . .) after the last capacitor

$$C_T = C_1 + C_2 + C_3 \ldots$$

Figure 8–26   Formula for parallel capacitors.

means you would continue to add additional capacitors to the formula in the same way as the first three capacitors are shown, so the formula would work for any number of capacitors connected in parallel.

8.8.7
Testing a Capacitor with an Analog-Type Ohmmeter

A capacitor can be tested for simple operation with an ohmmeter that has a needle movement as its indicator. The ohmmeter has a battery that will provide a small charge for the capacitor, and the needle-type meter movement will indicate current flow as the capacitor charges. You should set the meter to one of the mid-ranges for resistance such as 1 or 10 k. When you first place the meter leads on the capacitor terminals, the meter needle will deflect approximately half scale, and slowly (1–2 s) move back to zero. This shows that current flow starts out large when the capacitor begins to charge, and diminishes to zero as the capacitor reaches the full charge of the battery in the meter. If the meter needle does not move when you place the leads on the capacitor, change the resistance range and repeat the test. You can continue to change the resistance range until the meter needle moves and indicates the capacitor is charging.

After the capacitor charges, you will need to switch the polarity of the meter leads and touch the capacitor again. When you switch the meter leads, it will cause the positive battery voltage to be placed on the capacitor terminal that had just been charged with negative voltage. The capacitor should become charged again, and the meter needle will indicate current flow again. You must change the polarity of the voltage used to charge the capacitor each time you do the test by changing the leads of the ohmmeter.

It is important to understand that this test will not indicate the amount of capacitance the capacitor has. Rather, it will only indicate that the capacitor is operational and has the ability to charge and discharge. You will need to use a meter that has the ability to measure the amount of capacitance. Some digital voltmeters (DVMs) have a built-in circuit to measure the amount of capacitance a capacitor has.

Another way to determine if the capacitor is charging and discharging is to use a voltmeter and measure the voltage each time you allow the ohmmeter battery to charge the capacitor. If you used a lower resistance range, the capacitor will be charged with a small amount of voltage, approximately 1.5 V. If you used the 100-kΩ resistance range, the battery voltage will be between 9 and 30 V. The voltmeter should indicate the amount of DC voltage the capacitor has charged to. Since the voltmeter has extremely large resistance, the capacitor will not discharge, and the voltmeter should indicate the amount of the charge. Again this test does not indicate the amount of capacitance the capacitor has. Rather, it is a very quick test that indicates the capacitor's ability to charge or discharge

8.8.8
Capacitor Failure

Capacitors tend to fail in several ways including shorting and opening. Generally, you will not find a shorted capacitor when you are making tests in the field. The reason for this is that when a short circuit develops between the two conducting plates, it will very quickly draw excessive current, and an open will occur between one of the conducting plates and its terminal. By the time you are called to test the fault, the capacitor will have already gone from shorted to open. When you use the ohmmeter test to check a capacitor with this type of failure, it will not be able to charge or discharge, and the meter needle will not change, regardless of the polarity of the leads.

You may find a capacitor that is developing a small short circuit when you are troubleshooting. If the capacitor is mounted in a metal can or plastic case, it may be bulging slightly as gas builds inside the capacitor as its dielectric begins to break down and turns into a vapor. In this case, it is important to turn off all power to the capacitor and move away from it so it does not cause any bodily harm if it reaches the point where it explodes. After the power is turned off, you can remove the capacitor and replace it with a new one that has the same ratings.

8.9
RESISTANCE AND CAPACITANCE IN AN AC CIRCUIT

In a previous section we learned that when capacitors or inductors are used in an AC circuit, they will create an opposition that is similar to resistance. The opposition caused by a capacitor is called capacitive reactance, and the opposition caused by an inductor is called inductive reactance. The combined opposition caused by capacitive reactance, inductive reactance, and resistance in an AC circuit is called *impedance*. The main difference between capacitive reactance, induc-

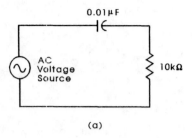

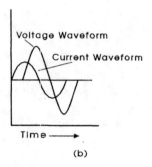

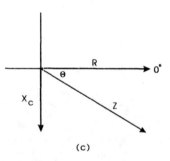

Figure 8–27   (a) Capacitor and resistor in an AC circuit. (b) Waveform of voltage and current for the AC circuit. Notice that the current waveform leads the voltage waveform. (c) Vector diagram that shows the relationship of voltage across the resistor and the voltage across the capacitor.

tive reactance, and resistance is that a phase shift occurs between the voltage waveform and current waveforms. Figure 8–27a shows a diagram of a capacitor in an AC circuit with a resistor. Figure 8–27b shows the voltage and current waveforms for this circuit, and Fig. 8–27c shows the vector diagram that is used to calculate the amount of phase shift for this diagram. A vector diagram is a graph derived from a trigonometry calculation used to determine phase angle.

When voltage encounters a capacitor in a circuit, it will take time to charge up the capacitor and then discharge it. This causes the reactance, which becomes an opposition to the voltage waveform. The capacitor does not bother the current waveform. In Fig. 8–27b you can see that the voltage waveform *lags*, or starts later than, the current waveform for this circuit.

8.10

CALCULATING CAPACITIVE REACTANCE

The total amount of opposition caused by the capacitor is called *capacitive reactance*, X_C. Even though you may never need to calculate capacitive reactance, you should understand the effects of changes in capacitance and frequency on the amount of capacitive reactance. Capacitive reactance can be calculated by the following formula:

$$X_C = \frac{1}{2\pi FC}$$

where $\pi = 3.14$
$F =$ frequency
$C =$ capacitance in microfarads

From this formula you can see that you must know the value of the capacitor and the frequency of the AC voltage. For example, if a 40-μF capacitor is used in a 60-Hz circuit, the amount of opposition (capacitive reactance) is 66.35Ω. Notice that because the capacitive reactance is an opposition, its units are ohms. This calculation is as follows:

$$X_C = \frac{1}{2\pi FC} = \frac{1}{2 \times 3.14 \times 60 \times 0.000040} = 66.35 \ \Omega$$

▶ *Example 8–4*

Calculate the capacitive reactance of a 60-Hz AC circuit that has a 90-μF capacitor.

Solution:

$$X_C = \frac{1}{2\pi FC} - \frac{1}{2 \times 3.14 \times 60 \times 0.000090} = 29.49 \ \Omega \ \blacktriangleleft$$

8.11

CALCULATING THE TOTAL OPPOSITION FOR A CAPACITIVE AND RESISTIVE CIRCUIT

The amount of total opposition (impedance) caused by a capacitor and resistor in an AC circuit can be calcu-

lated. Figure 8–28c shows the diagram that is used to determine the impedance for this type of circuit. For example, if a circuit has 70 Ω of resistance and 40 Ω due to capacitive reactance, the total impedance must be calculated using the distance formula, because the voltage and current in the circuit are out of phase. The formula for calculating impedance of this circuit is

$$Z = \sqrt{R^2 + X_c^2}$$

$$Z = \sqrt{70^2 + 40^2} \qquad Z = 80.62 \ \Omega$$

8.12

RESISTANCE AND INDUCTANCE IN AN AC CIRCUIT

When inductors (coils of wire) are used in an AC circuit, they will create an opposition that is similar to resistance. The opposition caused by an inductor is called *inductive reactance*. The main difference between capacitive reactance, inductive reactance, and resistance is that a phase shift between the voltage and current waveforms occurs. Figure 8–28a shows a diagram of an inductor in an AC circuit with a resistor. Figure 8–28b shows the voltage and current waveforms for this circuit, and Fig. 8–28c shows the vector diagram that is used to calculate the amount of phase shift for the circuit.

When current encounters an inductor in a circuit, it will take time to charge up the inductor and then discharge it. This causes the reactance, which becomes an opposition to the current waveform. The inductor does not bother the voltage waveform. In Fig. 8–28b you can see that the current waveform *lags*, or starts later than, the voltage waveform.

8.13

CALCULATING INDUCTIVE REACTANCE

The total amount of opposition caused by the inductor is called *inductive reactance*, X_L. As in the case of capacitive reactance, inductive reactance can be calculated, and you should remember that even though you do not calculate reactance in the field when you are troubleshooting, it is important that you understand the effect that changing the size of the inductor or changing the frequency has on the total amount of inductive

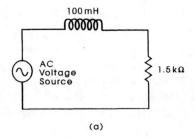

(a)

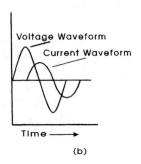

(b)

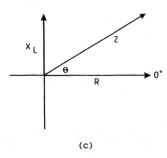

(c)

Figure 8–28   (a) An inductor and resistor in an AC circuit; (b) waveforms for the voltage and current in the inductive and resistive circuit; (c) vector diagram that shows voltage across the inductor leading the voltage across the resistor.

reactance. Inductive reactance can be calculated by the following formula:

$$X_L = 2\pi FL$$

where $\pi = 3.14$
F = frequency
L = inductance in Henries, or H

From this formula you can see that you must know the value of the inductor and the frequency of the AC voltage. For example, if an 80-H inductor is used in a 60-Hz circuit, the amount of opposition (inductive reactance) is 30,144 Ω (30.144 kΩ). Notice that because the inductive reactance is an opposition, its units are ohms. This calculation is as follows:

$$X_L = 2\pi FL = 2 \times 3.14 \times 60 \text{ Hz} \times 80 \text{ H} = 30,144\Omega$$

▶ *Example 8–5*

Calculate the inductive reactance of a 60-Hz AC circuit that has a 20-H inductor.

Solution:

$$X_L = 2\pi FL = 2 \times 3.14 \times 60 \times 20 = 7,536 \ \Omega \quad ◀$$

8.14
CALCULATING THE TOTAL OPPOSITION FOR AN INDUCTIVE AND RESISTIVE CIRCUIT

The amount of total opposition (impedance) caused by an inductor and resistor in an AC circuit can be calculated. Figure 8–28c shows the diagram that is used to determine the impedance for this type of circuit. For example, if a circuit has 70 Ω of resistance and 50 Ω due to inductive reactance, the total impedance must be calculated with a vector diagram, because the voltage and current in this circuit are out of phase. The formula for calculating impedance of this circuit is

$$Z = \sqrt{R^2 + X_L^2}$$
$$Z = \sqrt{70^2 + 50^2}$$
$$Z = 86.02 \ \Omega$$

8.15
TRUE POWER AND APPARENT POWER IN AN AC CIRCUIT

In a DC circuit you could simply multiply voltage times the current and determine the total power of the circuit. In an AC circuit you must account for the current caused by any resistors and calculate it separately from the power caused by resistors and capacitors or resistors and inductors. The reason for this is that when current is caused by a resistor, it is called *true power* (TP), and current caused by capacitive reactance and inductive reactance is called *apparent power* (AP). Apparent power does not take into account the phase shift caused by the capacitor or inductor. The power that occurs when current flows through a resistor is called true power because the resistor does not cause a phase shift between the voltage waveform and the current waveform.

The main point to remember is that true power can be used to determine the heating potential in an AC circuit. Thus if you have 1,000 W due to true power, you can determine that you can get 1,000 W of heating power from this circuit.

8.16

CALCULATING THE POWER FACTOR

The amount of true power and the amount of apparent power in an AC circuit combine to become a ratio called the *power factor* (PF). The formula for power factor is

$$PF = \frac{TP}{AP}$$

▶ **Example 8–6**

Calculate the power factor for a circuit that has 600 VA of apparent power and 500 W of true power. Note that when volts are multiplied by amperes in a reactive circuit, the units for apparent power are VA, which stands for voltamperes.

Solution:

$$PF = \frac{TP}{AP} = \frac{500 \text{ W}}{600 \text{ VA}} = 0.83, \quad \text{or} \quad 83\% \quad ◀$$

The true power in a circuit is always smaller than the apparent power, so the power factor is always less than 1. In a pure resistance circuit, the true power and the apparent power are the same, so the power factor is 1. When the power factor becomes too low, the power company adds a penalty to the electric bill that increases the bill substantially. When a factory has an electric bill of more than $20,000 per month, this penalty becomes important. In these types of applications, the power factor can be corrected by adding extra capacitance to an inductive circuit (having large or multiple motors) or by adding extra inductance to a capacitive circuit. Commercial power factor-correction systems are available for these applications.

QUESTIONS

Short Answer

1. Draw a diagram of an AC sine wave and identify its peak voltage and peak-to-peak voltage levels.

2. Explain the term *frequency* as it refers to AC electricity.

3. Discuss the difference between a load center and a fused disconnect.

4. Explain how you would test for voltage in a fusible disconnect on the line-side and load-side terminals.

5. Trace the path of a generated AC voltage from the generating station where it is produced to the point where you would connect it to a machine in a factory. Be sure to identify all the important parts of the system between the generation station and the disconnect on the machine.

True or False

1. Impedance is the total opposition to an AC circuit caused by inductive reactance, capacitive reactance, and resistance.

2. The units of impedance are ohms.

3. A load center is the point where incoming voltage is connected in a residence and where circuit breakers are mounted.

4. RMS voltage is the voltage your VOM voltmeter measures.

5. If the RMS voltage in an AC circuit is 110 V, the peak voltage is also 110 V.

Multiple Choice

1. If the size of a capacitor in an AC circuit increases from 10 μF to 20 μF and the frequency stays the same, the capacitive reactance will _____.
   a. increase
   b. decrease
   c. remain the same

2. When a capacitor is added to an AC circuit, the voltage waveform will _____ the current waveform, which provides additional torque to a motor if it is connected to the circuit.
   a. lag
   b. lead
   c. remain in phase with

3. If a circuit has two large compressor motors and a low power factor, the power factor can be raised by _____ to make the true power equal to the apparent power, which will avoid a penalty on the electric bill.
   a. adding capacitance
   b. removing capacitance
   c. adding inductance

4. The frequency of AC voltage is defined as _____.
   a. the period of the voltage
   b. the number of cycles per second
   c. the peak voltage of the voltage

5. The number of cycles per second in an AC voltage is called the _____ of the voltage.
   a. frequency
   b. period
   c. impedance

PROBLEMS

1. Calculate the RMS voltage for an AC sine wave that has 80-V peak.

2. Calculate the peak voltage for an AC sine wave that has 208-V RMS.

3. Calculate the period of a 60-cycle AC sine wave.

4. Calculate the frequency of an AC sine wave that has a period of 16 ms.

5. Calculate the impedance of an AC circuit that has 40 Ω of resistance and 30 Ω of capacitive reactance.

CHAPTER 9

Transformers and Three-Phase and Single-Phase Voltage

OBJECTIVES

After reading this chapter you will be able to:

1. Explain the operation of a transformer.

2. Discuss the difference between step-up and step-down transformers.

3. Explain how 480, 240, 120 VAC are developed from a delta-connected transformer.

4. Explain how a 440, 208, and 120 VAC are developed from a wye connected transformer.

9.0

OVERVIEW OF TRANSFORMERS

Transformers are needed to step up or step down voltage levels. For example, when voltage is generated, it needs to be stepped up to several hundred thousand volts so its current will be lower when it is transmitted. Because the current is lower, the size of wire used to transmit the power can be smaller, so the power can be transmitted over longer distances. When the voltage arrives at a city, it needs to be stepped down to approximately 40,000 V so that it is less dangerous. This level of voltage is sufficient to transmit power throughout a city. When the voltage arrives at a factory, it must be further stepped down to 480, 240, or 208 V and 120 V for all the power circuits in the factory, including the office. Transformers provide a means of stepping up or stepping down this voltage. Step-up transformers are used in some industrial applications, such as air cleaners, to increase the line voltage (120 VAC) to several thousand volts. Note that different parts of the country

use slightly different amounts of voltage, such as 440, 220, 208, or 110 V or other similar variations. These other voltage values are determined by the amount of the supply voltage from the power company and the different variations of transformers used. For this text we will try to standardize on 480, 240, 208, and 120 V to avoid confusion.

Another type of transformer is also used in factory electrical systems to step down 240 VAC to 120 VAC. The 120 VAC is used as control voltage for all the controls in the machine systems so that higher voltages are not used in start and stop buttons or other controls that people touch when they are setting them. The lower voltage provides a degree of safety against electrical shock, and it also allows the controls to last longer. This type of transformer is called a *control transformer*, and it is shown in Fig. 9–1.

As a technician you will encounter different voltages, such as 440, 220, 208, and 120 VAC, while you are

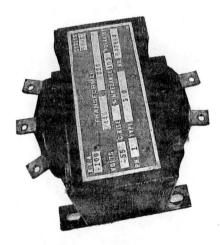

Figure 9–1   An example of a control transformer that is used in control panels of factory equipment.

101

working on electrical systems in factories. The equipment you are working with will require a specific amount of single-phase or three-phase voltage, and you will be responsible for going to the disconnect or load center where supply voltage is provided and making the proper connections to provide the correct voltage. When the proper amount of voltage is not present, you will need to know what transformers are required to step the voltage up or down to the proper level. This chapter explains how transformers operate and how the various levels of voltage are derived from connecting transformers to provide the proper amount of three-phase and single-phase voltage. The chapter also explains how to take voltage readings at the transformer and disconnect to troubleshoot the loss of a phase.

9.1

OPERATION OF A TRANSFORMER AND BASIC MAGNETIC THEORY

The transformer consists of two windings (coils of wire) that are wrapped around a laminated-steel core. The winding where voltage is supplied to a transformer is called the *primary winding*. H1 and H2 are connected to the primary terminals. The winding where voltages come out of the transformer is called the *secondary winding*. The terminals for the secondary winding are identified by the letters X1 and X2. Figure 9–2 shows a diagram of a typical transformer with the primary coil and secondary coil identified. From this figure you can see that the windings are also identified. If the transformer's secondary voltage is 120 V, then the secondary terminals will be identified as X1 and X2.

A transformer works on a principle called *induction*. Induction occurs when current flows through the primary winding and creates a magnetic field. The magnetic field produces flux lines that emanate from the wire in the coil as current flows through it. When

this current is interrupted or stopped, the flux lines collapse, and the action of the collapsing flux lines causes them to pass through the winding of the secondary coil that is placed adjacent to the primary coil. You should remember that the AC sine wave continually starts at 0 V and increases to a peak value and then returns to 0 V and repeats the waveform in the negative direction. The process of increasing to peak and returning to zero provides a means of creating flux lines in the wire as current passes through it and then allowing the flux lines to collapse when the voltage returns to zero. Because the AC voltage follows this pattern naturally 60 times a second, it makes the perfect type of voltage to operate the transformer.

When the AC voltage returns to zero during each half cycle, the flux lines that were created in the primary winding begin to collapse and start to cross the wire that forms the secondary coil of the transformer. When the flux lines from the primary collapse and cross the wire in the coil of the secondary winding, the electrons in the secondary winding begin to move, which creates a current. Because the current in the secondary winding begins to flow without any physical connection to the primary winding, the current in the secondary winding is called an *induced current*.

The following provides a more technical explanation of the relationship between the AC voltage and the windings of the transformer: The magnetic field in the primary winding of the transformer builds up when AC voltage is applied during the first half cycle of the sine wave (0° to 180°). When the sine voltage reaches its peak at 90°, the voltage has peaked in the positive direction, and it begins to return to 0 V by moving from the 90° point to the 180° point. When the voltage reaches the 180° point, the voltage is at 0 V, and it creates the interruption of current flow. When the sine-wave voltage is 0 V, the flux lines that have been built up in the primary winding collapse and cross the secondary coils, which creates a current flow in the secondary winding.

The sine wave continues from 180° to 360°, and the transformer winding is energized with the negative half-cycle of the sine wave. When the sine wave is between 180° and 270°, the flux lines are building again; and when the sine wave reaches the 360° point, the sine wave returns to 0 V, which again interrupts current and causes flux lines to cross the secondary winding. This means the transformer primary energizes a magnetic field and collapses it once in the positive direction and again in the negative direction during each sine wave. Because the sine wave in the secondary is not created until the voltage in the primary moves from 0° to 180°, the sine wave in the secondary is *out of phase* by 180° to the sine wave in the primary that created it. The voltage in the secondary winding is called *induced voltage* because it is created by induction.

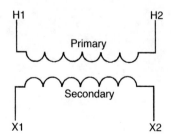

Figure 9–2   Diagram of a typical control transformer. The coil where incoming voltage is applied is called the primary, and the other coil is called the secondary.

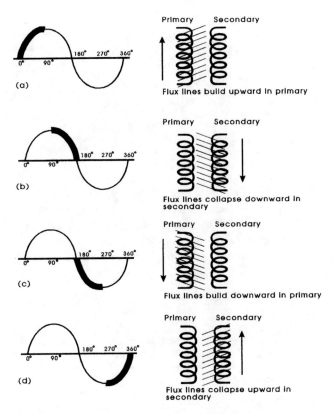

Figure 9–3   (a) AC voltage builds to a positive peak as the sine wave moves from 0° to 90°. Flux lines are shown building in the primary coil during this time. (b) AC voltage drops off from peak at 90° to 180°. During this time the flux lines in the primary coil begin to collapse and cut across the coils of the secondary. (c) AC voltage builds to a negative peak as the sine wave moves from 180° to 270°, and the flux lines build in the primary coil again. (d) When voltage decreases from the negative peak back to zero as the sine wave moves from 270° to 360°, the flux lines collapse.

The induced voltage is developed even though there is total isolation between the primary and secondary coils of the transformer. Figure 9–3 shows the four stages of voltage building up in the transformer and the flux lines building and collapsing at each point as the sine wave flows through the primary coil.

9.2

CONNECTING A TRANSFORMER TO A DISCONNECT FOR TESTING

An easy way to test a control transformer is to connect it directly to a disconnect and apply power to the

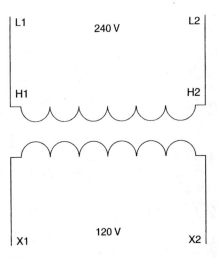

Figure 9–4   A control transformer connected to L1 and L2 terminals in a fusible disconnect. The primary voltage is 240 V, and the secondary is 120 V.

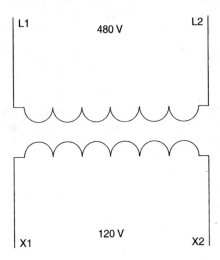

Figure 9–5   A control transformer connected to L1 and L2 in a disconnect for 480 V to test the transformer. The primary voltage is 480 VAC, and the secondary voltage is 120 VAC.

transformer primary circuit. When voltage is applied to the primary circuit, voltage should be available at the secondary. The amount of voltage at the secondary will depend on the rating of the transformer. Figure 9–4 shows a transformer connected to L1 and L2, which is the source for 240 V at the disconnect. The secondary voltage available at this transformer at terminals X1 and X2 is approximately 120 V. Figure 9–5 shows a similar transformer connected to L1 and L2 in a disconnect to provide 480 V to the primary circuit. Because this transformer is rated as 480/120, the secondary voltage is also 120 V.

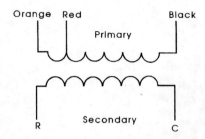

Figure 9–6 A control transformer that is made to operate with either 208-VAC or 240-VAC primary voltage and provide 24-VAC secondary voltage. 24 volts is primarily used in air conditioning and Furnaces.

9.3

TRANSFORMER RATED FOR 240 V AND 208 V PRIMARY

If the equipment is used in the office part of a factory application, such as heating or air conditioning, the supply voltage may be 240 or 208 V, depending on the transformer connections used by the utility company that supplies the power. The secondary voltage for the thermostat will be stepped down to 24 VAC. Because the equipment manufacturers do not know where their equipment will be used, they may provide a control transformer whose primary side can be wired to either 240 V or 208 V and whose secondary side will produce 24 V. This is accomplished by providing a second connection that is *tapped* in the primary winding. Figure 9–6 shows an example of this type of control transformer.

9.4

TRANSFORMER VOLTAGE, CURRENT, AND TURNS RATIOS

The amount of voltage a transformer will produce at its secondary winding for a given amount of voltage supplied to its primary is determined by the ratio of the number of turns in the primary winding to the number of turns in the secondary. This ratio is called the *turns ratio*. The amount of primary current and secondary current in a transformer also depends on the turns ratio.

Figure 9–7 shows a diagram that indicates the primary voltage (E_p), the primary current (I_p), the number of turns in the primary winding (T_p), the secondary voltage (E_s), the secondary current (I_s), and the number of turns in the secondary winding (T_s).

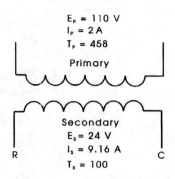

Figure 9–7 Diagram of a transformer that shows the primary voltage (E_p), the primary current (I_p), and the number of turns in the primary winding (T_p). This diagram also shows the secondary voltage (E_s), the secondary current (I_s), and the number of turns in the secondary (T_s).

Table 9–1

Turns Primary/ Secondary	Calculating Voltage	Calculating Current
$T_p = \dfrac{I_s \times T_s}{I_p}$	$E_p = \dfrac{E_s \times I_p}{T_s}$	$I_p = \dfrac{I_s \times T_s}{T_p}$
$T_p = \dfrac{E_p \times T_s}{E_s}$	$E_p = \dfrac{E_s \times I_s}{I_p}$	$I_p = \dfrac{E_s \times I_s}{E_p}$
$T_s = \dfrac{I_p \times I_p}{I_s}$	$E_s = \dfrac{E_p \times T_s}{T_p}$	$I_s = \dfrac{I_p \times T_p}{T_s}$
$T_s = \dfrac{E_s \times T_p}{E_p}$	$E_s = \dfrac{E_p \times I_p}{I_s}$	$I_s = \dfrac{E_p \times I_p}{E_s}$

The primary and secondary voltage, the primary and secondary current, and the turns ratio can all be calculated from formulas. See Table 9–1.

The turns ratio for a transformer is calculated from the following formulas:

$$\text{Turns ratio} = \frac{T_s}{T_p} \quad \text{or} \quad \frac{E_s}{E_p} \quad \text{or} \quad \frac{I_p}{I_s}$$

The ratio of primary voltage, secondary voltage, primary turns, and secondary turns is

$$\frac{E_p}{E_s} = \frac{T_p}{T_s}$$

From this ratio you can calculate the secondary voltage with the formula

$$E_s = \frac{E_p \times T_s}{T_p}$$

The ratio of primary voltage, secondary voltage, primary current, and secondary current is

$$\frac{E_p}{E_s} = \frac{I_s}{I_p}$$

(Notice that the ratio of voltage to current is an inverse ratio.) Using this ratio you can calculate the secondary current:

$$I_s = \frac{E_p \times I_p}{E_s}$$

If you know the primary voltage is 110 V, the primary current is 2 A, the primary turns are 458, and the secondary turns are 100, you can easily calculate the secondary voltage and the secondary current using these formulas:

$$E_s = \frac{E_p \times T_s}{T_p}$$

$$\frac{110 \text{ V} \times 110 \text{ turns}}{458 \text{ turns}} = 24 \text{ V}$$

$$I_s = \frac{E_p \times I_p}{E_s}$$

$$\frac{110 \text{ V} \times 2 \text{ A}}{24 \text{ V}} = 9.16 \text{ A}$$

▶ **Example 9–1**

Calculate the secondary current and secondary turns of the transformer shown in Fig. 9–8.

Solution:
Find the turns ratio:

$$\frac{E_p}{E_s} = \frac{240 \text{ V}}{24 \text{ V}} = 10{:}1 \qquad \blacktriangleleft$$

Because the number of turns in the primary is 458, the number of turns in the secondary will be 45.8.

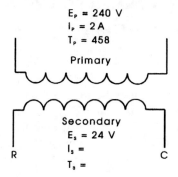

E_P = 240 V
I_P = 2 A
T_P = 458
Primary

Secondary
E_s = 24 V
I_s =
T_s =

R   C

Figure 9–8   Diagram for Example 9–1.

The secondary current can be calculated from the following formula:

$$I_s = \frac{E_p \times I_p}{E_s}$$

$$\frac{240 \text{ V} \times 2 \text{ A}}{24 \text{ V}} = 20 \text{ A}$$

9.5
STEP-UP AND STEP-DOWN TRANSFORMERS

If the secondary voltage is larger than the primary voltage, the transformer is known as a *step-up transformer*. If the secondary voltage is smaller than the primary voltage, the transformer is known as a *step-down transformer*. The control transformer that was presented in the previous sections is an example of a step-down transformer because the secondary voltage is smaller than the primary voltage. Step-up transformers are used to boost the secondary voltage, which also has the effect of lowering the secondary current so that more power can be transferred on smaller-size wire. Some industry applications where a step-up transformer is used include electrostatic air cleaners and the ignition system of an oil furnace. The oil furnace uses an ignition transformer that is a step-up transformer to increase the 120 VAC used to power the furnace to approximately 16,000 V. The high voltage is supplied to the igniter points so that the high voltage can jump the gap between the ignition points, creating a spark that ignites the fuel oil as it comes out of the nozzle.

9.6
VA RATINGS FOR TRANSFORMERS

The VA (voltampere) rating for a transformer is calculated by multiplying the primary voltage and the primary current or the secondary voltage and secondary current. In the example in Fig. 9–8 you can see that the primary voltage is 240 V and the primary current is 2 A. The VA rating for this transformer is 480 VA. The VA rating indicates how much power the transformer can provide. As a technician you can say that the primary VA is equal to the secondary VA. If you were a circuit designer, you would need to be more precise; you would find that the secondary VA is slightly less than the primary VA because of transformer losses. It is important that the VA rating of the transformer be large enough for an application. If the VA rating is too small, the transformer will be damaged and fail prematurely. Any time you need to replace a transformer, you must be sure that the voltage ratings match and that the VA size of the replacement transformer is equal to or larger than that of the original transformer.

9.7
THE 120-VAC CONTROL TRANSFORMER

The control transformer provides 120 VAC in the secondary for use in factory-control systems. You should remember that utility companies in different areas of the country provide voltage that is between 110 and 120 V. This means that one part of the country may refer to its lower voltage as 110, whereas another part calls its voltage 115 or 120 V. For this section the voltage will be referred to as 120 VAC. The 120 VAC secondary voltage is necessary because some of the control devices require 120 VAC, and the larger loads in the system, such as compressors and fan motors, require 240 VAC. The control transformer allows the larger loads in a system, such as compressor motors, pump motors, and fan motors, to be supplied with 240 V while providing the control circuit with 120 VAC.

The 240/120-VAC type of control transformer operates on a similar principle as the 120/24-VAC transformer. The diagram for this transformer is also provided in Fig. 9–9. From this diagram you can see that the primary windings of this transformer are identified with the letter H and the secondary windings are identified with the letter X. The primary winding for this transformer is constructed in two equal sections (windings). The terminals of the first winding are identified as

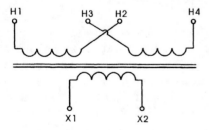

Figure 9–9 Diagram of a control transformer whose secondary voltage is 120 V. The primary windings of this transformer are identified as H1, H2, H3, and H4. The secondary of this transformer is identified as X1 and X2.

H1 and H2. The terminals of the second winding are identified as H3 and H4. The primary winding is broken into two separate windings so that the transformer primary side can be powered with 480 V or 240 V. If these two sections are connected in parallel with each other, the primary side of the transformer will be powered with 480 V; and if the two sections are connected in series with each other, the transformer will be powered with 240 V.

9.8
WIRING THE CONTROL TRANSFORMER FOR 480-VAC PRIMARY VOLTS

The control transformer can also be powered with 480 VAC by connecting its two primary windings in series. Figure 9–10 shows three diagrams of the control transformer. Figure 9–10(a) shows the transformer as you would see it on a wiring diagram without any jumpers connected. Notice that the H2 and H3 terminals are positioned so that they can be connected in either series or parallel. In this application the jumper is attached between terminals H2 and H3 to connect them in series. Figure 9–10b shows the physical location of the jumper when it is connected across these terminals. Figure 9–10c shows the equivalent electrical diagram. The two windings are connected in series because each winding is rated for 240 V, and the primary voltage is 480 VAC. Because the windings are connected in series, each of them will receive 240 V of the total 480 V.

9.9
WIRING THE CONTROL TRANSFORMER FOR 240-VAC PRIMARY VOLTS

The control transformer can also be connected to operate with 240-VAC primary volts and produce 120 VAC on the secondary terminals. Figure 9–11 shows a set of three diagrams for the same control transformer that is connected for 240-VAC primary. Figure 9–11a shows an electrical diagram of the control transformer as you would see it in a circuit diagram without any jumpers. The reason this diagram is shown is that some equipment manufacturers will show the control transformer

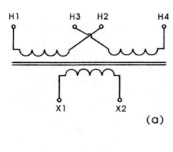

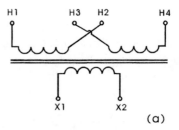

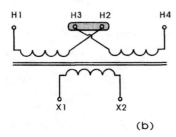

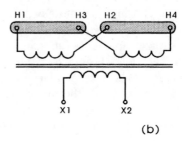

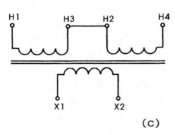

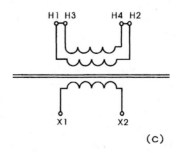

Figure 9–10 (a) Electrical diagram of the control transformer. Notice that the H2 and H3 terminals are located so that a jumper can be placed on them to connect them in series or parallel. (b) Diagram showing the jumper in place connecting the two primary windings in series for 480-VAC primary. (c) The equivalent electrical diagram that shows the two windings connected in series.

Figure 9–11 (a) Electrical diagram of a control transformer; (b) diagram that shows the jumpers in place for the transformer to have 240-VAC primary; (c) equivalent diagram that shows the two primary windings connected in parallel.

in the electrical diagram for the system since they are not sure whether the system voltage will be 480 or 240. This diagram indicates that the control transformer is to be connected by the technician when the unit is installed and the supply voltage can be verified. Figure 9–11b shows the diagram with two jumpers in place to connect the two primary windings in parallel with each other. This diagram will help you connect the jumpers to the correct terminals so that the windings are connected in parallel. Figure 9–11c shows the equivalent circuit of the two windings in parallel as you would see them in the system's electrical diagram provided by the equipment manufacturer. Sometimes it is difficult to see that the connections of the two jumpers actually connect the two windings in parallel, so the diagram in Fig. 9–11c shows how the windings look when they are in parallel.

As a technician you will be responsible for making the proper jumper connections based on the primary voltage that the system will have. If the jumpers on the transformer are connected for 480-VAC primary when the unit is shipped and the unit is connected to 240 V supply voltage, you will need to make the changes in the jumper so that the secondary winding of the transformer provides exactly 120 VAC.

9.10

TROUBLESHOOTING A TRANSFORMER

As a technician you will need to troubleshoot a variety of transformers. If the transformer you are testing is not connected to power, you can test each of the windings with an ohmmeter for continuity. Each of the coils should have some amount of resistance that indicates the amount of resistance in the wire in each coil. If you measure infinite resistance (∞), it indicates the winding has an open, and if you measure 0 Ω, it indicates one of the windings is shorted.

Another way that you can test a transformer is by applying power to the primary winding. It is important that you provide the continuity test first to detect any shorts before you apply power. If the transformer windings are not shorted, you can apply any amount of AC voltage that is equal to or less than the primary rating of the transformer. If the transformer is operational, a voltage will be present at the secondary terminals of the transformer. If no voltage is present at the secondary, be sure to check all the connections to ensure they are correct. If no voltage is present at the secondary, the transformer is defective and should be replaced. It is important to remember that the transformer has a primary winding and a secondary winding that are placed in close proximity to each other, when AC voltage is applied to one winding, induction will cause voltage to be available at the other winding.

The transformer may have a problem where its secondary voltage is higher or lower than its rating. If this occurs, it is possible that the amount of primary voltage is incorrect or that the transformer jumpers are not connected properly. For example, if the primary voltage is 208 V instead of 240 V, the secondary voltage will be less than 120 V, which will cause a problem. Be sure that the secondary is the proper amount before you use the transformer in a circuit.

9.11
NATURE OF THREE-PHASE VOLTAGE

Industrial equipment installed in factories may require three-phase AC voltage or single-phase voltage. Three-

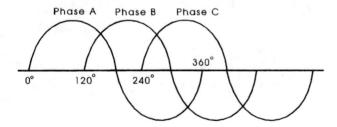

Figure 9–12   Sine waves for three-phase electricity. Notice that phase A is 120° out of phase with phase B, and phase B is 120° out of phase with phase C.

phase voltage is usually indicated on equipment data plates by the symbol 3φ voltage. Three-phase voltage is generated as three separate AC sine waves. Figure 9–12 shows an example of the three sine waves. From this figure you can see that the sine waves are identified as phase A, phase B, and phase C. All the phases are separated by 120° from each other. The 120° phase separation is caused by the generator windings being 120° out of phase with each other. Three-phase voltage is generated and transmitted because it is more efficient to produce than single-phase since the generator can have three separate windings.

9.12
WHY THREE-PHASE VOLTAGE IS GENERATED

Three-phase voltage is more useful than single-phase voltage for larger equipment and systems because it provides more power than single-phase voltage for an equal-size system. If a 10-hp motor is required as the hydraulic pump motor, a single-phase motor would need to be larger than the 10-hp three-phase motor. The three-phase motor can be smaller because it has three sets of windings and it receives three equal sources of voltage (L1, L2, and L3), whereas a single-phase motor receives only two sources of power. Because the power is shared by three circuits instead of two, the wire sizes, fuse sizes, and switch sizes are also smaller for a three-phase system.

For example, if a 10-hp motor is connected to a single-phase 240 V power source, each line (H1 and H2) would need to be capable of providing 50 A. A three-phase 10-hp motor connected to a three-phase power distribution would need only 28 A from each wire. This means that the three-phase system could use much smaller wire, which would be lighter and less expensive.

9.13
THREE-PHASE TRANSFORMERS

Figure 9–13 shows a picture of a typical three-phase transformer you will encounter on the job. The transformer may be mounted near the equipment, or it may be located in a transformer vault (a special room where transformers are mounted). In some cases the transformers are mounted on a utility pole just outside of the commercial site. Figure 9–14 shows the three-phase

Figure 9–13 A typical three-phase transformer with its cover in place. (Courtesy of Acme Electric Corp.)

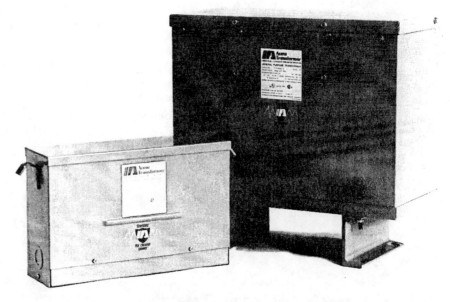

Figure 9–14 A typical three-phase transformer with its cover removed so that you can see the three independent transformers and connection terminals. (Courtesy of Acme Electric Corp.)

transformer with its cover removed so you can see three separate transformer windings. The three-phase transformer operates exactly like a single-phase transformer in that AC voltage is applied to the primary side of the transformer and induction causes voltage to be created in the secondary winding. In the case of the three-phase transformer, the 120° phase shift of the three-phase voltage applied to the primary windings will be maintained in the secondary side of the transformer.

When you are working on an industrial system, you may need to make connections between the transformer and the disconnect box or between the transformer and a load center. You may also need to make a voltage test at the terminals of the transformer. These connections will be made at the terminal strip of the transformer, as shown in Fig. 9–14. The three-phase transformer is essentially three single-phase transformers that are connected together in a *wye configuration* or *delta configuration*. The picture in Fig. 9–14 shows the three separate transformers. In most cases, you will not be requested to make the physical connections at the transformer, but you must understand the amount of voltage that will be available when a transformer is connected in a wye configuration or in a delta configuration. Because the primary voltage may be rather high (1,300–1,700 VAC), the electrical technicians for the electric utility company or technicians from a high-voltage service company will make the primary voltage connections for the transformer and connect the secondary to a disconnect box or a load center. You will be expected to make the connections for the equipment at the disconnect or at the load center.

9.14
THE WYE-CONNECTED THREE-PHASE TRANSFORMER

Figure 9–15 shows a diagram of a wye-connected three-phase transformer. Notice that the physical shape of the transformer windings looks like the letter *Y*, which gives this type of configuration its name. Also notice that this diagram shows only the secondary coils of the transformer. It is traditional to show only the primary-side connections or only the secondary-side connections when discussing a transformer power-distribution system, because showing both the primary and secondary connections in the same diagram tends to become confusing.

The amount of voltage measured at the L1–L2, L2–L3, and L3–L1 terminals on the secondary winding will be 208 V for the wye-connected transformer if its turns ratio is set for low voltage. If the turns ratio of the transformers is set for high voltage, the amount of volt-

age between each terminal will be 480 V. The voltage indicated between each of the windings for the transformer shown in Fig. 9–15 is 208 V, which indicates the turns ratio for this transformer will provide the lower voltage. If 480 V is needed, a transformer with a higher turns ratio must be used. The primary and secondary voltages for each transformer are provided on its data plate and can be specified when the transformer is purchased and installed.

9.15
THE DELTA-CONNECTED THREE-PHASE TRANSFORMER

Figure 9–16 shows a diagram of a delta-connected three-phase transformer. You should notice that the shape of the transformer windings in this diagram looks like a triangle (Δ). This shape is the Greek letter *D*, which is named *delta*. This diagram also shows only the secondary side of the three-phase transformers.

The amount of voltage measured at L1–L2, L2–L3, and L3–L1 is 240 V for the delta-connected transformer if it is wired for its lower voltage. If the delta-connected transformer is wired for its higher voltage, the voltage between L1 and L2, L2 and L3, and L3 and L1 is 480 V. If the delta-connected transformer is wired for its lower voltage, it is very easy to differentiate it from a wye-connected transformer. If the delta-connected transformer and wye-connected transformer are connected for their higher voltage, you cannot tell them apart, because both will provide 480 V between their terminals.

It is important to understand at this point that you will basically purchase the equipment so that its voltage requirements match the voltage supplied by the transformers at the commercial or industrial site. This means

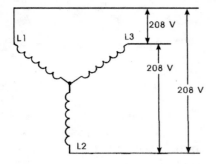

Figure 9–15   A diagram that shows the secondary windings of a three-phase transformer connected in a wye configuration. The voltage available at L1–L2, L2–L3, and L3–L1 is 208 V for this transformer.

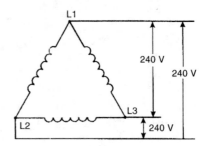

Figure 9–16   Electrical diagram of the secondary windings of a transformer connected in a delta configuration. Notice that the voltage between L1 and L2, L2 and L3, and L3 and L1 is 240 V.

that you do not need to determine if the transformer is connected as in a delta configuration or a wye configuration. You will need only to measure the amount of voltage at the secondary of the transformer, if it is 480 V, you will need to ensure that the system is rated for 480 V. If the secondary voltage is 240 V, the system must be rated for 240; and if it is 208 V, the equipment must be rated for 208.

If you need to know if the transformer windings are connected as delta or wye, you can check the physical connections, or you can make an additional voltage measurement between each line and the neutral terminal of the transformer if one is provided. The next section explains these measurements.

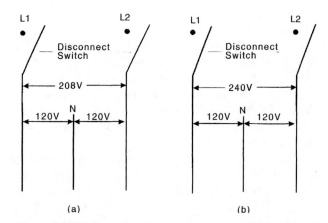

Figure 9–18   (a) Voltage from L1–L2 for a wye-connected transformer is 208 V and from L1–N or L2–N is 120 V; (b) voltage from L1–L2 for a delta-connected transformer is 240 V and from L1–N or L2–N is 120 V.

9.16

Delta- and Wye-Connected Transformers with a Neutral Terminal

Figure 9–17 shows the diagrams of a wye-connected transformer and delta-connected transformer, each with a neutral terminal. The neutral terminal on the wye-connected transformer is at the point where each of the three ends of each individual winding are connected together. This point is called the *wye point.*

Notice that the neutral point for the delta-connected transformer is actually the midpoint of the secondary winding connected between L1 and L2. This point is essentially the center tap of one of the transformer windings. Traditionally it will be the winding that is connected between L1 and L2 if a neutral connection is used with the three-phase transformer.

Figure 9–18a shows the amount of voltage that you would measure between terminals L1 and L2 of a wye-connected transformer and between terminals L1 and N and L2 and N. Notice that the voltage between L1 and L2 is 208 V, so this transformer is wired for the lower voltage. The voltage between L1 and N and L2 and N is shown as 120 V. Figure 9–18b shows the amount of voltage that you would measure between terminals L1 and L2 and between L1–N and L2–N for a delta-connected

transformer. The voltage between L1 and L2 is 240 V, so this transformer is connected for its lower voltage. The voltage between L1 and N and L2 and N is shown as 120 V. Because the neutral point for the delta-connected transformer is exactly halfway on the transformer winding, the voltages L1–N and L2–N will always be exactly half the voltage between L1 and L2.

This is the main difference between a wye-connected transformer and a delta-connected transformer. The voltage between L1 and N and L2 and N for any delta-connected transformer is always exactly half that of L1–L2, but the L1–N or L2–N voltage for a wye-connected transformer will always be more than half the voltage L1–L2. The exact amount of voltage L1–N can be calculated by dividing the voltage between L1 and L2 by 1.73, which is the square root of 3 ($\sqrt{3} = 1.73$). The square root of 3 is used because of the relationship between the phase shift of the three phases. Notice that 208 divided by 1.73 is 120 V.

The same relationship of voltage between L1 and L2 and L1 and N exists when the transformers are wired for their higher voltage. Figure 9–19a shows that a wye-connected voltage between L1 and L2 is 480 V and that between L1 and N is 277 V. The 277 V can be calculated by dividing 480 by 1.73. The 277 V that comes from L1–N

Figure 9–17   Electrical diagram of a wye-connected transformer with a neutral point and a delta-connected transformer with a neutral point.

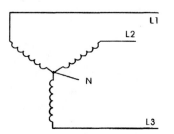

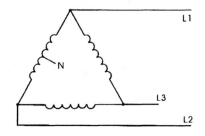

or L2–N is generally used for fluorescent lighting systems in commercial buildings. Because the supply voltage originates from a three-phase transformer, the lighting system in a commercial or industrial building will also use L3–N, so that voltage is used from all three legs of the transformer. This voltage is referred to as single-phase voltage because only one line of the three-phase transformer is used for each circuit. For example, L1–N is a single-phase circuit.

It is also important to understand at this time that L1–L2, L2–L3, and L3–L1 can each be used to supply complete power for a machine's electrical system. Even though this type of power supply uses two legs of the transformer, it is still called a single-phase power supply because only one phase of voltage is used at any instant in time. For example, if the system is powered with voltage from L1–L2, during any given half-cycle of AC voltage, the power source for the system would come from L1, and then during the next half-cycle it would come from L2. The source of power would continue to oscillate between L1 and L2, but only one phase is in use at a time.

Figure 9–19b shows that the higher voltage for the delta system between L1 and L2 is also 480 V, but this time the voltage between L1 and N or L2 and N is 240 V. Again the L1–N or L2–N voltage is exactly half of the supply voltage, because the neutral point on the transformer is the center tap of one of the transformers. It is important to understand that it is very easy to distinguish between a wye-connected power source and a delta-connected power source by measuring the voltage L1–L2 and L1–N. If the L1–N voltage is half the L1–L2 voltage, the system is a delta-connected system. If the

L1–N voltage is more than half, it is a wye-connected system. You should remember that the L1–N wye voltage can always be calculated by dividing the L1–L2 voltage by 1.73.

9.17
THE HIGH LEG DELTA SYSTEM

When a three-phase transformer system is used for the power source for an electrical system, it may have the neutral tap. It is important to remember that a three-phase system does not need to have a neutral to operate correctly. The neutral is added only if the lower voltage (120 V) is needed for some part of the system. In general, equipment manufacturers make all the components in a three-phase system a higher voltage, or they may supply a small transformer inside the equipment's power panel to drop the higher voltage to the necessary voltage level. For example, if the equipment needs 208-V three-phase for a pump motor, a control transformer can be provided in the power panel of the equipment to drop the 208 V between L1 and L2 to 120 V. The 120 V is used to provide power for the control circuit and relay coils. Because the primary side of the control transformer can be powered by L1–L2 (208 V), a neutral is not needed in this system. If the system has small pump motors, they may require 120 V for power, and so a neutral tap is needed; these components will be connected between L1 and N, L2 and N, or L3 and N.

If the transformer is connected as a delta transformer, it is important to understand that a different voltage becomes available between L2 and N. Figure 9–20 shows the diagram for this voltage. From this diagram you can see that the secondary winding of this three-phase transformer is connected as a delta transformer. The voltage from L1–N and L3–N is 120 V, so the voltage between L1 and L2, L2 and L3, or L3 and L1 is 240 V.

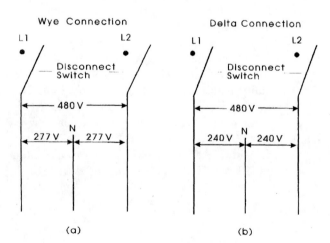

Figure 9–19   (a) Voltage from L1–L2 for a wye-connected transformer is 480 V and that from L1–N or L2–N is 277 V; (b) voltage from L1–L2 for a delta-connected transformer is 480 and that from L1–N or L2–N is 240 V.

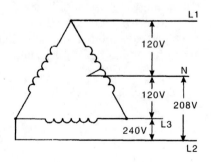

Figure 9–20   A high leg delta voltage occurs between L3 and N because the center tap that produces the neutral is the center tap of the L1–L2 winding.

The different voltage occurs between L2 and N. Because the winding between L1 and L3 has the center tap, it stands to reason that the voltage between L1 and N or L3 and N will be exactly half of the voltage L1–L3. Because the center tap is not between L2 and L1, the voltage for L2–N must come from one complete phase (240 V) and half of the next phase (120 V). This voltage uses two phases, so the 240 V and the 120 V are out of phase, and the resultant voltage from them is 208. This voltage is not the same as the voltage of 208 V that occurs between L1 and L2 or L2 and L3 or L3 and L1 of the wye-connected transformer, because the neutral point is not at a midpoint.

Because the 208 V between L2 and N comes from two phases, it will cause the transformer to overheat if it is used to power any components that require 208 V. For this reason the voltage is called the *high leg delta voltage*, to indicate that it is derived from L2–N of a delta-connected transformer and that it should not be used to power 208-V components. It is very important to understand that the L2 leg of the transformer poses no problem when it is used with L1–L3 or L2–L3 as part of the three-phase system or the 240-V single-phase system. The only problem occurs when the L3 terminal is used in conjunction with the neutral, which creates the L2–N voltage of 208 V.

The L2 terminal in any power distribution box for a delta-wired system should always be marked with an orange wire to identify it as the high leg delta. In some areas of the country, the high leg delta is also called the *wild leg*. The high leg delta voltage of 208 V may occur between L1 and N if the neutral point is produced by a center tap of the L2–L3 winding, or it may occur between L2 and N if the neutral point is produced by a center tap of the L1–L3 winding.

9.18
THREE-PHASE VOLTAGE ON SITE

The source for three-phase voltage is an electrical panel or a load center. As you know, the voltage actually comes from the three-phase transformers, but as stated before, as a maintenance technician you will not be expected to make connections right at the transformers. Instead technicians from the utility company will connect the wires between the transformers and the load centers or main disconnects. The load center may also be called a *circuit breaker panel*. Figure 9–21 shows an example of a three-phase fusible disconnect, and Fig. 9–22 shows an example of a three-phase load center. In the picture of the three-phase disconnect, you can see the three fuses. Incoming voltage is connected to the

Figure 9–21 A three-phase fusible disconnect that is used to provide three-phase voltage to equipment in commercial and industrial applications.

line-side terminals at the top of the disconnect. The line-side terminals are identified as L1, L2, and L3. The disconnect has a handle on the right side that is used to turn its switch on and off. If the disconnect switch is depressed at the top, the switch will close and the disconnect will have voltage at the load-side terminals at the bottom of the disconnect. The load-side terminals are identified as T1, T2, and T3. It is important to understand that the line-side terminals will always have power even when the disconnect switch is open (in the off position).

The three-phase load center is similar to the circuit breaker panel for a single-phase system, except it has three individual circuits. In smaller shops that have only a few pieces of equipment, the equipment may be connected directly to the circuit breakers in the load center. This is also generally the case for loads such as exhaust fans and lighting, because they need to be individually protected. The load center is specifically designed to mount three-phase circuit breakers or each individual circuit that will be used. Thus if you have four exhaust fans mounted in the ceiling and they are connected to the load center, each one will be connected to its own circuit breaker. When you are installing a new fan or servicing existing equipment, you must locate the load center and circuit breakers for the circuit you intend to use, or you must locate the fusible disconnect if one is installed near the unit location.

In larger factories, the power is distributed through bus ducts throughout all areas of the factory. The bus duct has a number of openings that allow the technician to connect bus box disconnects directly onto the bus. These disconnects allow power to be tapped from the three-phase supply as close to the machine as possible. Most bus ducts are suspended approximately 8 to 15 ft above the floor, and service to the disconnect on the machine is made by using flexible cable or fixed conduit.

Figure 9–22 A three-phase load center that contains the circuit breakers for the electrical system. (Courtesy of Eaton Electrical)

9.19

INSTALLING WIRING IN A THREE-PHASE DISCONNECT

When you are ready to connect the wires between the disconnect and the equipment you are installing, you will need to ensure that the source of the three-phase voltage is turned off and a padlock is used to lock out the system. The padlock should also have a tag attached to it that has your picture and name on it to identify you as the technician who has locked out the system. The lock-out and tag-out procedure ensures that you can safely work on the system without anyone accidentally turning the power on when you are still working in the panel.

Figure 9–23 shows a diagram of the locations in the disconnect where you will connect the equipment wires. The utility company will have the supply voltage

wires previously installed at the three terminals at the top of the disconnect box. These terminals are the line-side terminals, and they are identified as L1, L2, and L3. You will make your connections at the bottom terminals, which are identified as T1, T2, and T3. These terminals are the *load-side terminals*. The other end of each wire will be connected in the power panel for the equipment. This process is called *connecting the field wiring*.

When you need to make connections for field wiring in a circuit breaker panel, you will connect to its *main breaker* at the top of the panel. The individual circuit breakers will receive power any time the main breaker is in the *on* position. You need to add a three-phase circuit breaker for each piece of equipment for which you are providing power. Again, you should notice that the wire from the circuit breaker is connected at the incoming power terminals in the equipment.

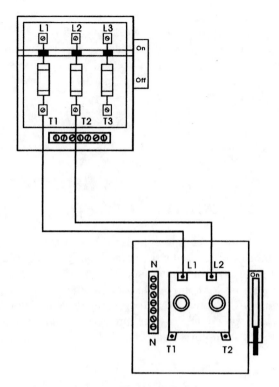

Figure 9–23 A single-phase disconnect is connected to the load-side terminals of a three-phase disconnect to provide a single-phase power source.

9.20
TESTING FOR A BAD FUSE IN A DISCONNECT

At times you will be working on a system that receives its power from a three-phase disconnect. The system will not run, and you may suspect one or more of the fuses are bad. You can check for a blown fuse by measuring the voltage or by removing the fuses and testing them for continuity with an ohmmeter. If you select to check the fuses with a voltage test, you will need to test the incoming voltage at L1, L2, and L3 to determine that the supply voltage is good. Check the voltage between terminals L1 and L2, L2 and L3, and L3 and L1. The voltage at each of these sets of terminals should be the rated voltage for the system, such as 480 V. If you do not get the same voltage reading at each set of terminals, the incoming power has a problem, and you will need to determine where the voltage has been interrupted.

The next test should be at the bottom terminals of the disconnect at T1–T2, T2–T3, and T3–T1. Each of these readings should be the same as the voltage read-

ings at the line side of the disconnect. If any of the voltage readings are lower than the line-side readings or 0 V, you have an open fuse. At this point the simplest way to find the bad fuse is to turn the disconnect off and remove all the fuses and test them for continuity. Be sure to use a fuse puller when you are removing or installing fuses in the disconnect. It is also important to remember that the line-side terminals of the disconnect are "hot" (powered) even though the disconnect switch is open.

The reason the load-side voltage may be lower than the line-side voltage rather than 0 V is because voltage may feed back through two of the windings in any three-phase motor or three-phase transformer that is connected to the load side of the disconnect. It is difficult to determine the exact amount of voltage that is dropped when voltage feeds back through other windings, so you must test for an open fuse if the line-side voltage is not the same as the load-side voltage at each set of terminals where you take a reading. If you find an open fuse, you will need to replace it with one that has the same current rating and same voltage rating.

Remember that you can also measure for voltage across each fuse; if you measure full voltage across any fuse, the fuse is open and the voltage is actually *back-feed voltage*.

9.21
SINGLE-PHASE VOLTAGE FROM A THREE-PHASE SUPPLY

At times you will need to provide a single-phase source of voltage for a small motor, pump, or shop tools such as grinders or drill presses. The single-phase voltage can be either 480, 208, or 240 V L1–L2. In Fig. 9–23 you can see the connections between the load-side terminals of a three-phase disconnect and the line-side terminals of a single-phase disconnect. If the single-phase circuit requires a neutral, a neutral wire can be connected between the neutral terminal of the three-phase disconnect and the single-phase disconnect and between the three-phase disconnect and the transformer.

This type of circuit may also be used in applications inside a factory or commercial location where three-phase voltage is supplied to the system and you need to provide a single-phase voltage source to power a window-type air conditioner or a small refrigerator. Once you understand that you can get single-phase voltage from the three-phase system, you will begin to see how the AC power-distribution system works in industrial, commercial, and residential applications.

The remaining chapters of this book will continue to refer to the supply voltage as you would find it connected to the equipment at the line-side terminals. As a technician you will need to be aware of where this voltage comes from and how it is distributed so that you can trace the circuits if no voltage is present at the unit.

9.22
Bus Duct and Bus Disconnect Boxes

In most factories, the power is distributed throughout the factory by a series of bus ducts that are suspended in the ceiling. In some cases, the bus duct is mounted on the wall or in the floor. The bus box is a disconnect switch that is designed specifically to mount on the bus duct and provide a disconnect and fuse combination. Because the bus duct has multiple bus boxes, one or more can be turned off while others remain powered, so that you can turn the power off to the machine you are working on while allowing power to be connected to a machine right beside you that you are not working on. The front of the bus box is a door that can be opened to be tested or serviced. A safety latch is provided to keep you from opening the door while power is turned on. You can use a screwdriver to override the latch to check the fuses in the box while power is supplied. It is rec-

ommended that you turn off the disconnect switch on the box any time you are working on the system. The only time that it is permissible to override the latch is when you need to take a voltage reading across the fuses while power is applied. Figure 9–24 shows a typical bus duct and bus box.

9.23
Wiring a Duplex Receptacle as a Utility Outlet

In most factory equipment it is necessary to provide one or more duplex receptacles so that you can plug in 120-V service equipment, such as service lamps or soldering irons, or hand power tools, such as electric drills. Figure 9–25 shows a duplex receptacle. When you hold a duplex receptacle in your hand, you should notice that it has gold-colored screws on one side and silver-

Figure 9–25   A duplex receptacle that provides 120 V for tools and equipment.

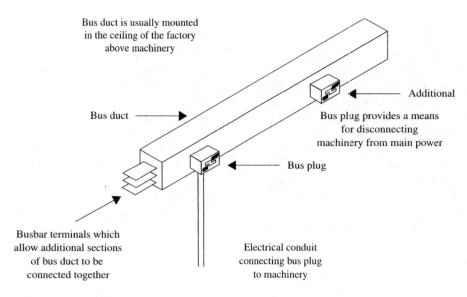

Figure 9–24   Typical bus duct with bus boxes shown connected in the bus.

colored screws on the other side. The L1 wire should be connected to a gold-colored screw and the neutral wire should be connected to the silver-colored screw. The ground wire should be connected to the green screw on the receptacle.

The openings in the face of the receptacle are different sizes. The larger opening on the left side of the receptacle face is connected to the neutral terminal, and the smaller opening on the right side of the receptacle face is connected to the L1 terminal. The round opening at the top is connected to the ground terminal. Because the openings in the face of the receptacle are different sizes, any plugs that are connected to it have to be connected correctly. If a tool such as an electric drill is made of plastic, its plug will be manufactured so that it can be plugged into the receptacle either way. If the electric drill has a metal case, the plug will have a ground terminal, and the plug will connect to the receptacle only one way.

When you are ready to connect L1 and N wires to the receptacle and disconnect, be sure the disconnect switch is open. It is also important to test the voltage from L1–N at the line side of the disconnect to ensure that you have 120 V instead of the 208 V of the high leg delta voltage. When all the connections have been made, you can close the main disconnect switch and test the two terminals of the receptacle with your voltmeter to ensure that 120 V is present.

QUESTIONS

Short Answer

1. Explain how inductance is used in the operation of a transformer.

2. Provide an example of a step-up transformer in an industrial system.

3. Discuss how it is possible to have a control transformer whose primary winding can be connected to 220 or 208 V and provide 120 V at its secondary.

4. Explain how you would test for 240 and 120 VAC in a fusible disconnect on the line-side and load-side terminals.

5. Explain why a control transformer is required in some electrical equipment.

True or False

1. If voltage L1–L2 is 240 V and voltage L1–N is 120 V, the transformer is connected as a delta transformer.

2. The secondary voltage is larger than the primary voltage in a step-up transformer.

3. Voltage in a transformer moves from the primary winding to the secondary winding by conductance.

4. The VA rating for a transformer is determined by multiplying the voltage at the secondary winding by the current flow in the secondary winding.

5. The 208 V from L3 to N for a high leg delta transformer can be used for motor loads that require 208 V.

Multiple Choice

1. The voltage at the secondary terminals of a _____ transformer will be larger than the applied voltage to the primary terminals.
   a. step-up
   b. step-down
   c. isolation

2. A transformer that is wired as a delta transformer has _____.
   a. 440, 208, and 120 V
   b. 480, 240, and 120 V
   c. 480, 277, and 120 V

3. A transformer that is wired as a wye transformer has _____.
   a. 480, 208, and 120 V
   b. 480, 240, and 120 V
   c. 500, 300, and 100 V

4. If a control transformer has a 2:1 turns ratio and it has 240 V applied to its primary winding, the voltage at the secondary winding will be _____.
   a. 480 V
   b. 240 V because the secondary voltage is strictly controlled to equal the primary voltage
   c. 120 V

5. The pumps and motors in a system are connected to the _____ of a fused disconnect.
   a. line-side terminals
   b. load-side terminals
   c. power company terminals

PROBLEMS

1. Calculate the secondary voltage of a transformer that has a 4:1 turns ratio and 120 VAC applied to the primary terminals.

2. Calculate the VA of a 120-V control transformer that requires 5 A.

3. Determine the turns ratio of a transformer that has 4,000 turns in its primary windings and 800 turns in its secondary winding.

4. If the transformer in Problem 3 has 480 V applied to its primary winding, how many volts will be measured at its secondary?

5. Calculate the amount of secondary amps for a control transformer that is rated for 800 VA and 120 V at the secondary.

CHAPTER 10

Relays, Contactors, and Solenoids

OBJECTIVES

After reading this chapter you will be able to:

1. Identify the contacts and coil in a relay and contactor.

2. Explain the operation of a relay and contactor.

3. Identify the components in a control circuit.

4. Select the proper size of contactor from a NEMA table.

5. Identify the basic parts of a solenoid and explain their functions.

10.0

OVERVIEW OF RELAYS AND CONTACTORS IN THE CONTROL CIRCUIT

The *control circuit* in any industrial electrical system is the heart of the system. It consists of the control devices, such as start-stop switches, limit switches, or pressure switches that determine when to energize or de-energize each system. Each control circuit uses relays or contactors to control the various motors, such as pump motors and fan motors. This chapter introduces the main components that are used in the control circuits of these systems and their theory of operation. The components in the control circuit are energized, and in turn they energize components such as compressors and motors in the *load circuit*. Figure 10–1 shows a typical control circuit for a system. Notice that it is the bottom part of the circuit; the load circuit is the top part of the electrical diagram.

10.1

THE CONTROL TRANSFORMER

The control transformer was introduced in Chapter 9. In that chapter you learned that the transformer converts line voltage to lower voltage power for the control circuit. The lower voltage (120 volts) is used in each case to provide power for the control circuit.

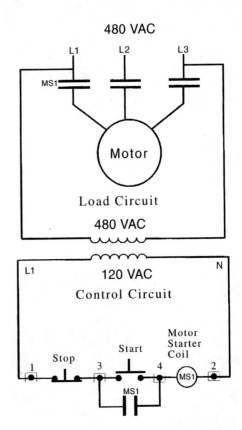

Figure 10–1 A typical control circuit at the bottom of the diagram. This start-stop button controls a three-phase motor through a motor starter.

In the diagram in Fig. 10–1, the control circuit is shown as the bottom part of the diagram, and it starts at the secondary side of the transformer, where it is identified by the letters L1 and N. The terminal that is identified as the L1 terminal is considered the source of power; it is called the *hot side of the circuit*. The right side of the diagram, identified by the transformer terminal N, is considered to be the *common side of the circuit*. It is important to understand that because the voltage in the transformer is AC voltage, both sides of the circuit have the same potential during every other half-cycle. The concepts of a hot side and a common side are defined for people to make it easier to understand the operation of the control circuit.

The coil becomes an electromagnet when it is energized, and its magnetic field causes each set of normally open contacts to close and each set of normally closed contacts to open. The contacts are basically a switch that is operated by magnetic force. The part of the relay that moves and causes the contacts to move is called the *armature*. Power is applied to the coil of the relay, and the magnetic flux causes the armature to move, which in turn causes the contacts to change position. The coil

Figure 10–2   A typical relay. (Courtesy of Honeywell Sensing and Control, Honeywell International, Inc.)

10.2

THE THEORY AND OPERATION OF A RELAY

A *relay* is a magnetically controlled switch that is the main control component in a typical electrical system. Figure 10–2 shows a picture of a typical relay, and Fig. 10–3 shows a cutaway diagram of a typical relay, which consists of a *coil* and a number of *sets of contacts*.

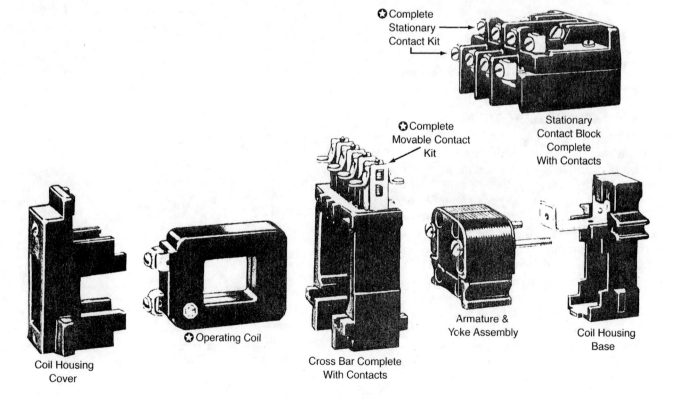

Figure 10–3   A cutaway diagram of a relay. Notice that the coil and contacts are mounted so that the magnetic field of the coil will make the contacts move. (Courtesy of Rockwell Automation's Allen-Bradley Business)

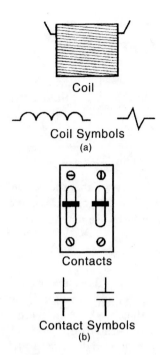

Coil

Coil Symbols
(a)

Contacts

Contact Symbols
(b)

Figure 10–4 Electrical symbols (a) and diagram (b) for the coil and contacts of a relay.

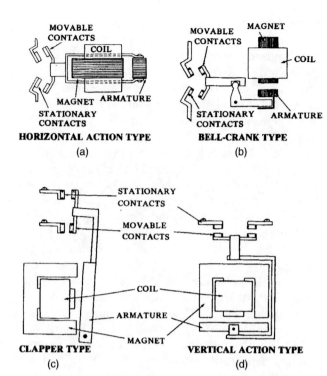

Figure 10–5 (a) Horizontal-action armature assembly; (b) bell-crank armature assembly; (c) clapper armature assembly; (d) vertical-action armature assembly. (Courtesy of Square D/Schneider Electric.)

is part of the control circuit, and the contacts are part of the load circuit.

Figure 10–4 shows the electrical symbol for the coil and contacts of a relay. The armature is not shown in the electrical symbols because it is considered part of the contacts. When you encounter a relay in the control panel of a system, you must relate the physical parts with the electrical symbols that are used to identify the components. You must also learn to envision its operation as two separate pieces, the coil and contacts, even though they are mounted near each other and operate almost simultaneously. It is important to understand that the coil must be energized first, and a split second later the magnetic field that is built up in the coil will cause the contacts to move.

10.3
TYPES OF ARMATURE ASSEMBLIES FOR RELAYS

You will encounter many different types and brands of relays that look like they all have different theories of operation. In reality the operation of all relays can be categorized into four basic methods, because one of four basic armature assemblies is used in all relays. Figure 10–5 shows an example of these four basic armature assemblies. A few relays will use hybrid armatures that involve some features of more than one type of armature assembly.

Figure 10–5a shows an example of a *horizontal-action armature* for a relay. In this example a set of stationary contacts is mounted in the horizontal position. A set of movable contacts is mounted on the armature assembly directly across from the stationary set. The armature is mounted inside the coil in such a way that when the coil is energized, its magnetic field will cause the mass of the armature to center in the middle of the coil. Because the armature is slightly offset when it is placed in the coil, it will move to the left to center itself directly in the middle of the coil when the magnetic field is developed in the coil. The movement of the armature to the left will shift the movable contacts to the left until they come into contact with the stationary set of contacts. When the movable contacts and stationary contacts touch each other, they complete a circuit that allows large amounts of current and voltage to pass through them, just as a normal switch does when it is in the closed position. When the current to the coil is de-energized, a small spring causes the armature to shift back to the left to its original position.

The diagram in Fig. 10–5b shows an example of a *bell-crank armature assembly*. In this example the coil is mounted above the armature, and when it is energized, it produces a magnetic field that pulls the armature upward. The movable contacts are connected to an arm that is bent at a right angle. When the end of the arm is pulled upward by the magnet, the other end of the arm is moved to the left. This action shifts the movable contacts against the stationary contacts to complete the electrical circuit. When the coil is de-energized, gravity causes the armature to drop downward away from the coil. This movement causes one end of the arm also to move downward, which makes the left end of the arm shift back to the left. This movement causes the contacts to move to the open position again.

The diagram in Fig. 10–5c shows an example of a *clapper armature assembly.* The armature is a large arm on the right side of the coil. The armature (arm) has a pin through the bottom to act as an axis. The movable contacts are mounted at the top of the armature on the left side. When the coil is energized, it creates a magnetic field that pulls the armature to the left toward the coil. This action causes a small amount of travel at the bottom of the arm and a large amount of travel at the top of the arm, because the bottom of the arm is held in place at the axis. The movable contacts are mounted near the top of the arm, so they will move to the left a significant distance until they come into contact with stationary contacts. When the coil is de-energized, a small spring pulls the armature to the left, which causes the contacts to return to their open position.

Figure 10–5d shows an example of a *vertical-action armature assembly.* The armature is mounted on a bracket that is shaped like the letter C. The bottom part of the C-shaped bracket is mounted directly on the armature, and a set of movable contacts is mounted directly on the top of the C-shaped bracket. When the coil is energized, it creates a magnetic field that pulls the entire C bracket upward until the armature is pulled tight against the coil. This movement causes the contacts mounted on the top of the bracket to shift upward until they touch the stationary contacts. When the coil is de-energized, gravity causes the complete bracket to drop downward and move away from the coil, which makes the movable contacts move away from the stationary contacts.

10.4

PULL-IN AND HOLD-IN CURRENT

When voltage is first applied to a coil of a relay, it draws excessive current. This occurs because the coil of wire presents resistance to the circuit only when current first starts to flow. As the flow of current increases in the coil, inductive reactance begins to build, which causes current to become lower. When the current is at its maximum, it creates a strong magnetic field around

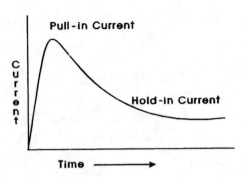

Figure 10–6   A diagram that shows pull-in and hold-in current for relay coils. Notice that the pull-in current is approximately three to five times as large as the hold-in current.

the coil, which causes the armature to move. When the armature moves, it makes the induction in the magnetic coil change so that less current is required to maintain the position of the armature.

Figure 10–6 shows a diagram of the *pull-in current* and the *hold-in current.* The pull-in current is also called the *inrush current*, and the hold-in current is also called the *seal-in current.* The pull-in current is typically three to five times as large as the hold-in current. At times the supply voltage is slightly lower in the summer. This condition is called a brownout, and it may cause a problem with activating relays. If the voltage is too low, it may not be able to supply sufficient current to pull the armature into place when the relay is first energized because the pull-in current is too small.

If the system is running when the brownout condition occurs, the relay coil will probably remain energized, because the amount of hold-in current is sufficient to keep the armature in place even though the voltage is low. Because the brownout condition occurs when it is very hot outside, it will most likely affect air conditioners or other systems that cycle on and off frequently through this condition. If the air conditioner ever gets the room cool enough so it can cycle off during a brownout condition, it will probably not energize again because of the low voltage and the large amount of pull-in current required.

10.5

NORMALLY OPEN AND
NORMALLY CLOSED CONTACTS

A relay can have normally open or normally closed contacts. The electrical symbol in Fig. 10–4 shows exam-

ADDING or CONVERTING CONTACT CARTRIDGES HAVING "SWINGAROUND" TERMINALS
General Instructions (Specific cases below.)
1.1 Adding a contact cartridge:
As received, accessory cartridges are in the normally open mode with terminal screws adjacent to N.O. symbols. If normally closed mode is desired, convert contact as indicated in Step 1.2 below. When cartridges are inserted, the terminal screws must face the front. The clear cover may face either side. Do not install more than 8 N.C. contacts per relay. When installing one cartridge, locate it at an inner pole position. When installing 2 cartridges, locate both in inner or outer (balanced) positions.
1.2 Converting a contact to its alternate mode (N.O. ⇄ N.C.):
Withdraw an assembled cartridge for replacement or conversion by inserting the blade of a suitably-sized screwdriver under a terminal screw pressure plate. Slide cartridge out. See Figure 2. Back the terminal screws out of the cylindrical nuts a sufficient amount (approximately 2 turns for a fully-tightened screw) to permit rotation of each screw and nut assembly to its alternate position. See Figure 3.
Screwdriver Blade
Pressure Plate of Terminal Screw
Cartridge
Direction of Withdrawal
Contact Block
FIGURE 2
Terminal Screw
Pressure Plate
Cartridge Housing
N.C. Contact Symbol
Cylindrical Nut
N.O. Contact Symbol
FIGURE 3

Figure 10–7 Example of changing normally open contacts to normally closed contacts in the field by turning them over. Normally closed contacts can also be converted to normally open contacts in the same manner. (Courtesy of Rockwell Automation's Allen-Bradley Business)

ples of normally open and normally closed contacts. It is important to understand that the word *normal* for contacts indicates the position the contacts are in when no voltage is applied to the coil. The contacts can be held in their normal position by a spring or by gravity. The contacts will move from their normal position to their energized position when power is applied to the coil.

Some types of contacts can be changed or converted from normally open to normally closed in the field. Other types are manufactured in such a way that they cannot be changed. Figure 10–7 shows examples of converting normally open contacts to normally closed contacts while the relay is installed. The contacts can be converted in the field by the technician by simply removing them from the relay and turning them upside down. When normally open contacts are inverted, they become normally closed; and when normally closed contacts are inverted, they become normally open. This means that as a technician in the field, you can change the contacts in a relay to get the exact number of normally open or normally closed contacts needed for the application.

10.6 Ratings for Relay Contacts and Relay Coils

When you must change a relay that is worn or broken, you need to ensure that the coil of the new relay matches the voltage of the control circuit exactly. This means that if the voltage for the control circuit is 120 VAC, the coil must be rated for 120 VAC. If the control voltage is 240 VAC, the coil must be rated for 240 VAC. The voltage rating for a relay coil is stamped directly on the coil. If the coil is rated for 24 VAC, it will be color-coded black. If the coil is rated for 120 VAC, it should be color-coded red or have a red-colored stamp on the coil. If the relay coil is rated for 208 or 240 VAC, it will be color-coded green or identified with a green stamp or green printing on the coil. DC coils are color-coded blue. It is important to understand that the current rating for a relay coil is seldom listed on the component. If it is important to know the current rating for the coil, you can look for it in the catalog or on the specification sheet that is shipped with the new relay. If you change a relay, you must also make sure that the rating for the contacts meets or exceeds the current rating and the voltage rating of the load to which it will be connected. For example, if the contacts of the relay are used to energize a 240-VAC fan motor that draws 3 A, the contacts must be rated for at least 240 V and 3 A. It is permissible to have the contacts in this example rated for more voltage and current, such as 600 V and 10 A.

Contact ratings are grouped by voltage and by current. The voltage ratings are generally broken into two groups, 300 V and 600 V. This means that if you are using the contacts to control 240 or 208 VAC, you would use contacts that have a 300-V rating. If you are using contacts to control a 480-VAC motor, you would need to use contacts with a 600-V rating.

The current ratings of contacts are listed in amperes or horsepower. The current rating or horsepower rating must exceed the amount of current the relay is controlling. This means that if the relay is controlling 12 A, the contacts need to be rated for more than 12 A of current. The current rating of the contacts and the voltage rating for the contacts are printed directly on the contacts or on the side of the relay.

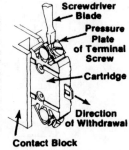

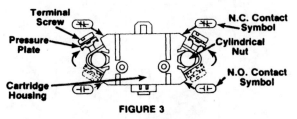

10.7

IDENTIFYING RELAYS BY THE ARRANGEMENT OF THEIR CONTACTS

Some types of contact arrangements for relays have become standardized so that they are easier to recognize when they are ordered for replacement or when you are trying to troubleshoot them. The diagrams in Fig. 10–8 show examples of some of the standard types of relay arrangements. Figure 10–8a shows a relay with a set of normally open contacts. This type of relay could also have a single set of normally closed contacts instead of normally open contacts. Because this relay has only one contact and it can only close or open, this type of relay is called a *single-pole, single-throw (SPST) relay.* The word *pole* in this identification refers to the number of contacts, and the word *throw* refers to the number of terminals to which the input contacts can be switched. Because the contact in this relay has one input and it can be switched only to a single output terminal, it is said to have a single throw.

Figure 10–8b shows a relay with two sets of normally open contacts. Because this relay has two sets of single contacts, it is called a *double-pole, single-throw relay (DPST).* The double-pole part of the name comes from the fact that the relay has two individual sets of normally open contacts, and the single-throw part of the name comes from the fact that each contact has only one output terminal. When the coil is energized, both sets of contacts move from their normally open position to the normally closed position.

Figure 10–8c shows a relay with a set of normally open and a set of normally closed contacts that are connected on the left side. The point where this connection is made is called the *common terminal,* and it is identified with the letter C. When the relay coil is energized, the normally open part of the contacts will close, and the normally closed part of the contact, will open. Because these contacts basically have a common point as the input terminal and two output terminals, one normally open (NO) and one normally closed (NC), it is called a *single-pole, double-throw relay (SPDT).* The most important part of this relay is that the contacts have two terminals on the output side, so it is called a double-throw relay. The single-pole, double-throw relay is used where two exclusive conditions exist and you do not ever want them both to occur at the same time. For example, if this relay is controlling the cooling (air conditioning) and the heating (furnace) for the cooling and heating system to the offices of a factory, you would never want them both to be on at the same time. By connecting the air-conditioning system to the normally open terminal on the right side of the relay and the furnace to the normally closed terminal,

you create the conditions so that the furnace and air-conditioning system cannot be on at the same time.

The relay in Fig. 10–8d has two sets of single-pole, double-throw contacts, so it is called a *double-pole, double-throw relay (DPDT).* In this case the term *double throw* is used because two sets of normally open/normally closed contacts are provided. Each set has a common point on the left side (input side) and a terminal that is connected to the normally open (NO) set and

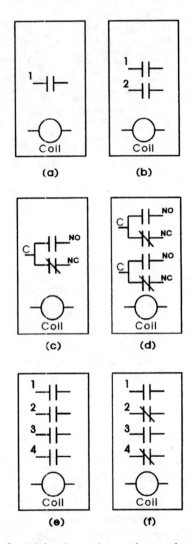

Figure 10–8   (a) A relay with a single set of normally open contacts. This type of relay is called a single-pole, single-throw (SPST) relay. (b) A relay with two individual sets of contacts. This relay is called a double-pole, single-throw (DPST) relay. (c) A relay with two sets of contacts that are connected at one side at a point called the common terminal (C). The output terminals are identified as normally open (NO) and normally closed (NC). This type of relay is called single-pole, double-throw (SPDT). (d) A relay with two sets of SPDT contacts. This relay is called a double-pole, double-throw (DPDT) relay. (e) A relay with multiple sets of normally open contacts. (f) A relay with a combination of normally open and normally closed contacts.

a terminal that is connected to the normally closed (NC) set on the right side. This type of relay is used where exclusion is needed and the loads of 208 VAC or 240 VAC need power from both L1 and L2. In this type of application, $L1$ is connected to the common terminal (C) of one set of contacts, and $L2$ is connected to the common terminal (C) of the other set. This causes $L1$ and $L2$ to be switched the same way in both conditions.

Figure 10–8e shows multiple sets of normally open contacts. This type of relay can have any number of sets of normally open contacts. The additional sets of contacts can be added to original contacts in some types of relays. If the original relay is manufactured with this provision, you can purchase the additional contacts and add them to the original relay by placing them on top of the original relay and tightening the mounting screws to make the additional contacts operate with the relay armature. The contacts for this type of relay can all be normally closed if the application requires it. The main feature of this type of relay is that it can have any number of contacts.

Figure 10–8f shows a relay with multiple sets of individual normally open and normally closed contacts. The combination of normally open and normally closed contacts can be any mixture of the two. This type of relay is similar to the one shown in Fig. 10–8e, except in this type of relay the contacts can be any combination of normally open or normally closed sets. In most cases, the contacts in this type of relay are convertible in the field, and as a technician, you can

add sets of contacts and change them from normally open to normally closed, or vice versa, as needed. In most original installations, the relays are provided in the original equipment, and you will need only to identify them for installation and troubleshooting purposes. Later, if the equipment is modified or if additional machinery is added, you may need to locate additional contacts on a relay to connect the add-on equipment so that it will operate correctly with the original system.

10.8

EXAMPLES OF RELAYS USED IN INDUSTRIAL ELECTRICAL SYSTEMS

Relays are used in a variety of applications in industrial electrical systems. For example, one of the most common relays is the fan relay used in heating systems. Another common relay is the sequencing relays used in some electric furnaces. Figure 10–9 shows examples of several types of relays that you will find in these systems. In some newer systems, the relays use plug-in terminals so that a relay can be removed and replaced without changing any wiring. The relay is simply

Figure 10–9 Variety of relays used in industrial electrical applications. (Courtesy of Honeywell Sensing and Control, Honeywell International, Inc.)

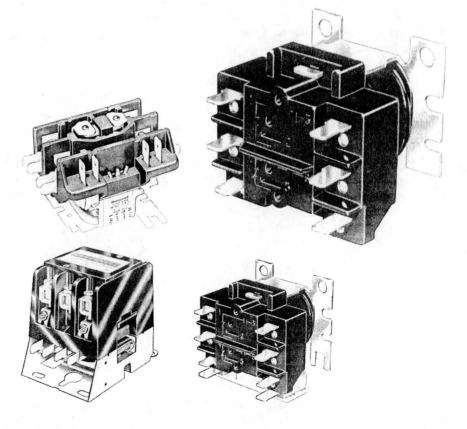

plugged into a socket-type base. You will begin to become familiar with relays and recognize them in the control panels of the equipment you work on.

10.9

CURRENT RELAYS AND POTENTIAL RELAYS FOR STARTING SINGLE-PHASE COMPRESSORS

Current relays and potential relays are special relays that are designed to energize the start winding of single-phase compressor motors during starting and then immediately de-energize as soon as the motor is running. Single-phase air-conditioning and refrigeration compressor motors have a special problem in that they are motors that are sealed inside a metal case, yet they need a device to operate as a start switch to switch the start winding of the motor into the circuit during starting and then disconnect the winding during the run phase. Because the motor uses lubrication, an explosion is possible if a switch is inside the sealed case. This means that the starting device or relay must be mounted outside the compressor.

The current relay has a low-impedance coil that takes advantage of pull-in and hold-in current to energize a set of normally open contacts. The coil for the current relay is connected in series with the run winding of a single-phase motor, and the contacts of the relay are connected in series with the start winding of the motor. In some cases a capacitor is also connected in series with the contacts and start winding. When voltage is applied to the motor, current can flow only through the run winding, which causes very large current to be drawn. This current is large enough to exceed the pull-in current level of the contacts, which makes them close. When the current relay contacts close, voltage is applied to the start winding of the motor, which provides additional power that allows the motor to start and continue running. After the motor has started and begins to run, the large initial current that is flowing through the run winding and current relay coil diminishes to a minimal level,

which is below the hold-in current level. When this occurs, the current relay coil is not able to keep its contacts closed, so they open and de-energize the start winding of the motor. This action is exactly what the single-phase motor needs, because its start winding can be energized only a few seconds while the motor is starting. Figure 10–10 shows the current relay and electrical diagram. The current relay is also available as an electronic-type relay.

The *potential relay* uses a high-impedance coil with a set of normally closed contacts to energize a start capacitor for a capacitor-start, capacitor-run (CSCR) single-phase compressor motor. Figure 10–11 shows a picture and diagram of a typical potential relay. When power is applied to the compressor motor, current flows through the run winding and through the normally closed contacts of the potential relay to the start winding. When the motor begins to run, its windings create a back voltage called back EMF. The back EMF is large enough to cause the coil of the potential relay to energize and pull its contacts open, which de-energizes the start winding. When main power is turned off to the motor, it stops running, and the normally closed contacts of the potential relay return to their normally closed position so the relay is ready to start the motor again.

10.10

THE DIFFERENCE BETWEEN A RELAY AND A CONTACTOR

A contactor is similar to a relay in that it has a coil and a number of contacts. The main difference is that the contactor is larger and its contacts can carry more current. A *relay* is generally defined as magnetically controlled contacts that carry current less than 15 A. A *contactor* is defined as having contacts that are rated for 15 A or more. Some manufacturers do not follow the 15-A rating, so sometimes you will find a relay that has a current rating for its contacts in excess of 15 A, and you may also find a contactor with contact ratings less than 15 A. In general the main difference is that a contactor is specifically designed so its contacts can carry larger amounts of current, up to 2,250 A. Contactors are rated

Figure 10–10 (a) A typical current relay; (b) an electrical diagram of a current relay.

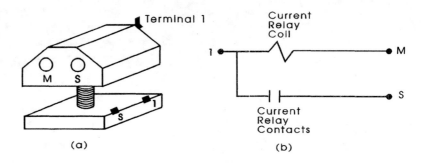

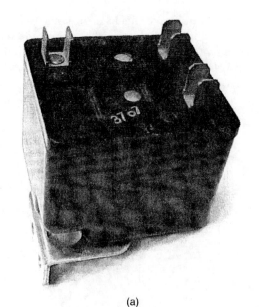

(a)

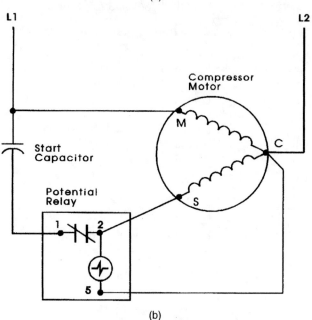

(b)

Figure 10–11   (a) A potential relay used to start single-phase compressors; (b) a potential relay connected to a Capacitor-Start, Capacitor-Run (CSCR) compressor motor.

by the National Electrical Manufacturers Association (NEMA), and their sizes range from size 00 to size 9.

An example of a place in which a contactor is used instead of a relay is the compressor motor of an air-conditioning system. In this type of system the motor is controlled by a contactor instead of a relay because the amount of current the compressor motor draws is usually in excess of 15 A. When you look at the compressor contactor, you will see that it looks like a very large relay. Because its contacts can carry current larger than 15 A, it will technically be called a contactor instead of a relay.

10.11
NEMA RATINGS FOR CONTACTORS

Figure 10–12 shows a table of NEMA ratings for contactors. This table shows that a size 00 contactor is rated to safely carry up to 9 A for a continuous load. You should also notice that the contacts are rated for up to 600 V. You should remember that the current rating for contacts depends on the size of the contacts, and the voltage rating of the contacts depends on whether the relay is manufactured so that arcs do not jump between different terminals. This means that contacts rated for a higher voltage have more plastic or insulating material between the sets of contacts so arcs do not jump from the contacts to other parts of the relay or other sets of contacts. This table also identifies the load as a maximum horsepower rating. Thus, you may identify the load from its current rating or its horsepower rating and select the proper contactor size to safely control the load.

Notice from this table that the next-larger size of contactor is a size 0. This contactor is rated for up to 18 A on a continuous basis. The largest contactor is the size 9 contactor, its current rating is 2,250 A.

Industrial electrical systems usually use contactors up to a size 4. One application for contactors is to control electric heating elements for electric furnaces. These contactors are called *heating contactors*. Because heating elements may draw up to 100 A, the size 4 contactor can easily carry this current. If larger heating elements are used, a larger contactor is used. Figure 10–13 shows two examples of contactors, a size 3 and a size 5. The size 3 and size 5 contactors are approximately 5 and 12 in. tall, respectively.

10.12
THE FUNCTION OF THE CONTROL CIRCUIT

As you have seen in this section, the function of the control circuit in industrial electrical systems is to control the conditions that cause the equipment to become energized or de-energized. The components in the control circuit of typical industrial electrical systems are generally low-voltage (120-VAC) devices. The relay coil is located in the control section of the electrical system, and the relay contacts are located in the load section, and they are typically connected in series with the load it is controlling. It is important to remember that the function of the control circuit is to energize the relay coil so power can be sent through the relay contacts to the load. The action in the control circuit must always occur before the action in the load circuit.

If you must troubleshoot an electrical system for an industrial electrical system, you should make two checks. First you should see if the loads are energized and operating correctly. If the loads are not energized, you should suspect the control circuit is not energized, move on to the second test, and make checks in the control circuit. As you become more familiar with the control circuit and load circuit, you will find it is almost automatic to check the control circuit to ensure that the relay coil is energized so its contacts are energized. The next sections of this book present in-depth material concerning the operation of typical loads, such as compressor motors and fan motors.

Figure 10–12 NEMA ratings for contactors. (Courtesy of Rockwell Automation's Allen-Bradley Business)

NEMA Size	Continuous Ampere Rating	Maximum Horsepower Rating *Full Load Current Must Not Exceed "Continuous Ampere Rating"*			
		Motor Voltage			
				50 Hz	
		200V	230V	380V-415V	460V-575V
		3 Ø • 4 Power Poles • 600V AC Maxi			
00	9	1-1/2	1-1/2	2	2
0	18	3	3	5	5
1	27	7-1/2	7-1/2	10	10
2	45	10	15	25	25
3	90	25	30	50	50
4	135	40	50	75	100
5	270	75	100	150	200
6	540	150	200	300	400
7	810	-	300	600	600
8	1215	-	450	900	900
9	2250	-	800	1600	1600

Figure 10–13 Examples of NEMA size 3 and size 5 contactors. The size 3 is approximately 5 in. tall, and the size 5 is approximately 12 in. tall. (Courtesy of Rockwell Automation's Allen-Bradley Business)

Size 0, 3-Pole,
Open Type without Enclosure
Top Wiring Construction

(a)

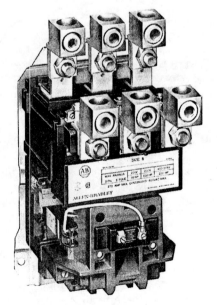

(b)

10.13
SOLENOIDS USED IN INDUSTRIAL SYSTEMS

A *solenoid* is a *magnetically controlled valve*. You learned in earlier chapters about the power of current flowing through a coil of wire and causing a strong magnetic field to develop. The strong magnetic field in the solenoid valve is used to make the valve plunger move from the open to the closed position or from the closed to the open position. The valve is part of the plumbing system, such as an inlet valve for a water system. Solenoid valves allow the plumbing system to be interfaced with the electrical system, which means the plumbing system can be automated.

10.14
BASIC PARTS OF A SOLENOID VALVE

Figure 10–14 shows a typical solenoid valve in the closed position and in the open position. From this figure you can see that the solenoid valve is a plumbing-type valve that has a movable plunger used to close off the valve. The valve either can be designed so that the plunger is held in the open position by spring pressure or can be designed to be held in the closed position by spring pressure. When electric current is applied to the coil in the valve, the strong magnetic field causes the plunger to move upward. If the valve is the type where spring pressure is causing the plunger to stay in the closed position, the magnetic field will cause the plunger to open the valve passage. If the valve is the type where spring pressure is causing the plunger to stay in the open position, the magnetic field will cause the plunger to close the valve passage.

The coil for the solenoid is exactly like the coils that you learned about earlier in the chapter. Because the

coil is basically a long wire that is coiled on a plastic spool, it will have two ends. You can test the coil for continuity, and you should notice that each type of solenoid coil has some amount of resistance that will vary from approximately 50 to 1,500 Ω. Remember that the exact amount of resistance does not matter; because the coil is made from one long piece of wire, it is doubtful that the amount of resistance will change. The main problem a solenoid will have is an open somewhere along the length of the wire in its coil. If the coil is open, its resistance will be infinite.

10.15
TROUBLESHOOTING A SOLENOID VALVE

At times you will be working on a system that has one or more solenoid valves, and you will suspect the solenoid valve is not operating correctly. The first test you should make when troubleshooting a solenoid valve is to read its data plate and determine the amount of voltage the valve should have; then measure the amount of voltage that is available across the two wires that are connected to the coil. It is important to remember that the coil of any solenoid valve must have the correct amount and type of voltage applied to its two wires if it is to conduct sufficient current to cause the plunger to move. This means that if the valve is rated for 120 VAC, you should measure 120 VAC across the two coil wires. If the voltage is missing or is an incorrect amount, the coil will not operate properly. If the right amount of voltage is not present, you will need to use troubleshooting procedures to locate the open in the circuit that is causing the loss of voltage before it reaches the coil.

If the proper amount of voltage is present and the coil will not activate, you must make several additional tests. You can test to see if the coil has developed a magnetic field by placing a metal screwdriver near the coil to see if it is attracted by the magnetic field when voltage is applied. If the screwdriver is attracted to the coil

Figure 10–14 Example of a solenoid valve in the closed position (de-energized) and the open position (energized).

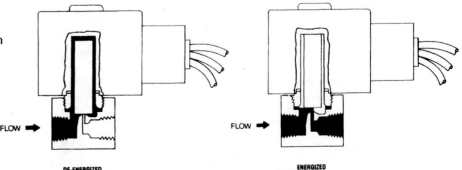

DE-ENERGIZED

ENERGIZED

FLOW

FLOW

by the magnetic field but the valve is not opening or closing, the valve seat may be frozen in place; in that case the valve will need to be replaced.

The other problem that the coil may have if the correct amount of voltage is present at the coil wires is that the coil may have an open in its wire. You can test the coil wire with a continuity test. Be sure to turn off all voltage to the solenoid and disconnect the coil wires so that the coil is isolated from backfeeding to the remainder of the circuit. If the continuity test indicates infinite (∞) resistance at the highest resistance range, you can suspect the coil has an open and should be replaced. You should notice that the coil can be replaced without unsoldering the valve.

QUESTIONS

Short Answer

1. Identify the main parts of a relay and explain its operation.

2. Explain the operation of the four types of armature assemblies presented in this chapter.

3. Explain the operation of a current relay and a potential relay.

4. Discuss the differences between a contactor and a relay.

5. Identify the main parts of a solenoid and explain its operation.

True or False

1. Plug-in relays are used in industrial electrical systems because they are easy to change out when they become faulty.

2. The main difference between a relay and a contactor is the contactor usually has more contacts.

3. A solenoid is similar to a relay in that it has a coil that becomes magnetic and moves the plunger of a valve.

4. It is possible to change normally open contacts to normally closed contacts in the field for some relays.

5. For a relay to operate correctly, its contacts must close first to provide current to its coil.

Multiple Choice

1. The normally closed contacts of a relay _____.
   a. pass current when the coil is not energized
   b. pass current when the coil is energized
   c. pass current at all times, whether the coil is energized or not energized

2. A current relay is _____.
   a. a special relay designed to control only current
   b. a relay that is designed to be used as a relay or a contactor
   c. a relay that is designed to start single-phase compressors

3. A solenoid is _____.
   a. a valve that is controlled (opened and closed) by gravity and springs
   b. a reversing relay that is used to start heat pumps
   c. a valve that is controlled (opened and closed) by springs and a magnetic field produced by a coil

4. A contactor is _____.
   a. a special type of solenoid used to start a pump motor
   b. similar to a large relay in that it has contacts and a coil, but its contacts are rated for more than 15 A
   c. a relay that has both normally open and normally closed contacts

5. An SPDT relay has _____.
   a. one set of contacts that connect to two output terminal points
   b. two sets of contacts that connect to one terminal point
   c. one set of contacts and one output terminal point

PROBLEMS

1. Draw a waveform that shows pull-in and hold-in current for a relay coil.

2. Use the NEMA table provided in Fig. 10–12 to select a NEMA starter size that can control 45 continuous amps.

3. Draw an example of single-pole, double-throw (SPDT) contacts and explain their operation.

4. Draw an example of a double-pole, single-throw (DPST) contact and explain its operation.

5. Draw an example of a double-pole, double-throw (DPDT) contact.

CHAPTER 11

Motor Starters and Over-Current Controls

OBJECTIVES

After reading this chapter you will be able to:

1. Explain the operation of a motor starter.
2. Identify the main parts of a motor starter.
3. Identify the main parts of an overload and explain their operation.
4. Explain the operation of single-element and dual-element fuses.
5. Identify single-pole, two-pole, and three-pole circuit breakers and explain their operation.

11.0
OVERVIEW OF MOTOR STARTERS

Motor starters are magnetically controlled contacts that are usually used in industrial applications. The motor starter is a larger version of a relay or contactor, and it is used to control larger motors. The motor starter also has *over-current protection* for motors built into it, but relays and contactors do not. This over-current protection is called an *overload*, and it is sized to trip if the amount of current drawn by the motor exceeds the designated limit.

11.1
WHY MOTOR STARTERS ARE USED IN INDUSTRIAL ELECTRICAL SYSTEMS

In Chapter 10 you learned about relays and contactors. Relays and contactors are designed to close or open their contacts and controls current to small loads in the system.

In most industrial systems the motors are large and expensive. These motors must be protected by fuses or circuit breakers in the disconnect for short-circuit protection and by the overloads in motor starters to protect against slow over-currents. If an open motor has a problem such as the loss of lubrication or if the motor is overloaded, it will draw extra current, which will damage the motor if the overload current is allowed to continue for any length of time. The overloads in the motor starter sense the excess current and trip the motor starter coil so that its contacts open and stop all current flow to the motor until a technician manually resets the overloads. A special contactor that has integral over-current protection built in is called a *motor starter*. This chapter explains the basic parts and operation of a motor starter.

11.2

THE BASIC PARTS OF A MOTOR STARTER

Figure 11–1 shows a typical motor starter with all the parts identified. From this picture you can see that it is a three-pole starter which is used in three-phase circuits. The incoming voltage is connected on the top at the terminals identified as L1, L2, and L3, and the motor leads are connected to the bottom terminals identified as T1, T2, and T3. The three major sets of contacts are located in the top part of the motor starter, and the overload assembly is mounted in the lower part of the motor starter.

The coil is located in the middle of the motor starter, and it has an indicator that shows the word *ON* when the coil is energized and *OFF* when the coil is de-energized.

Figure 11–2 shows a wiring diagram of a motor starter connected to a three-phase motor at the top of the diagram and a ladder diagram of just the control circuit, which consists of the start push button, the stop push button, the motor-starter coil, and the motor-starter overloads, in the bottom diagram. It is easy to see the parts of the motor starter in the wiring diagram because they are all in the shaded area. You can see that the coil circuit uses smaller wire than the contact circuit. The motor starter has three sets of contacts that have a heater in series with it. This ensures that all the current

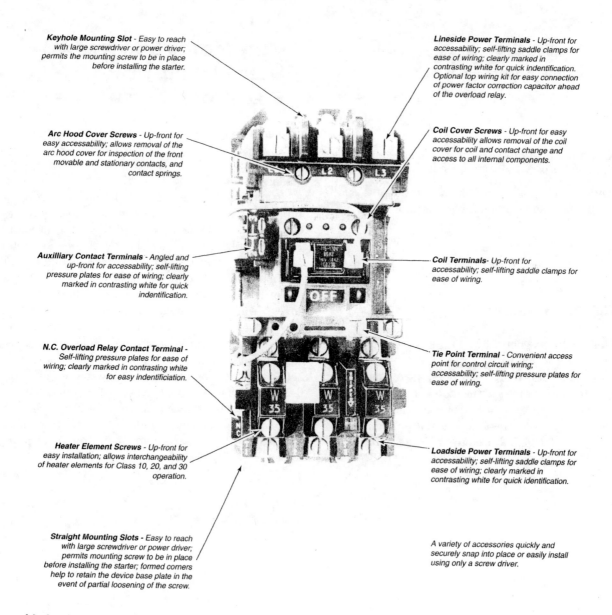

Keyhole Mounting Slot - *Easy to reach with large screwdriver or power driver; permits the mounting screw to be in place before installing the starter.*

Arc Hood Cover Screws - *Up-front for easy accessability; allows removal of the arc hood cover for inspection of the front movable and stationary contacts, and contact springs.*

Auxilliary Contact Terminals - *Angled and up-front for accessability; self-lifting pressure plates for ease of wiring; clearly marked in contrasting white for quick indentification.*

N.C. Overload Relay Contact Terminal - *Self-lifting pressure plates for ease of wiring; clearly marked in contrasting white for easy indentificiation.*

Heater Element Screws - *Up-front for easy installation; allows interchangeability of heater elements for Class 10, 20, and 30 operation.*

Straight Mounting Slots - *Easy to reach with large screwdriver or power driver; permits mounting screw to be in place before installing the starter; formed corners help to retain the device base plate in the event of partial loosening of the screw.*

Lineside Power Terminals - *Up-front for accessability; self-lifting saddle clamps for ease of wiring; clearly marked in contrasting white for quick indentification. Optional top wiring kit for easy connection of power factor correction capacitor ahead of the overload relay.*

Coil Cover Screws - *Up-front for easy accessability allows removal of the coil cover for coil and contact change and access to all internal components.*

Coil Terminals - *Up-front for accessability; self-lifting saddle clamps for ease of wiring.*

Tie Point Terminal - *Convenient access point for control circuit wiring; accessability; self-lifting pressure plates for ease of wiring.*

Loadside Power Terminals - *Up-front for accessability; self-lifting saddle clamps for ease of wiring; clearly marked in contrasting white for quick identification.*

A variety of accessories quickly and securely snap into place or easily install using only a screw driver.

Figure 11–1   A typical motor starter with all its parts identified. (Courtesy of Rockwell Automation's Allen-Bradley Business)

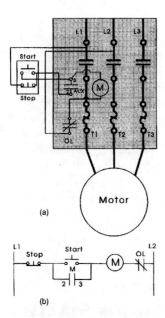

Figure 11–2 (a) Electrical diagram of a motor starter and control circuit. (b) A ladder diagram of the control circuit.

that flows to the motor must pass through a heater. If the current flowing to the motor is normal, the heater does not provide sufficient heat to cause the overload to trip. If the motor draws excess current, the current flowing through the heater will cause it to create excess heat that will trip the normally closed overload contacts. Because the normally closed overload contacts are connected in series with the motor-starter coil, current to the motor-starter coil will be interrupted when the overload contacts open. A reset button on the motor starter must be manually reset to set the overload contacts back to their normally closed position.

As mentioned above, the control circuit is shown as a ladder diagram at the bottom of this figure. In this diagram you can see that the motor-starter coil is energized when the start push button is depressed.

The motor starter also has one or more additional sets of normally open contacts called *auxiliary contacts*. The auxiliary contacts are connected in parallel with the start push button. These contacts serve as a seal-in circuit after the motor-starter coil is energized. The start push button is a momentary-type switch, which means that it is spring-loaded in the normally open position. When the start push button is depressed, current flows from L1 through the normally closed stop push-button contacts and through the start push-button contacts to the motor-starter coil. This current causes the coil to become magnetized so that it pulls in the three major sets of load contacts and the auxiliary set of contacts. When the auxiliary contacts close, they create a parallel path around the start push-button contacts so that current still flows around the start push-button

contacts to the coil when the push button is released. Because the stop push button is connected in series in this circuit, the current to the coil is de-energized, and all the contacts drop out when the stop push button is depressed.

11.3
THE OPERATION OF THE OVERLOAD

The overload for a motor starter consists of two parts. The heater is the element that is connected in series with the motor, and all the motor current passes through it. The heater is actually a heating element that converts electrical current to heat. The second part of the overload device is the trip mechanism and overload contacts. The trip mechanism is sensitive to heat, and if it detects excess heat from the heater, it will trip and cause the normally closed overload contacts to open. Because the motor-starter coil is connected in series with the normally closed overload contacts, all current to the coil will be interrupted, and the coil will become de-energized when the overload contacts are tripped to their open position. When the coil becomes de-energized, the motor-starter contacts will return to their open position and all current to the motor will be interrupted. When the overload contacts open, they remain open until the overload is reset manually. This ensures that the overloaded motor stops running and cools down until someone comes to investigate the problem and resets the overloads.

Figure 11–3 shows a typical heating element for a motor-starter overload device, which is called the *heater*. In the cutaway diagram for this figure you can see that the heater is actually a heating element that converts electrical current into heat as it passes through the heating element. You should also notice the knob protruding from the bottom of the heater. This knob has a shaft that is held in position inside the heater and teeth machined into the part that protrudes from the heater. The teeth are called the *ratchet mechanism.*

Figure 11–4 shows the heating element mounted into the trip mechanism. The trip mechanism consists of the ratchet from the heater and a pawl. The ratchet is actually the knob that protrudes from the bottom of the heater that has teeth in it. The pawl has spring pressure, which tries to rotate the ratchet. Because the heater holds the shaft of the ratchet tight with solder, the pawl cannot move. When the heater becomes overheated, it melts the solder that holds the ratchet in place and allows it to spin freely. When the ratchet spins, it allows

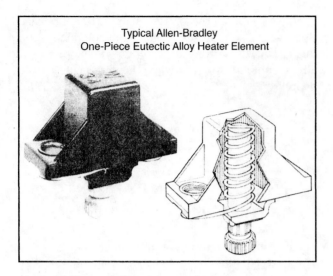

Figure 11–3   A typical heater assembly. Notice the ratchet that protrudes from the bottom of the heater. (Courtesy of Rockwell Automation's Allen-Bradley Business)

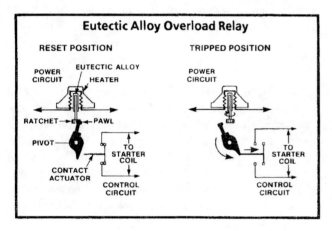

Figure 11–4   Diagram of the ratchet and pawl mechanism that trips and allows the normally closed overload contacts to open. (Courtesy of Rockwell Automation's Allen-Bradley Business)

the pawl to move past it, which in turn allows the normally closed contacts to move to their open position.

After the overload condition occurs and the overload contacts open, the motor starter de-energizes and the motor stops running. When the motor stops running, the heating element is allowed to cool down. After the heating element cools down for several seconds, the reset button can be depressed, which moves the pawl back to its original position, and the normally closed overload contacts move back to their closed position. If the motor continues to draw excess current when it is restarted, the excess current will cause the heater to trip the overload mechanism again. If the motor current is within specification, the heaters will not produce enough heat to cause the overload mechanism to trip.

Because the over-current condition must last for several minutes for the overload mechanism to trip, the overloads allow the motor to draw high locked-rotor amperage (LRA) during the few seconds the motor is trying to start without tripping. If the motor has an over-current condition while it is running, the overloads allow the condition to last several minutes to see if it clears up on its own before the motor is de-energized. If the problem continues, the overloads sense the over-current and trip to protect the motor. Figure 11–4 shows the overload contacts in their normally closed position and in their tripped, or open, condition.

11.4
EXPLODED VIEW OF A MOTOR STARTER

It may be easier to see all the parts of the motor starter in an exploded view. Figure 11–5 shows an exploded view of the motor starter. At the far left in this figure, you can see that the main contacts of the motor starter are much larger than those in a traditional relay. The coil is shown in the middle of the picture. The coil has two square holes in it, which allow the magnetic yoke to be mounted through it. The magnetic yoke and coil are mounted in the movable contact carrier. When the coil is energized, it pulls the magnetic yoke upward, which causes the movable contacts to move upward until they make contact with the stationary contacts that are shown at the far left. The overload mechanism is shown at the far right in this figure. It is mounted at the lower part of the motor starter, and all current that flows through the contacts must also flow through the overload mechanism.

11.5
SIZING MOTOR STARTERS

At times you will need to select the proper-size motor starter for an application. The size of motor starters is determined by the National Electrical Manufacturers Association (NEMA). The ratings refer to the amount of current the motor-starter contact can safely handle. The sizes are shown in the table provided in Fig. 11–6, and you can see the smallest-size starter is a size 00, which is rated for 9 A, which is sufficient for a 2-hp three-phase motor connected to 480 V or for a 1-hp single-phase motor connected to 240 V. You can see that the size 1 motor starter is rated for 27 A, which is sufficient for a 10-hp, three-phase motor connected to 480 V or a 3-hp single-phase motor connected to 240 V. A size 00 motor starter

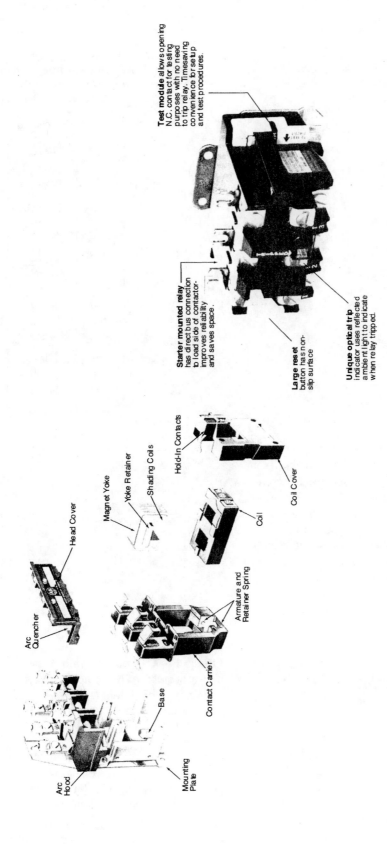

Figure 11-5 Exploded view of a motor starter. The stationary contacts are shown at the far left, and the movable contacts are shown to the right of the stationary contacts. The coil and magnetic yoke are shown in the middle of the picture. The overload mechanism is shown at the far right side of the figure. (Courtesy of Rockwell Automation's Allen-Bradley Business)

Arc Hood

Mounting Plate

Base

Contact Carrier

Arc Quencher

Head Cover

Magnet Yoke

Yoke Retainer

Shading Coils

Armature and Retainer Spring

Hold-In Contacts

Coil

Coil Cover

Test module allows opening N.C. contact for testing purposes with no need to trip relay. Timesaving convenience for setup and test procedures.

Starter mounted relay has direct bus connection to load side of contactor—improves reliability and saves space.

Large reset button has non-slip surface

Unique optical trip indicator uses reflected ambient light to indicate when relay tripped.

Figure 11–6 The NEMA sizes for motor starters. Notice the smallest motor starter is a size 00, and the largest is a size 9. (Courtesy of Rockwell Automation's Allen-Bradley Business)

		Maximum Horsepower Rating Full Load Current Must Not Exceed "Continuous Ampere Rating"			
Ratings for NEMA Full Voltage Starters					
		Motor Voltage			
				50 Hz	
NEMA Size	*Continuous Ampere Rating*	*200V*	*230V*	*380V-415V*	*460V-575V*
00	9	1-1/2	1-1/2	2	2
0	18	3	3	5	5
1	27	7-1/2	7-1/2	10	10
2	45	10	15	25	25
3	90	25	30	50	50
4	135	40	50	75	100
5	270	75	100	150	200
6	540	150	200	300	400
7	810	–	300	600	600
8	1215	–	450	900	900
9	2250	–	800	1600	1600

is about 4 in. high, and a size 2 motor starter is approximately 8 in. high. You will typically use up to a size 3 or 4 motor starter to protect compressor motors for commercial air-conditioning and refrigeration systems. The size 4 starter will protect motors up to 100 hp. It is important to understand that the overload heaters for the motor starter can also be purchased for a specific current rating. This means that each motor starter can have a heater that is rated specifically to the amount of current drawn by the motor that is connected to it.

Figure 11–7 Model 460 motor saver is a solid-state motor protector. This device protects the motor against low voltage, loss of phase, and short cycling. (Courtesy of SymCom, Inc.)

11.6

SOLID-STATE MOTOR PROTECTORS

Solid-state motor protectors provide a feature similar to motor starters in that they protect motors against problems. For example, when the motor is supplied with three-phase voltage, the solid-state motor protector provides protection against the loss of any phase of voltage, low voltage on any phase, high voltage on any phase, reversal of any phase, unbalanced voltage between any phase, and short cycling. In some systems, such as hydraulic pumps, air compressors, and refrigeration compressors, short cycling is a common problem that occurs when a motor is turned off and turned on again after a few seconds before the pressure in the system has a chance to

equalize. This can occur if someone sets the controls incorrectly or if power is interrupted due to a power outage such as one that occurs during a lightning storm and then immediately comes back on. If the motor is allowed to try and start before its pressure is equalized, it will draw large locked-rotor current and possibly not start.

Figure 11–7 shows a solid-state motor-protection device, and Fig. 11–8 shows a diagram of the location of a motor-protection device in a compressor-starting circuit. In the diagram you can see that the motor saver has a set of normally open contacts that are connected in series

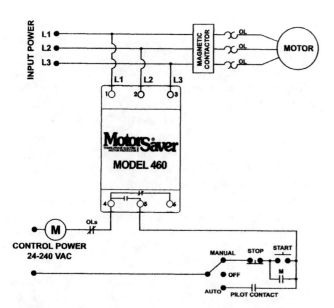

Figure 11–8 Electrical diagram of the model 460 motor saver that is connected to a coil of the motor starter or contactor. (Courtesy of SymCom, Inc.)

with the motor-starter coil. The motor starter samples the voltage at L1, L2, and L3 from the three-phase voltage supplied to the compressor. If the supply voltage has a problem (low voltage, phase loss, etc.), the motor-saver circuit sensing the problem opens its contacts and causes current to the motor starter to be interrupted. A second circuit provides contacts for an audible alarm. When the motor saver trips, the alarm indicates the problem exists.

11.7
OVER-CURRENT CONTROLS

Over-current controls include thermal overloads used in motor starters, thermal overloads used internally in motor windings, circuit breakers, and fuses. In this section you will learn about thermal overloads, solid-state overloads, circuit breakers, and fuses used to protect compressors and other types of motors from drawing too much current and overheating. You will find that the heat that damages motors can come from over-currents or from other conditions such as insufficient air flow that cause the motor to overheat when the current is within normal limits.

11.8
THERMAL OVERLOADS

Thermal overload is specifically designed to detect excessive heat build up in a motor that comes from over-current or from other physical conditions. One appli-

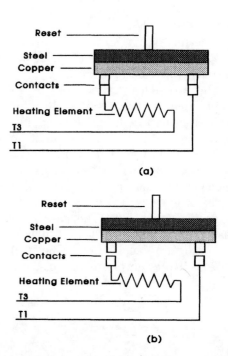

Figure 11–9 (a) Electrical diagram of thermal overload with contacts closed; (b) electrical diagram of thermal overload with contacts open.

cation of a thermal overload is where it is mounted directly on the dome of a compressor to sense the temperature in the compressor. It is also connected in series with the compressor run and start windings to detect excessive current. Figure 11–9 contains a series of diagrams that show how the thermal overload operates. In part a of this figure, you can see the physical shape of the overload and the location of all the parts inside the overload. Part b shows the contacts when the overload senses too much current and the overload contacts are open.

All the current flowing through the compressor motor windings flows in terminal T1 at the bottom of the diagram, through the heating element, and through the contacts and out terminal T3. Because all the current for the motor must flow through the heating element, the overload must be sized for the exact amount of current the motor will draw safely. When the motor draws current in excess of this limit, the heater produces extra heat, which causes the bimetal contacts to warp to the open position. When the overload is allowed to cool, the bimetal will snap the contacts to the closed position.

11.9
FUSES

A fuse performs a similar function to an overload, except the fuse uses an element that is destroyed when the over-current occurs. The fuse provides a thermal-sensing element that is capable of carrying current. When the amount of current becomes excessive, the heat that is generated is sensed by the fuse element, which melts when the temperature is high enough.

Fuses are available in a variety of sizes and shapes for different applications. Figure 11–10 shows a plug-type fuse and a cartridge-type fuse. Each fuse is sized for the amount of current it will limit. When the amount of current is exceeded, the fuse link melts and opens the fuse. Figure 11–11 shows examples of a single-element fuse and a dual-element fuse that are in the process of blowing. The single-element fuse provides protection at one level. This type of fuse is generally used for noninductive loads, such as heating elements or lighting applications. The dual-element fuse provides two levels of protection. The first level is called *slow over-current protection*, and it consists

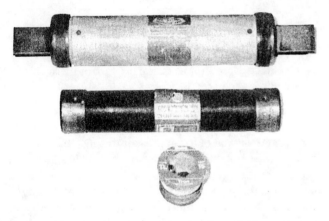

Figure 11–10   Plug-type and cartridge-type fuses.

of a fusible link that is soldered to a contact point and attached to a spring. When a motor is started, it draws locked-rotor amperage (LRA) for several seconds. This excess current causes heat to build in the fuse, and this heat is absorbed in the slow over-current link. If the motor starts and the current drops to the FLA level, the link will cool off and the fuse will not open. If the LRA current continues for 30 s, the amount of heat generated will cause the solder that holds the slow over-current link to melt. When the solder melts, the spring pulls the link open and interrupts all current flowing through the fuse. The slow over-current link allows the fuse to sustain over-current conditions for short periods of time, and if the condition clears, the fuse will not open. If the condition continues to exist, the fuse will open.

The second type of element is called the short-circuit element. This element opens immediately when the amount of current exceeds the level of current the link is designed to handle. In the dual-element fuse, the short-circuit link is sized to be approximately five times the rating of the fuse. A short circuit is by definition any current that exceeds the full-load current rating by five times. (Some manufacturers use a factor of 10 times the rating.)

The single-element fuse has only a short-circuit element in it. These types of fuses are generally not used in circuits to start motors because the motor draws LRA. If a single-element fuse is used to protect a motor circuit, it must be sized large enough to allow the motor to start, and then it is generally too large to protect the motor against an over-current condition of 20%, which will eventually damage the motor if it is allowed to occur for several hours.

Figure 11–11   (a) Single-element fuses in the process of blowing; (b) dual-element fuses in the process of blowing. (Courtesy of Cooper Bussmann)

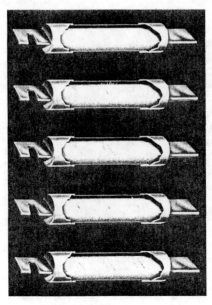
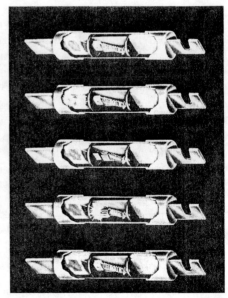

(a)   (b)

11.10
FUSED DISCONNECT PANELS

The cartridge-type and screw-base-type fuses are generally mounted in a panel called a *fused disconnect*. The fused disconnect is normally mounted near the equipment it is protecting, and it serves two purposes: First, it provides a location to mount the fuses; second, it provides a means of disconnecting the electrical supply voltage to a circuit. A three-phase disconnect is shown in Fig. 11–12. The fuse disconnect has a switch handle that is used to disconnect the main power from the circuit and the fuses so that you can safely remove and replace or test the fuses.

SAFETY NOTICE!

It is recommended that you always use plastic fuse pullers to remove and replace fuses from a disconnect to protect you from electrical shock even when the fuse-disconnect switch is in the off position. It is important to remember that even though the switch handle is open, line voltage is still present at the top terminal lugs in the disconnect. You should never use metal pliers or screwdrivers to remove fuses.

11.11
CIRCUIT BREAKERS AND LOAD CENTERS

The *circuit breaker* is similar to the thermal overload in that it is an electromechanical device that senses both over-current and excess heat. Some circuit breakers also sense magnetic forces. Circuit breakers are mounted in electrical panels called *load centers*. The load center can be designed for three-phase circuits or for single-phase circuits. The single-phase panel provides 240-VAC and 120-VAC circuits. Figure 11–13 shows a typical load center without any circuit breakers mounted in it.

The circuit breakers are manufactured in three basic configurations for single-phase, 120-VAC applications that require one supply wire; 240-VAC, single-phase applications that require two supply wires; and three-phase applications that require three wires. The three-phase circuit breakers can be mounted only in a load center that is specifically manufactured for three-phase circuits. The two-pole and single-pole breakers can be mounted in a single-phase or three-phase load center. Figure 11–14 shows examples of circuit breakers.

The operation of the circuit breaker is similar to the thermal overload in that it senses excessive current that will trip its circuit after a specific amount of time.

Figure 11–12 A three-phase fused disconnect. (Courtesy of Eaton Electrical)

Figure 11–13 A load center that is used to mount circuit breakers. The load center is sometimes called a circuit breaker panel. (Courtesy of Eaton Electrical)

Figure 11–14 Single-pole, two-pole, and three-pole circuit breakers for load centers. (Courtesy of Eaton Electrical)

The main difference between the circuit breaker and the thermal overload is that the circuit breaker is mounted in the load center to protect both the circuit wires and the load. The circuit breakers are sized to the total current rating for the wire and all the loads that are connected to the wire. In some cases this means that the circuit breaker is sized for the current flowing to several motors that are connected on the circuit. For instance, if a system has a hydraulic pump that draws 20 A and two conveyor motors that each draw 5 A, the circuit breaker would be rated for 30 A. A problem can occur if the hydraulic pump is turned off and the conveyors are running. If one of the conveyors has an overload that doubles its current to 10 A, the circuit breaker will not be able to detect 10 A in the conveyor motor as an overload because the total circuit current is below the 30-A total. In these circuits it may be necessary to use a circuit breaker to protect the entire circuit and overloads at each motor to protect them individually from overheating. This means that the main job of the circuit breaker is to protect the circuit against short-circuit conditions and protect the entire circuit against over-current conditions rather than protecting individual motors against over-current. This is why many circuits have a combination of circuit breakers and overload devices.

QUESTIONS

Short Answer

1. Explain the differences and similarities between a motor starter and a contactor.

2. An overload on a motor starter has contacts and a heater. Describe the function of each.

3. Identify the parts of thermal overload used to protect a compressor and explain each part's operation.

4. Discuss the differences between a single-element fuse and a dual-element fuse.

5. Explain the difference between slow over-current and short-circuit current.

True or False

1. The heater is the part of the motor-starter overload that opens and interrupts current flow.

2. The overload contacts on a motor starter are normally open and interrupt current when an overload condition is detected.

3. The main job of the circuit breaker is to protect the circuit against short circuits and the entire circuit against over-current rather than to protect individual motors against over-current.

4. A fused disconnect provides a location for mounting fuses for the system and also provides a means of disconnecting main power from the circuit.

5. A thermal overload device may be mounted internally in a motor such as an air-conditioning system's sealed compressor to detect heat inside the compressor shell.

Multiple Choice

1. A motor starter _____.
   a. is just a larger version of a relay, and it has only a coil and contacts
   b. has a coil and contacts like a relay or contactor and also has overloads to provide over-current protection
   c. is more like a circuit breaker than a relay in that it can detect excessive heat buildup on the outside of the motor through its heaters

2. A dual-element fuse _____.
   a. can detect slow over-current and large short-circuit currents
   b. can be used two times before it must be replaced
   c. has two elements so it can be used in single-phase circuits to protect both L1 and L2 supply-voltage lines

3. The overload device on a motor starter _____.
   a. has a heater to detect excess current and overload contacts to interrupt the control circuit current
   b. has a heater to detect excessive current and overload contacts that directly interrupt the large load current to the motor
   c. has a heater that interrupts control-circuit current and overload contacts that directly interrupt the large load current to the motor

4. The motor starter has auxiliary contacts that _____.
   a. are connected in parallel with a start switch to seal in the circuit when it is operating correctly and ensure the coil remains de-energized if the overload contacts trip
   b. are connected in series with a start switch to seal in the circuit when it is operating correctly and ensure the coil remains de-energized if the overload contacts trip
   c. can switch current to an auxiliary motor such as a conveyor motor at the same time the main contacts supply current to a hydraulic pump motor

5. Heaters for motor starters are available in different sizes so that they can _____.
   a. provide the correct amount of current protection to match the motor connected to the motor starter
   b. match the voltage rating of the fuses used in the circuit
   c. match the temperature rating of the ambient air in which the motor will operate

PROBLEMS

1. Draw a sketch of a motor starter that has its contacts connected to a three-phase motor and its coil controlled by start and stop push buttons. Be sure to show the heaters and the overload contacts in the proper circuits.

2. Draw a sketch of a thermal-type overload and explain the function of each of the parts.

3. Draw a sketch of a single-element fuse and a dual-element fuse and explain how they are different.

4. Draw a sketch of a thermal overload device connected to the outside of a compressor and explain how it protects the compressor against overheating.

5. Use the table shown in Fig. 11–6 to select a motor starter for a single-phase, 3-hp motor that draws 27 A at 240 V.

CHAPTER 12

Motor-Control Devices and Circuits

OBJECTIVES

After reading this chapter you will be able to:

1. Identify pilot devices in a control circuit and their symbols in electrical diagrams.

2. Identify the load circuit and control circuit.

3. Explain the operation of a push-button switch, limit switch, flow switch, level switch, and pressure switch as they are used as pilot devices.

4. Identify a two-wire control circuit.

5. Identify a three-wire control circuit.

6. Explain the operation of a drum switch as it is used to reverse the operation of an AC or DC motor.

7. Explain the operation of a single-element and a dual-element fuse.

8. Identify the proper enclosure for specific industrial applications.

12.0

INTRODUCTION

In this chapter we explain basic motor-control circuits commonly in use on the factory floor. It is very important to understand that many of the electronic circuits are used to supplement or replace standard motor-control circuits that have been in operation for the last 30 years. These motor-control circuits have been named and are easily identified so that when you hear about them on the job or when you see them on the job you will easily recognize them. You must also understand

that in the past few years, companies have changed their concept of an electrical or maintenance technician, and they now expect to hire each person to troubleshoot and repair the motor-control systems as well as the electronic systems. Traditionally the motor-control systems have been tested by electricians and the electronics technician. In the past 10 years, companies have streamlined their personnel so that they expect the maintenance technician to work on motor-control circuits. For this reason it is vitally important that you learn as much as you can about standard motor-control circuits so that you can compete for the best-paying jobs.

The circuits in this chapter are identified so that you can come back and study them when you are on the job if you need to review them. This will be extremely useful when you find machinery that incorporates four or five of these basic circuits with additional complex circuits. In these circuits you will be able to gain knowledge that you can transfer to more difficult circuits for troubleshooting and repair. These circuits will include pilot devices, motor starters, and different types of motors. If you do not understand the function or operation of relays and motor starters, you may need to review the chapters where they were introduced.

12.1

PILOT DEVICES

Pilot devices are a group of components that include push-button switches, limit switches, and other switches that are commonly found in motor-control circuits. These switches are called pilot devices because they are rated for control-circuit voltage and current. That is, these switches normally have 120 VAC and less than 1 A current flowing through their contacts. The

Figure 12–1 A typical set of push-button switches and selector switches used to operate a machine. (Courtesy of Rockwell Automation's Allen-Bradley Business)

control circuits use pilot duty switches to energize or de-energize the coil of a relay, motor starter, solenoid, or indicator lamp. All the loads in these circuits use low current. Again, you must understand that the L1-N voltage in your part of the country may be any value between 110 and 125 VAC. In this chapter all these voltages are referred to as 120 VAC.

Figure 12–1 shows a set of multiple push-button switches and selector switches that are pilot devices used to control a large industrial machine. These push buttons are the most commonly used switches to start and stop a machine, and selector switches are used to select the operation of the machine for such functions as manual or automatic operation. Figure 12–2 shows an exploded view of a typical push-button switch installation. From this diagram you can see that the switch is actually an assembly of a number of parts, including the push-button actuator on which the operator presses to

activate the switch, a number of seals to keep moisture and other contaminants out of the switch assembly, and the contact block and contacts where the wires are actually connected. When the push-button activator is depressed, the contacts in the bottom of the switch transition. If the switch has normally open contacts, they will go closed; and if the switch has normally closed contacts, they will go open. Some switches will have one set of contacts, whereas other switches have multiple sets of contacts. The important point to recognize is that all wires are connected to the contact block, and the other parts of the switch can be changed without removing any wiring. This allows switches to be changed in the field with a minimum of downtime for rewiring.

12.1.1
Limit Switches and Other Pilot Switches

A variety of pilot switches are available for use in control circuits. One of the most popular is called a *limit switch*. Figure 12–3 shows examples of several types of limit switches. These switches are physically mounted in the machine so that the motion of the machine or the parts around the switch will cause the switch to activate. Figure 12–3a shows a yoke-type limit switch. The yoke has a roller on each section of its activator arm. This arrangement is used where machine motion is in two directions. When the machine moves in one direction, it strikes the right-hand roller and causes it to switch to the right. This action causes the right roller to be pressed downward, and the left roller snaps to the up position, where it is in position to detect machine motion when the machine moves back to the left. The yoke arm on the switch will continue to detect motion to the right and then to the left. This type of switch is very useful for surface grinders, where the part being finished is moved back and forth under the grinding wheel. Each time the part moves to the end of its stroke, the switch is activated in the other direction.

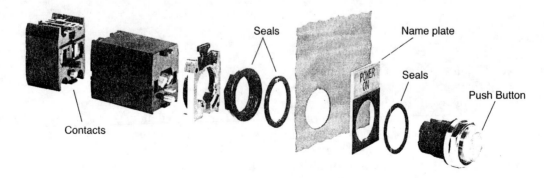

Figure 12–2 An exploded view of a typical switch. Notice that the switch-activation push button is shown above the panel where it is accessible to the operator, and the remainder of the switch, including rubber seals, and the switch contacts are shown below the panel. (Courtesy of Rockwell Automation's Allen-Bradley Business)

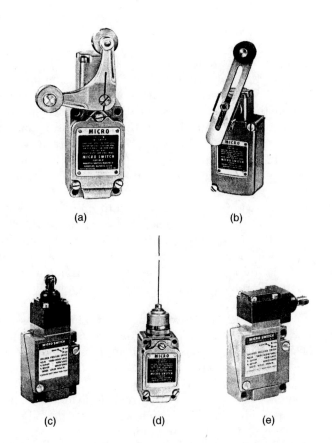

Figure 12–3 (a) Fork-lever-roller yoke-type limit switch; (b) roller-arm-type limit switch; (c) top-roller-type limit switch; (d) wobble-lever-actuated cat whisker limit switch; (e) side-roller-type limit switch. (Courtesy of Honeywell Sensing and Control, Honeywell International, Inc.)

LIMIT SWITCHES	
Normally Open	Normally Closed
⌇⌇	⌇⌇
⌇⌇	⌇⌇
Held Closed	Held Open

Figure 12–4 The electrical symbol for the normally open limit switch, the normally open, held-closed limit switch, the normally closed limit switch, and the normally closed, held-open limit switch.

switch is also much more sensitive than other types of limit switches. Figure 12–3e shows a side-roller-type limit switch. This type of limit switch is mounted in such a way as to detect end of travel. When the machine reaches the end of its travel, it depresses the plunger and causes the switch to activate.

12.1.2
Symbols for Limit Switches and Other Motor-Control Devices

The limit switch can be connected as a normally open or a normally closed switch. It can also be used on a machine in such a way that the machine travel will hold the switch in the normally open or normally closed position before the machine starts its travel. Figure 12–4 shows the four electrical symbols for a limit switch. The top left switch symbol is for a *normally open limit switch*. In this symbol notice that the switch's contact arm is shown below the terminal on the right side of the switch and that the switch is shown as a normally open switch. The *normally open, held-closed limit switch* is shown in the bottom left of the figure. You should notice this is similar to the normally open switch in that the switch contact arm is shown below the switch's output terminal on the right side. The major difference with this switch is that it is shown with normally closed contacts. The reason it is shown with normally closed contacts is because the machine motion will keep this switch in its normally closed position.

Figure 12–3b shows a limit switch with a single roller arm. This type of switch detects motion in only one direction. When the machine moves past the roller arm, it moves the arm to the right so that the switch is activated. Figure 12–3c shows a top-roller-type limit switch. This type of switch has a small roller mounted on a plunger-type arm. When the machine moves past the roller, it causes the plunger to depress to activate the switch. When the machine moves past the switch, a spring causes the plunger to move back upward into place to be activated again. This type of switch can detect motion in either direction. Another variation of this type of switch uses a blunt end on the plunger, which means the switch can detect only motion that is directed down on the plunger. This type of switch is generally mounted at the very end of machine travel, and when the machine reaches the end of its stroke, it depresses the plunger.

Figure 12–3d shows a wobble-lever-actuated limit switch that is sometimes called a *cat whisker*–type limit switch. This type of limit switch can detect motion in any direction. This means that any movement of the cat whisker can cause the switch to activate. This type of

NOTICE!
It is important to remember that when you are purchasing limit switches, the switch is available only as a normally open or as a normally closed switch. The held-open and held-closed conditions occur when the switch is mounted in place on machinery, and the location of the switch causes the machine to hold the switch in the activated position.

The switch in the top right section of this figure is a *normally closed limit switch*. This symbol is different in that the switch contact is shown on top of the switch's output terminal. When the machine motion activates the switch, it moves the contact arm upward to its open position. When this switch is wired in the field, you should select the normally closed contacts.

The symbol for the *normally closed, held-open limit switch* is shown in the bottom right corner of this figure. You can see this symbol shows the switch's contact arm in the open position, and it is shown above the output terminal. When you wire this switch in the field, you use the normally closed set of contacts, and the position of the machine when it is at rest keeps the switch held in the normally open position.

12.1.3
Electrical Symbols for Other Pilot Devices

Other pilot devices, such as pressure switches, flow switches, float switches, temperature-activated switches, time-delay relays, and lamps, are also used in motor-control circuits. The symbols for these devices are shown in Fig. 12–5. These symbols are a standard for pilot devices used in motor-control circuits. When you encounter a typical motor-control diagram on a blueprint, you can use the symbols in this figure to identify each type of switch. You should notice that the symbol for the part that causes the switch to activate is shown at the bottom of each symbol. For example the symbol for a liquid-level switch is a circle, which represents the ball-type float typically found on this type of switch, and the symbol for the pressure-type switch looks like a pressure dome. The symbol for each switch should look very similar to the actuator for that switch. You should also notice that the contact arm for each switch is shown in the normally open or normally closed position.

When you examine each symbol carefully, observe whether its contacts are normally open or normally closed. You should also understand that the switch arm can be drawn on top of the terminals or below the terminals. Each of these conditions allows four combinations of switches to be identified. It is important to understand these subtle differences. For example, Fig. 12–6 shows a pressure switch drawn with its contact arm in each of the four ways described previously, so each switch is different from the other. Figure 12–6a shows the switch arm normally closed and above the contact terminal; the switch is activated by pressure. This indicates that the

switch is a normally closed switch, and an increase in pressure causes the switch to open. It may help to remember the switch symbol by the type of application for which the switch would be used. For example, this type of switch is used as a high-pressure safety switch. When the pressure gets too large in a system, the switch opens and shuts off a compressor that is adding pressure to the system. The pressure switch in Fig. 12–6b shows a normally open switch with the switch arm above the contact terminal. This type of switch activates a circuit when pressure falls. This type of switch can be used to turn on a compressor when air pressure drops below a minimum so the system will sustain the pressure. Because the switch arm is drawn above both these switches, they may also be called *high-pressure switches* when equipment manufacturers refer to them, because they typically sense high pressure.

Figure 12–6c shows a normally closed pressure switch with the switch arm below the contact terminal. This symbol indicates that when pressure decreases, the switch contacts will fall to the open position. This type of switch can be used as a low-pressure safety switch. For example, if a system is to maintain a minimum oil pressure, the switch will open if the oil pressure drops below the setpoint. Figure 12–6d shows a normally open pressure switch with the switch arm below the contact terminal. When pressure increases, it causes the switch arm to go up and close. This type of switch can be used to start a motor when the pressure gets too high. Because the switch arm is shown below the contact on both these switches, they may be referred to as *low-pressure switches*.

It is important to remember that the pressure switches are sold only with normally open or normally closed contacts or the combination of one set of normally open and one set of normally closed contacts. Again, as with the limit switches, the conditions in which the switches are used when they are mounted on machinery may be referred to in the name the machine manufacturer gives the switch. For example, the switch may be called the compressor high-pressure cutout safety, or it may be called the low-oil-pressure protection switch. In both cases, the switch is still a pressure switch, and a set of normally open or normally closed contacts is used to control circuit current when pressure increases or decreases.

You should now understand that whichever way the contact arm is drawn, a decrease in the physical parameter the switch is measuring will cause the switch arm to drop, and an increase will cause the arm to raise. The next section shows how all the pilot devices are used together to make common motor-control circuits.

STANDARD ELEMENTARY DIAGRAM SYMBOLS

The diagram symbols shown below have been adopted by the Square D Company and conform where applicable to standards established by the National Electrical Manufacturers Association (NEMA).

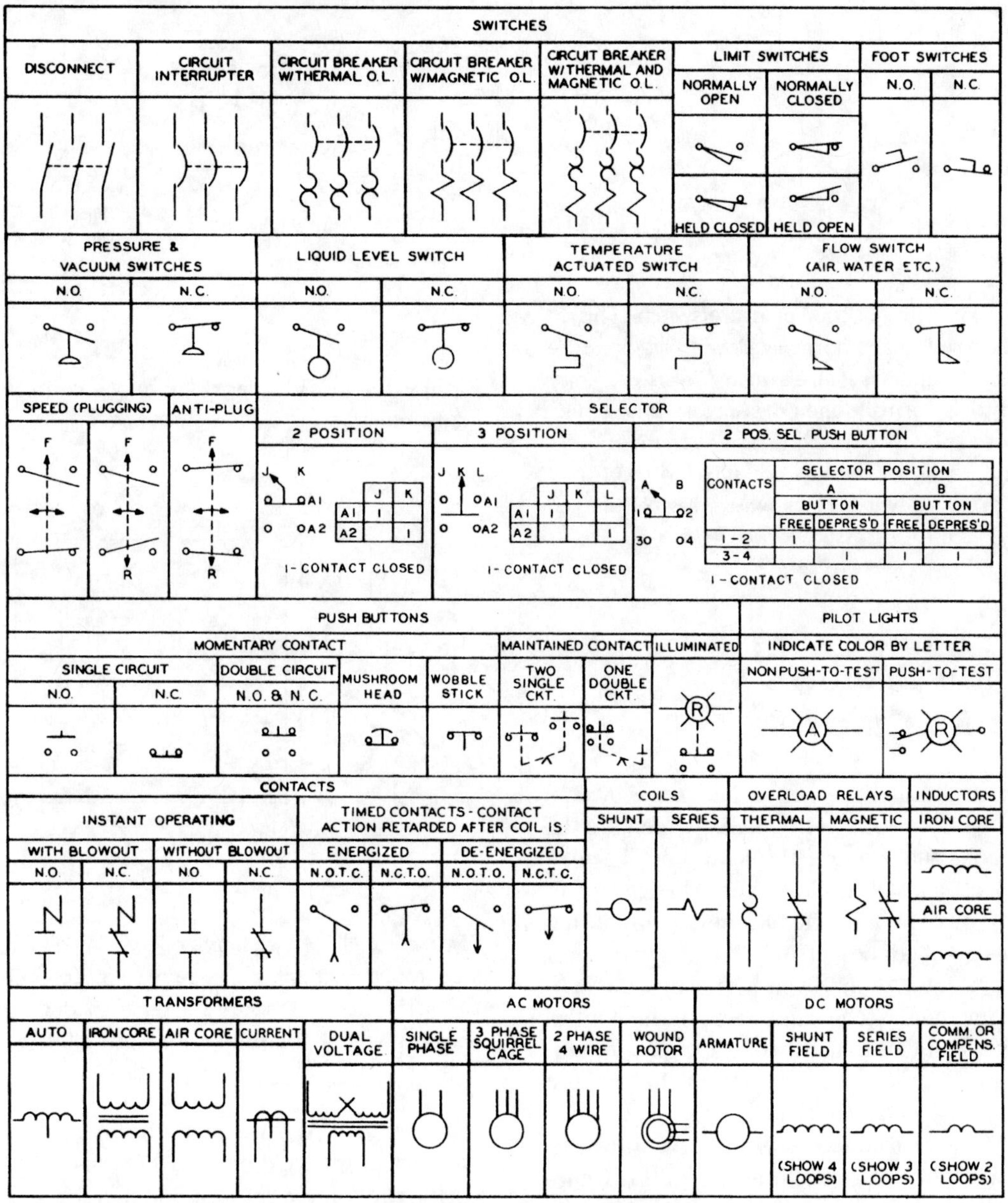

SUPPLEMENTARY CONTACT SYMBOLS

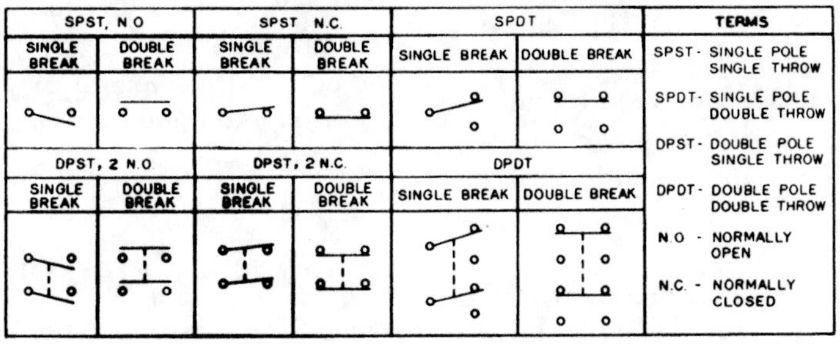

Figure 12–5 Electrical symbols for pilot devices and other motor-control devices. (Courtesy of Square D/Schneider Electric)

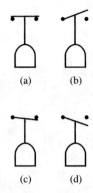

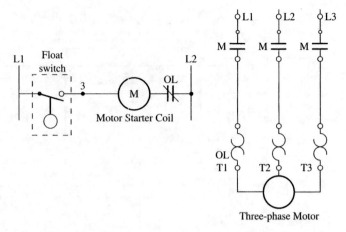

Figure 12–6 (a) Normally closed pressure switch. (This switch may be referred to as a normally closed, high-pressure switch.) (b) Normally open pressure switch. (This switch may also be referred to as a normally open, high-pressure switch.) (c) Normally closed pressure switch. (This switch may be referred to as a normally closed, low-pressure switch.) (d) Normally open pressure switch. (This switch may be referred to as a normally open, low-pressure switch.)

12.2
CONTROL CIRCUITS AND LOAD CIRCUITS

In motor-control circuits, the components are broken into two groups: the *control components*, which are the switches and coils that control the load, and the *load components*, such as motors that cause the system to perform some type of action. The part of the circuit that has the controls is called the *control circuit*, and the part of the circuit that has the motor in it is called the *load circuit*. Typically, the switches and wires in the control circuit use lower voltages than the load and smaller currents than the load. In fact, in most circuits, the load is a single-phase or three-phase motor.

Figure 12–7 shows a ladder diagram of a control circuit and a load circuit. You should notice in this figure that the control circuit is shown on the left, and it has a float switch connected to the relay coil. A normally closed set of overload (OL) contacts is shown connected in series with the motor-starter coil. The load circuit is shown on the right side of the figure, and you can see that it consists of a three-phase motor starter and a three-phase motor. (*It is important to note that the relay coil is technically a load, but it is considered part of the control circuit because of the function it performs.*)

The same control and load circuits are shown as a wiring diagram in Fig. 12–8. When this circuit is shown as a wiring diagram, you will notice that some of the control-circuit components, such as the motor-starter coil, are drawn where they are physically located, on

Figure 12–7 A ladder diagram of a control circuit and a load circuit.

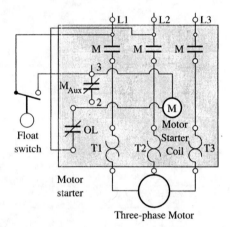

Figure 12–8 An example of a wiring diagram that shows a control circuit and a three-phase motor in the load circuit. Notice the motor starter is shaded in the wiring diagram so that you can locate it and all its terminals.

the motor starter. Because the motor-starter contacts are part of the load circuit, it is hard to determine in a wiring diagram what components are part of the load circuit and what components are part of the control circuit. In some diagrams the load circuit is usually drawn with darker and wider lines than the wires of the control circuit to indicate the wires are larger. The remaining part of this chapter explains the different types of control-circuit and load-circuit configurations.

12.2.1
Two-Wire Control Circuits

The two-wire control circuit is commonly used in applications where the operation of a system is automatic and basically two wires are used to provide voltage to the load. This may include such applications as sump pumps, tank pumps, electric heating, and air compressors. In these systems you typically close a disconnect switch or circuit breaker to energize the circuit, and the actual energizing of the motor in the system is controlled by the operation of the pilot device.

The circuit is called a two-wire control because only two wires are needed to energize the motor-starter coil. The circuit controls that are located in front of the coil will provide the operational and safety features, and the overload contacts that are located after the coil are used to protect the circuit against over-currents.

Figure 12–7 shows a ladder diagram of the two-wire control circuit for a pumping station, and Fig. 12–8 shows the control circuit as a wiring diagram. You will notice that it is difficult to see the sequence of operation in the wiring diagram, so you generally use the ladder diagram to determine the sequence of operation. The wiring diagram shows the location of all wires and components.

Voltage to the pumping station is provided through a fused disconnect (not shown). After the disconnect is closed, the float switch is in complete control of the motor starter. Notice that the float control is identified as a float switch. When the level of the sump rises past the set point on the control, its contacts energize, and the motor-starter coil becomes energized. The coil of the motor starter energizes and pulls in the motor-starter contacts, which energize the motor. As the motor operates, it pumps the liquid, which reduces the level in the sump. When the level of the sump is lowered, the level switch senses the change in level and switches its contacts off, which de-energizes the motor-starter coil. When the motor-starter coil is de-energized, the motor-starter contacts are opened, and the motor is turned off.

This on-off sequence continues automatically until the disconnect is switched off or unless the motor starter overloads are tripped. If the motor draws too much current when it is pumping, the heaters trip the overload contacts and open the control circuit, which interrupts current flow to the motor-starter coil. Because the overloads must be reset manually, the circuit does not become energized automatically when the overload is cleared. In this case, someone must physically come to the motor starter and press the reset button. At this time, the system should be thoroughly inspected for the cause of the overload. It is also important to use a clamp-on ammeter to determine the full-load amperage (FLA) of the motor and cross-reference this to the size of the heaters in the motor starter to see that they match. The FLA should be checked against the motor's data plate. You should also take into account the service factor listed on the data plate when you are trying to determine if the pump motor is operating correctly.

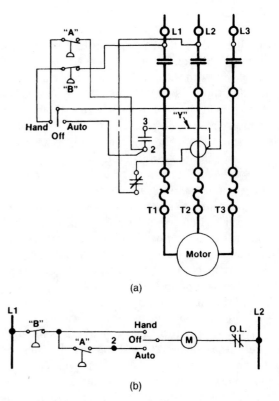

Figure 12–9 (a) A wiring diagram of a two-wire control circuit for a compressor motor. The compressor motor is controlled by a motor starter that is energized by one of two pressure switches. (b) A ladder diagram of the same circuit shown in the wiring diagram. In this circuit a selector switch provides hand or auto control. Pressure switch A and B are both in the circuit when the selector switch is in auto, and pressure switch B is only in the circuit when the selector switch is in manual. (Courtesy of Rockwell Automation's Allen-Bradley Business)

Figure 12–9 shows a more complex two-wire control circuit that is used to control an air compressor. Notice that the wiring diagram for the control and load circuit is shown in Fig. 12–9a, and the ladder diagram of the control circuit is shown in Fig. 12–9b. This circuit is still a two-wire control circuit, and it is used to turn the air compressor on automatically when the pressure drops below 30 psi and to turn off the compressor when the pressure reaches 90 psi. Pressure switch A in this circuit controls the operation of the control at these pressures. Its high and low pressures are adjustable so that the system can be energized and de-energized at other pressures if the need arises. A hand switch is provided in this circuit that allows the system to be pumped to a predetermined pressure. The hand switch is intended to be used when the automatic switch is not functioning properly or when you need to test the system.

Pressure switch B in this diagram acts as a safety for this circuit. This pressure switch is set at 120 psi and is not adjustable. It is in the circuit to prevent the tank pressure from rising too high. This can occur if the operational pressure switch becomes faulty and will not open when the pressure reaches 90 psi. It can also be used to protect the system against too much pressure when the switch is in the hand position. Generally, the safety switch is meant to protect the system against component failure or control failure.

If a pressure control were to fail, the compressor would continue pumping air into the tank, which would allow its pressure to rise to an unsafe level. Because the operational switch would still be closed, the air pressure in the tank could be increased to a level where the pump would cause the motor to stall if the safety switch were not in the circuit. This could build up pressures to several hundred pounds, which could cause lines and fittings in the air system to explode.

The nonadjustable pressure switch acts as a backup to the operational switch and trips off any time the pressure reaches 120 psi. This switch is also different in that it is interlocked when it trips, so that it must be reset manually by having someone pressing the reset button. If you find the reset button activated on the safety switch, you must test the system thoroughly to determine why the operational switch did not control the circuit.

In normal operation, the motor starter cycles the air compressor on and off to keep the pressure in the tank between the high and low set points on the operational control. The motor-starter overloads could also trip this circuit and require manual reset. This could occur if the motor were incurring overcurrent problems. The over-current could occur due to bad bearings or because someone has physically depressed the switch. When this push button is released, the normally open contacts return to their normally open state.

The normally open auxiliary contacts from the motor starter close when the motor-starter coil is energized. When they close, they provide an alternative path around the push-button contacts for current to get to the coil. These contacts are called *seal* or *seal-in* contacts when they are used in this manner. Sometimes the seal contacts are said to have memory, because they maintain the last state of the push buttons in the circuit.

12.2.2
The Three-Wire Control Circuit

The *three-wire control circuit* is the most widely used motor-control circuit. This circuit is similar to the two-

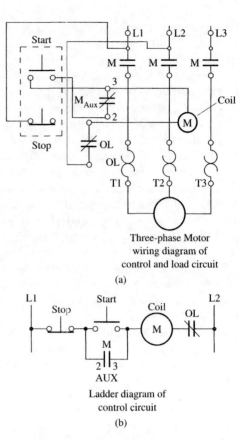

Three-phase Motor
wiring diagram of
control and load circuit

(a)

Ladder diagram of
control circuit

(b)

Figure 12–10   (a) A wiring diagram of a typical three-wire control system. This control circuit gets its name because of the auxiliary contacts that are connected in parallel with the start button. The auxiliary contacts seal in the circuit to keep the coil energized after the start push button is released. (b) A ladder diagram of a three-wire control circuit. (Courtesy of Rockwell Automation's Allen-Bradley Business)

wire circuit except it has an extra set of contacts that are connected in parallel around one of the original pilot switches to seal it in. The extra set of parallel contacts provides the third wire, which also gives this circuit its name. You should fully understand this circuit and learn to recognize it when it is shown as a ladder diagram and as a wiring diagram. In this way you will be able to understand the operation of the circuit wherever it appears.

One variation of the three-wire circuit is shown in Fig. 12–10. In this figure, a stop push button and a start push button are used as the pilot devices for control. Figure 12–10a shows the wiring diagram for the three-wire circuit, and Fig. 12–10b shows the ladder diagram for the three-wire control circuit. In the ladder diagram you can see that the stop push button is wired normally closed. When the start push button is depressed, voltage flows through its contacts to energize the motor-starter coil. When the coil becomes energized, it pulls its main contacts closed to cause the motor to become energized, and it also pulls its normally open auxiliary contacts that are connected across the start push button to their closed positions. When the auxiliary contacts close, they provide an alternative route for voltage to travel around the start push-button contacts, which will return to their normally open position when the switch is no longer depressed. The auxiliary contacts are called the *seal-in* circuit or, in some cases, the *memory circuit*. If the auxiliary contacts are not used, you need to keep your finger on the start button to keep the circuit energized.

You should notice that in the three-wire circuit, the stop button is used to unseal the circuit and de-energize the motor-starter coil. Any time the stop button is depressed, the coil becomes de-energized and the seal-in circuit drops out. This means that the start push button must be depressed again to energize the circuit.

It is also important to understand in the three-wire circuit that the stop push button traditionally is always the first switch in the circuit. Most people assume this has something to do with safety. In reality, it has to do with economics. When you look at a motor starter, you will notice that the auxiliary contacts are physically located near the coil. If the start button is connected in the circuit next to the coil, the wire that is connected to the auxiliary contact on the right side (terminal 3) can actually be a short jumper wire that is connected directly to the left side of the coil. If the start button is not located next to the coil, the wires that connect the auxiliary contacts in parallel with the start push button need to be as long as the distance between the start push button and the motor starter. This may be a distance of several hundred feet.

12.2.3
A Three-Wire Start-Stop Circuit with Multiple Start-Stop Push Buttons

Additional start and stop push buttons may be needed on a system that is very large. For example, if the system is installed over a large area, such as for a long conveyor system that may be more than several hundred feet long, it would be inconvenient and unsafe to have one start and stop button at only one end of the system. To make the system more safe and to make it more convenient to start and stop the system, push-button switches can be installed every 40 ft. Each additional start button should be connected in parallel with the first start button, and all additional stop buttons should

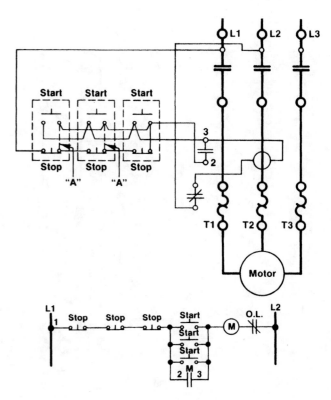

Figure 12–11   The wiring diagram for a three-wire control circuit with additional start and stop push buttons added to the circuit is shown at the top of this figure, and the ladder diagram of just the control circuit is shown at the bottom. (Courtesy of Rockwell Automation's Allen-Bradley Business)

be connected in series with the original stop button. Figure 12–11 shows examples of the additional start and stop push buttons in a three-wire circuit.

12.2.4
Three-Wire Control Circuit with an Indicating Lamp

An indicator lamp can be added to a three-wire circuit to show when the coil in the circuit is energized or de-energized. The indicator lamp can be green to show when the system is energized and red to show when the circuit is de-energized. The lamp is usually mounted where personnel can easily see it at a distance. Sometimes the indicators are used where one operator must watch four or five large machines. After a machine has been set up, the operator will move on to the next machine. Because the installation is very large and spread out over a distance, the operator can watch for the indicator lamps to see if the machine is still in operation.

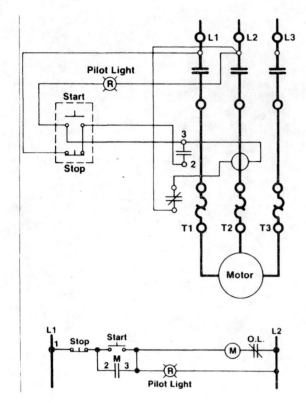

Figure 12-12 A wiring diagram and a ladder diagram of a three-wire control circuit with an indicator lamp added to show when the motor-starter coil is energized. (Courtesy of Rockwell Automation's Allen-Bradley Business)

The indicator lamps can also be used by maintenance personnel when a machine has more than one motor starter. In this type of application, each motor starter has an indicator lamp to indicate its coil is energized. This helps the maintenance personnel begin testing for faults in the correct part of the circuit when the system has stopped or is not operating correctly. This is especially useful if the motor starter is mounted in a NEMA (National Electrical Manufacturers Association) enclosure, where the technician cannot see through the enclosure door to verify if the starter is energized or de-energized.

Figure 12-12 shows a wiring diagram and a ladder diagram of a circuit with an indicator lamp connected in the control circuit. The lamp is called a *pilot light*, and it is connected in parallel with the seal-in contacts on the motor starter. When the motor starter closes, the lamp is energized to indicate that the motor-starter contacts are closed. Any time the lamp is de-energized, the operator and maintenance personnel know that the motor starter is not energized. A press-to-test lamp can also be used in this circuit. The press-to-test lamp allows the operator and maintenance personnel to put their fingers on the lamp and depress the lens at any time to test it to see if it is operational. When the lamp

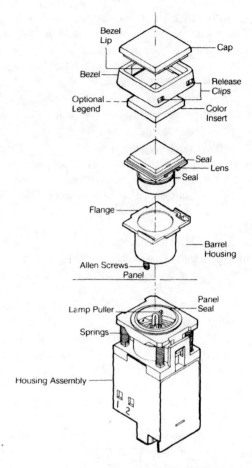

Figure 12-13 An exploded view of a typical indicator lamp. Notice that the lens is replaceable, so different-color lenses can be used. The field wiring is connected to a terminal section that allows the lamp to be changed when it is damaged without removing and replacing wiring. (Courtesy of Rockwell Automation's Allen-Bradley Business)

lens is pressed, it causes a special set of contacts in the base of the lamp holder to provide voltage instantly to the lamp and illuminate it. If the bulb is burned out, the lamp does not illuminate, and the maintenance personnel or the operator can change the bulb. If the indicator is energized most of the time, such as in a continuous operation, the press-to-test lamp may not be necessary.

Indicator lamps are available for 120 V, 240 V, 480 V, and 600 VAC, which means they are available for any control-circuit voltage. Various colored lenses are also available to indicate other conditions with the machine. These indicators can be connected across different individual motor starters in the machine to provide other information, such as hydraulic pumps running, heaters energized, conveyor in operation, and other conditions that are vital to the machine. Figure 12-13 shows an exploded-view diagram of a

Figure 12–14 A wiring diagram and ladder diagram of a forward and reversing motor starter. Notice that you can clearly see the interlock system in the ladder diagram.
(Courtesy of Rockwell Automation's Allen-Bradley Business)

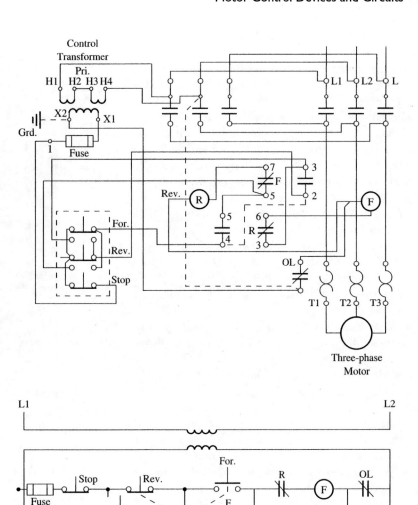

typical indicator lamp. In this diagram you can see that the lens is replaceable, and the part of the lamp that has wires connected to it can easily be removed from the socket part of the lamp in case the socket needs to be replaced.

12.2.5
Reversing Motor Starters

In previous chapters you have found that DC and AC motors can be reversed. These chapters provided the terminal connections for each type of DC motor, AC single-phase motors, and AC three-phase motors. The circuit shown in Fig. 12–14 shows a forward and reversing motor starter for an AC three-phase motor.

When you must install or troubleshoot this circuit, you must be sure to provide interlocks so that the motor cannot be energized in both the forward and reverse directions at the same time.

From this diagram you can see that the control circuit has forward and reverse push buttons. The forward and reverse push buttons each have a normally open and a normally closed set of contacts. In the ladder diagram each button switch has a dashed line, which indicates that both sets of contacts are activated by the same button. The forward and reverse push buttons are better defined in the wiring diagram, which shows that each switch has an open set and a closed set of contacts. The stop button is wired in series with the open contacts of both of these switches, so the motor can be stopped when it is operating in either the forward or the reverse direction.

The operation of the push buttons is best understood through the use of a ladder diagram. The ladder diagram is used to show the sequence of the control circuit, whereas the wiring diagram shows the operation of the load circuit, which includes the motor and heaters for the overloads. The overload contacts are connected in the control circuit, where they will de-energize both the forward and reverse circuit if the motor is pulling too much current.

From the wiring diagram you can also see that two separate motor starters are used in the circuit. The forward motor starter is shown on the right side of the reverse motor starter. Each starter has its own coil and auxiliary contacts, which are used as interlocks. The location of the auxiliary contacts is shown in the wiring diagram, but their operation is difficult to determine there. When you see the auxiliary contacts in the ladder diagram, their function can be more clearly understood.

This comparison of the ladder diagram and the wiring diagram should help you understand that you need both diagrams to work on the equipment. The ladder diagram will be useful in determining what should be tested, and the wiring diagram is useful in showing where the contacts you want to test are located.

You should also notice that the control circuit is powered from a control transformer that is connected across L1 and L2. The secondary side of the transformer is fused to protect the transformer from a short circuit that may occur in either coil.

12.2.6
Reversing Motors with a Drum Switch

The drum switch is a manual switch that allows you to manually reverse the direction in which a motor is turning. The switch contacts are open and closed manually by moving the drum switch from the *off* position to the *forward* or *reverse* position. Figure 12–15 shows a picture of the drum switch, and Fig. 12–16 shows a diagram of the drum-switch contacts. Figure 12–16a shows the drum-switch contacts when the switch is in the reverse position, Fig. 12–16b shows the contacts when the switch is in the off position, and Fig. 12–16c shows the contacts when the switch is in the forward position. When the switch is in the *reverse position*, you should notice that terminal 1 is connected to terminal 2, terminal 3 is connected to terminal 4, and terminal 5 is connected to terminal 6. In the *off* position, all contacts are isolated from all other contacts. In the *forward posi-*

Figure 12–15   A drum switch. Notice the handle requires the operator to manually change the position of the switch from forward to reverse, or to the off position. (Courtesy of Eaton Electrical)

Handle end		
Reverse	Off	Forward
1 o——o 2	1 o o 2	1 o o 2
3 o——o 4	3 o o 4	3 o o 4
5 o——o 6	5 o o 6	5 o——o 6
(a)	(b)	(c)

Figure 12–16   (a) Contacts of a drum switch when it is switched to the reverse position; (b) contacts of a drum switch when it is switched to the off position; (c) contacts of a drum switch when it is switched to the forward position.

tion, terminal 1 is connected to terminal 3, terminal 2 is connected to terminal 4, and terminal 5 is connected to terminal 6.

After you understand the operation of the drum switch in its three positions and methods of reversing each type of motor, these concepts can be combined to develop manual reversing circuits for any motor in the factory as long as its full-load and locked-rotor amperage (FLA and LRA) do not exceed the rating of the

Figure 12–17 (a) A single-phase AC motor connected to a drum switch; (b) a three-phase AC motor connected to a drum switch; (c) a DC motor connected to a drum switch.

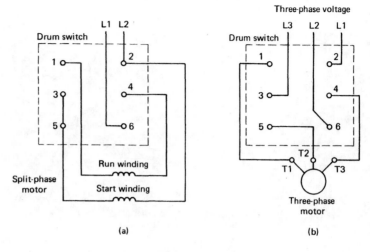

(a)

(b)

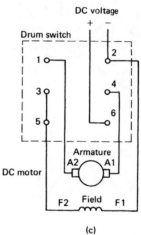

(c)

drum switch. Figure 12–17a shows an AC single-phase motor connected to a drum switch. Notice that the start winding must be reversed for the motor to run in the reverse direction, so the start winding is connected to terminals 3 and 2.

Figure 12–17b shows a three-phase motor connected to a drum switch. You should remember that with this switch, any two lines to the motor can be swapped so the motor will change direction of rotation. Figure 12–17c shows a DC motor connected to a drum switch. You should notice with the DC motor that the direction of current flow through the field is reversed to make the motor run in the opposite direction. You may need to study the connections inside the drum switch when it is in the forward or reverse position to fully understand how the drum switch is used with each of these motors to reverse the direction of their rotation.

These diagrams are especially useful for installation and troubleshooting of these circuits. The drum switch can be tested by itself or as part of the reversing circuit. The motors can also be disconnected from the drum switch and operated in the forward and reverse directions for testing or troubleshooting if you suspect the switch or motor is malfunctioning.

12.2.7
Industrial Timers

Timers that are used in industrial applications are designed to be easy to operate and provide a wide range of functions that fall under the categories of *time delay on* and *time delay off*. The time delay for these timers may be provided by air escaping a pneumatic bellows or by a motor-driven timer. More advanced types of solid-state timers use some type of electronic time-delay circuit, such as the 555 chip, to generate their time base. The main feature of the industrial timer is that it must be designed to be easily adjusted by a machine operator who does not understand anything about electronics. The second feature of industrial timers is that they should be able to handle the

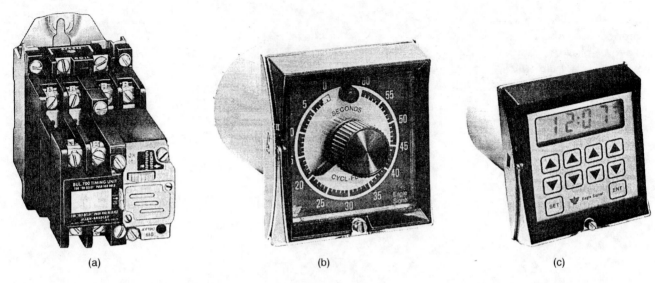

(a)   (b)   (c)

Figure 12–18   (a) A pneumatic-type industrial timer; (b) a motor-driven-type industrial timer; (c) a solid-state-type industrial timer. ((a) Courtesy of Rockwell Automation's Allen-Bradley Business; (b) and (c) Courtesy of Danaher Industrial Controls)

TIME DELAY-ON   TIME DELAY-OFF

Figure 12–19   Examples of time-delay-on and time-delay-off symbols.

high temperatures and strong vibrations that are present on the factory floor. Industrial timers can be interfaced to relay or solid-state logic circuits, and they use one of four basic types of operators that provide the amount of time delay. The time-delay operators can be classified as *pneumatic* (air chamber), *motor and cam, solid state* (time constant), or *programmable* (microprocessor). The operation of each type of timer is explained in this section, and examples of common applications are given. Figure 12–18 shows examples of industrial-type timers.

The electrical symbol for time-delay devices uses the *head* or *tail* of an arrow. The tail of the arrow is used to indicate a time-delay-on function, and the head of the arrow pointing down is used to indicate a time-delay-off function. Figure 12–19 shows examples of the timer symbols.

12.2.8
On-Delay Timers

Industrial timers have contacts and an operator that causes the time delay. On-delay timers change the position of contacts after power is energized to the operator of the timer. One of the largest uses of timers in industrial applications is to control the starting time of larger motors so

that they do not all try to start at the same time. When a motor starts, it draws a larger current than when it is running at full speed. This starting current is called *locked-rotor amperage* (LRA) or *inrush current*, and it is caused by the armature not turning when power is first applied to the motor. As the motor shaft begins to rotate, the current drops to normal levels as the motor comes up to full speed. If a machine has several large hydraulic pump motors, they will all try to start at the same time when power is turned on. The starting cycle of the motors can be staggered by using on-delay timers so that the effects of inrush current are minimized.

The on-delay timer may also be called a *delay-on* timer. When power is applied to the motors, the first motor is allowed to start normally, but power to the second motor starter is controlled through the on-delay timer. If the timer is set for 10 s, the timer contacts will close after a 10-s delay. The time-delay period starts when power is applied to the time-delay operator. Figure 12–20 shows an example of this time-delay-on circuit.

12.2.9
Off-Delay Timers

In some applications it is necessary to ensure that a motor continue to operate for several minutes after power is turned off to the main system. For example, in large industrial heating systems, the fan may need to run for up to 2 min after the heating element or gas valve has been de-energized. The additional time the fan is allowed to run after the heat source is turned off allows the system to capture all the heat that has built up in the heating chamber and use it. This provides a degree of ef-

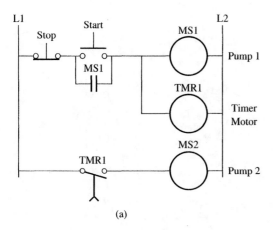

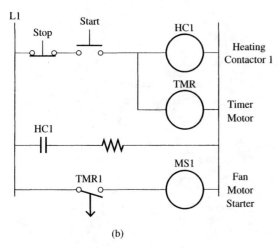

Figure 12–20 (a) Time-delay-on-type timer used to delay the time two pump motors start to limit the effects of inrush current; (b) time-delay-off-type timer used to delay the time a furnace fan stays on after the heating element is turned off. The additional time the fan runs allows extra heat to be removed from the heating chamber.

ficiency, because this heat would be lost if the fan were turned off at the same time as the heating element. The time delay also prevents the heating system from *over-shooting* the temperature set point on the controller. If the temperature set point is set for 150°F and the controller de-energizes the heating element when the temperature reaches 150°, the heat that remains in the heating chamber would cause the system to overheat. The additional heat that remains in the chamber acts like a flywheel and continues to add heat to the application, which causes the over-temperature condition.

The off-delay timer allows the heating control to de-energize the heating element several degrees prior to the set point and allows the fan to continue to operate for several additional minutes to dissipate the remaining heat. This type of timer control may also be called a *time-delay-off timer*. Figure 12–21 shows an example of the time-delay-off timer used to control a furnace fan.

12.2.10
Normally Open and Normally Closed Time-Delay Contacts

Time-delay-on and time-delay-off timers can have normally open, normally closed, or both types of contacts. Figure 12–21 shows the symbol for each of the four possible timers. This figure also shows a timing diagram for each type of timer contact.

An on-delay-type timer can have normally open and normally closed contacts. The symbol for an on-delay timer with normally open contacts is called *normally open, timed close* (NOTC). The contacts for this type of timer start out open, and once the timing element is energized, the contacts close after the delay time has elapsed. The symbol for the on-delay-type timer with normally closed contacts is called *normally closed, timed open* (NCTO). In Fig. 12–21 you should notice that the NOTC-type timer shows the contacts opening downward, and the tail of the arrow is used to indicate it is an on-delay-type timer. The NCTO-type timer symbol shows the contact closed.

Timing diagrams for the four types of timers are also shown in Fig. 12–21. The timing diagram shows the condition of the contacts as either open or closed during the four periods of the timer operation. The first timer period shows the contacts prior to timing. This represents the time *before* the timer operator is energized. The second period represents the time *during timing*, and the third period represents the time *after timing* but when power is still applied to the timer operator. The fourth period represents the time after timing when power is turned off to the operator. This condition is called *reset*, and it actually occurs twice during the cycle, once at the very beginning of the cycle during the first timing period before timing and again at the very end of the cycle. This means the very first part of the timing cycle may be called the time before timing, or reset. You should also notice that the timing diagrams show a diagram for the timing operator as well as a diagram for each type of contact used. The diagram that shows power applied to the operator is provided as a reference to indicate when the timing cycle actually begins. The time-delay period for each timer should be counted from the point where power is applied to the time-delay operator. Additional symbols are provided below each timing segment in the form of a triangle with an "x" or an "o" inside. The "x" indicates the timing contacts are closed during that part of the time cycle and an "o" indicated the timing contacts are open during that timing cycle.

An off-delay-type timer can also have normally open and normally closed contacts. If an off-delay timer uses normally open contacts, it is called *normally open, timed open* (NOTO). The first question that comes to

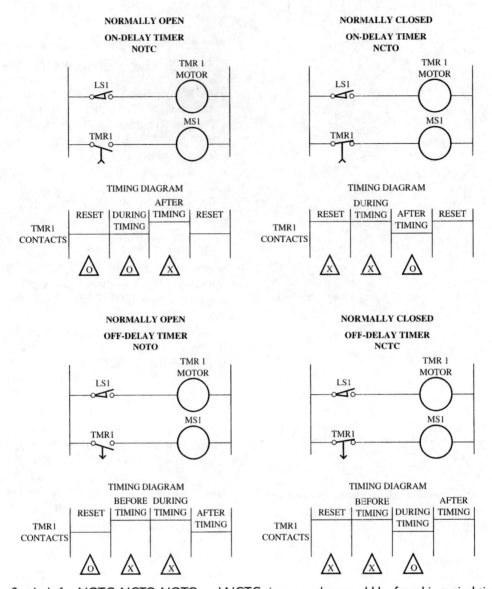

Figure 12–21 Symbols for NOTC, NCTO, NOTO, and NCTC timers as they would be found in typical timing circuits. Timing diagrams are provided for each type of timer.

mind about this type of timer is how it can be normally open and then also timed open. This situation occurs because one of the states the off-delay timer goes through is omitted. The actual timing cycle for the off-delay timer includes three distinct steps—normally open, *energized closed*, and then timed open. The timing diagram for this timer shows the contacts are normally open during the time prior to timing (reset). When power is initially applied to the circuit, the time-delay operator is powered, and the normally open contacts move to the closed position. When power to the time-delay operator is de-energized, it begins the time-delay period, and the contacts open at the end of the timing period. Thus the name *normally open, timed open* fits.

If an off-delay timer uses a closed set of contacts, it is called *normally closed, timed closed* (NCTC). Again, the name seems to be contradictory because the con-

tacts start out closed and end the cycle being timed closed. The part that is missing is the middle part of the cycle, where the contacts are opened when power is applied to the time-delay element. On-delay and off-delay timers can have a single set of normally open or normally closed contacts, or they can have multiple sets of both normally open and normally closed contacts.

12.2.11
Pneumatic-Type Timer Operators

Several different types of operators are used in industrial-type timers. The simplest type of operator is called a *pneumatic* operator. This type of time-delay operator is also called a time-delay element. The pneumatic operator or element uses a bellows that fills with air. Prior to the start of the timing cycle the bellows is completely filled with air; when the time delay begins, the air is released through a *needle valve*. The needle valve can be adjusted to regulate the time it takes for the air to move out of the bellows. A spring is mounted against the bellows to apply pressure to it to help force the air out. The amount of time it takes for the air to empty out of the bellows determines the amount of time delay. When sufficient air has emptied out of the bellows, the spring pressure is strong enough to trip the mechanism that ends the timing cycle. The trip mechanism also actuates the timed contacts.

The pneumatic-type time-delay element is an *add-on device* that can be added to a traditional relay. Figure 12–22 shows a picture of this type of add-on timing device. Because the pneumatic time-delay element is added to a normal control relay, it is possible to convert any control relay that is used in a circuit into a time-delay relay. Because the pneumatic element is added to the relay, it is also possible to take advantage of the original relay contacts that still operate with the relay coil. These contacts open or close instantly when the coil is activated, so they are called *instantaneous contacts*.

The pneumatic time-delay element can be converted from time delay on to time delay off right in the field. The element is designed so that when it is mounted on the relay in the upright position, it will provide time-delay-on functions. If the element is turned upside down when it is mounted on the relay, it will provide time-delay-off functions. You can determine if the element is mounted for delay-on or delay-off functions by looking closely at the lettering on the element. If the element is mounted for delay-on functions, the words *on-delay* will be right side up and the words *off-delay* will be upside down. When the element is mounted for off-delay functions, the words *off-delay* will be right side up.

The time-delay element generally has two sets of contacts mounted on it. One set is usually normally open, and one set is normally closed. These contacts can be changed from normally open to normally closed, or vice versa, by loosening a mounting screw and turning the contact set upside down. Thus the contacts can be changed in the field. You can determine if the set of contacts is normally open or normally closed by checking them with an ohmmeter.

12.2.12
Motor-Driven-Type Timers

Motor-driven-type timers are widely used in industrial applications. These timers are also called *synchronous motor-type* timers or *electromechanical*-type timers. The timer uses a motor to turn a shaft on which cams are mounted. The shaft has a clutch that engages it to or disengages it from the motor. When the shaft is engaged

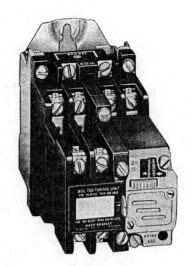

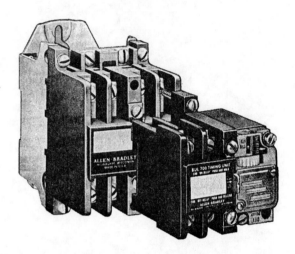

Figure 12–22 Pneumatic-type time-delay element. This element is added to a traditional relay to create a time-delay relay. Notice that the element can be mounted right side up for time-delay-on operation, or it can be mounted upside down for time-delay-off operation. The delay contacts can also be mounted upside down to change from normally open to normally closed. (Courtesy of Rockwell Automation's Allen-Bradley Business)

(energized), the motor turns the shaft and the cams actuate several set of contacts. When the time-delay period has elapsed, the time-delay contacts change from open to closed or from closed to open, and the timer motor resets for the next cycle. Figure 12–23 shows a picture and diagram of the motor-driven-type timer.

From the diagram in Fig. 12–23 you can see that the timer has a clutch coil and a motor that are connected in parallel with a common point identified as terminal 2. The timer has four sets of contacts. In the diagram contacts 9–10–C and 6–7–8 are located directly under the clutch coil to indicate they are energized instantaneously by the

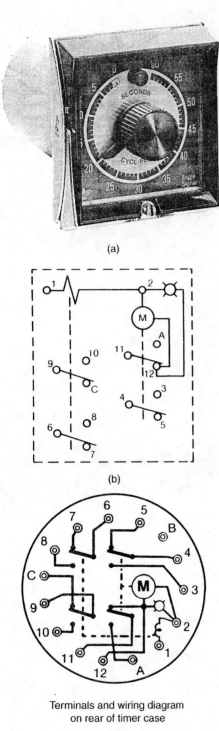

(a)

(b)

(c)

Terminals and wiring diagram
on rear of timer case

Figure 12–23 Synchronous motor-driven-type timer
with a diagram of the internal circuit of the electric
contacts. (Courtesy of Danaher Industrial Controls)

clutch when power is applied to the timer. Sets 11–12–A
and 3–4–5 are located in the diagram directly under the
timer motor to indicate they are controlled by the timer
motor. These contacts are called the *delay contacts.*

One typical application uses a limit switch to control a timer. When the limit switch is closed, power is

applied through it to the clutch coil. This same power
energizes the motor, but it must go through normally
closed time-delay contacts 11–12. When power reaches
the clutch coil, it immediately activates the instantaneous contacts and engages the clutch. Power also
reaches the timer motor, and it begins to rotate its
shaft. Because the clutch is engaged, the motor shaft
turns the time-delay shaft; and when the correct
amount of time delay has passed, the cams located on
the shaft move with the shaft and activate both sets of
time-delay contacts. Because the timer motor is connected through the normally closed set of time-delay
contacts 11–12, the timer motor becomes de-energized
when the contacts open, and the timer is reset and becomes ready for the next cycle. The instantaneous sets
of contacts controlled by the clutch coil remain energized until the limit switch is de-energized.

12.3
JOGGING CONTROL CIRCUITS

In some applications, such as motion control, machine
tooling, and material handling, you must be able to turn
the motor on for a few seconds to move the load slightly
in a forward or reverse direction. This type of motor control is called *jogging.* The jogging circuit utilizes a reversing motor starter to allow the motor to be moved slightly
when the forward or reverse push button is depressed.

Another requirement of the jogging circuit is that the
motor starters do not seal in when the push buttons are
depressed to energize the motor when it is in the jog
mode, yet operate as a normal motor starter when the
motor controls are switched to the run mode. A diagram
of a jogging circuit is provided in Fig. 12–24. The load circuit and control circuit are shown as electrical wiring diagrams, which give the location of each component, and
the control circuit is shown again as a ladder diagram so
you can see the sequence of operation for the forward
and reversing motor starter with the jog function.

The wiring diagram in this figure gives you a very
good idea of the way the jog-run switch operates. This
switch is shown to the left of the motor starter in the diagram. You can see that it is part of the start-stop station. The jog-run button is a selector switch that is
mounted above the forward-reverse-stop buttons.

When the switch is in the jog mode, the selector
switch is in the open position. From the ladder diagram,
you can see that the jog switch is in series with both the
seal-in circuits, which prevents them from sealing the
forward or reverse push buttons when they are depressed. This means that the motor operates in the forward direction for as long as the forward push button is
depressed. As soon as the push button is released, the

Figure 12–24 A ladder diagram of a forward and reversing jogging circuit. Notice the interlock between the forward push buttons and the reverse push buttons so that you cannot energize the forward and reverse motor starters at the same time. (Courtesy of Rockwell Automation's Allen-Bradley Business)

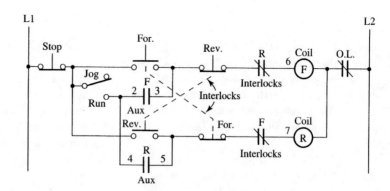

motor starter is de-energized. This jog switch also allows the motor to be jogged from one direction directly to the other direction without having to use the stop button.

The motor is protected by the overloads that are connected in series with the forward and reverse motor-starter coils. If the overload trips, the overload contacts in the control circuit open, and neither coil can be energized until it is reset.

These two diagrams allow you to understand the operation of the jog circuit. You can make a forward and reversing motor-starter circuit into a jogging circuit by adding a jog switch, but you must be sure that the motor and the motor starters are rated for jogging duty. *Remember:* Some motor starters and motors cannot take

the heat that builds up when a motor is started and stopped continually during a jogging operation. The motor and the motor starters will be rated for jogging or plugging if they can withstand the extra current and heat.

12.3.1
Sequence Controls for Motor Starters

Sequence control allows a motor starter to be utilized as part of a complex motor-control circuit that uses one set of conditions to determine the operation of another circuit. Figure 12–25 shows an example of this type of

Figure 12–25 Wiring diagram and Ladder diagram of Motor starters that are sequenced so that motor starter 1 must be on before motor starter 2 is started. (Courtesy of Rockwell Automation's Allen-Bradley Business)

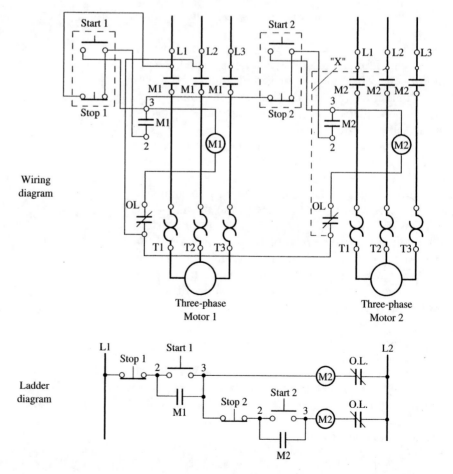

circuit. The circuits in this figure are presented in wiring-diagram and ladder-diagram form. You can really begin to see the importance of the ladder diagram, because it shows the sequence of operation, which would be very difficult to determine from the wiring diagram. The wiring diagram is still very useful, because it shows the field wiring connections and the locations of all terminals that will need to be used during troubleshooting tests.

The circuit in Fig. 12–25 shows two conveyors that are controlled by two separate motor starters. Conveyor 1 must be operating prior to conveyor 2 being started. This is required because conveyor 2 feeds material onto conveyor 1, and material would back up on conveyor 2 if conveyor 1 were not operating and carrying it away. The ladder diagram shows a typical start-stop circuit with an auxiliary contact being used as a seal-in around the start button. When the first start button is depressed, M1 is energized, which starts the first conveyor in operation. M1 auxiliary contacts seal the start button and provide the circuit power to the second start-stop circuit.

Because the second circuit has power at all times after M1 is energized, its start and stop buttons can be operated at any time to turn the second conveyor on and off as often as required without bothering the first conveyor's motor starter. Remember, this circuit requires conveyor 1 to be operating prior to conveyor 2, because conveyor 2 feeds material onto conveyor 1.

The circuit also protects the sequence if conveyor 1 is stopped for any reason. When it is stopped, the M1 motor starter becomes de-energized, and the M1 auxiliary contacts return to their open condition, which also de-energizes power to the second conveyor's start-stop circuit.

You should also notice that the power for this control circuit comes from L1 and L2 of the first motor starter. This means that if supply voltage for the first conveyor motor is lost for any reason, such as a blown fuse or opened disconnect, the power to the control circuit is also lost, and both motor starters are de-energized, which stops both conveyors. If the first motor draws too much current and trips its overloads, it causes an open in the motor starter's coil circuit, which causes the auxiliary contacts of the first motor starter to open and de-energizes both motor starters.

If you need additional confirmation that the belt on the first conveyor is actually moving, a motion switch can be installed on the conveyor; its contacts would be connected in series between the first start button and the M1 coil. This would cause the first motor-starter coil to become de-energized any time the conveyor belt was broken or slipping too much.

12.4
OTHER TYPES OF PILOT DEVICES

A wide variety of other pilot devices are commonly used in motor-control circuits that are interfaced to industrial electronic circuits. These switches include pressure, level, and temperature switches, as well as other types of switches. Figure 12–26 shows pictures of three

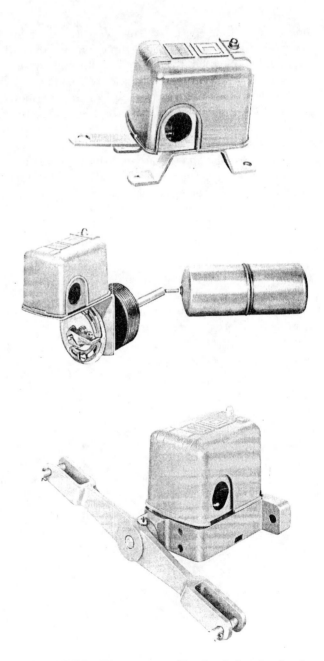

Figure 12–26 Three types of level switches. Applications and electrical diagrams of these switches are shown in Figure 12–27. (Courtesy of Square D/Schneider Electric)

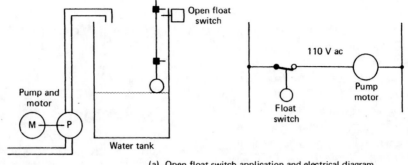

(a) Open float switch application and electrical diagram

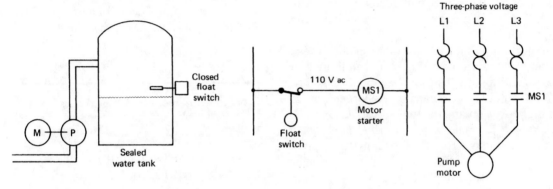

(b) Closed float switch application and electrical diagram

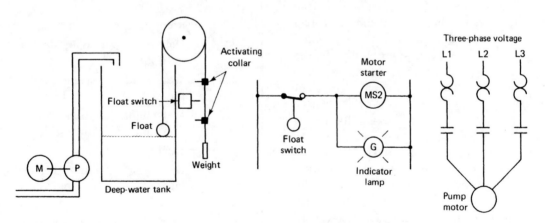

(c) Rod and chain float switch application and electrical diagram

Figure 12–27   (a) A float and rod are used with the level switch to turn it on and off. (b) A closed-type flow switch is mounted at the specific level that is to be controlled. (c) A float is attached to one end of a cable and a weight is attached to the other end. When the float moves up or down, the switch is activated. (Courtesy of Square D/Schneider Electric)

types of level switches, and Fig. 12–27 shows a diagram of each switch. Level switches are sometimes called *float switches*.

The first switch, in Fig. 12–26a, is used with a ball-type float that is attached to a long rod. The rod has adjustable cams attached at the bottom and the top of the rod. When the level in the sump raises, the float lifts the rod, and the cam that is attached to the lower part of the rod moves high enough to trip the switch handle to the up position, which closes the level-switch contacts. When

the pump runs, it pumps the sump level down and allows the float to drop. When the float drops low enough, the cam that is attached to the top of the rod causes the switch handle to move to the down position, which turns the pump off.

The switch in Fig. 12–26b uses a sealed-type float to activate the level switch. In this application, the switch is mounted in the tank at the precise level desired. When the level of the liquid in the tank is above this level, the float lifts and causes the switch contacts to close and turn on a pump. When the pump runs, the level of the tank is decreased, which allows the float to drop and turn the switch off. It is important to understand that the length of the switch arm determines the amount of travel the float must make to turn the switch on and off. This distance is called the *control band*, or *dead band*.

The float switch in Fig. 12–26c is a pivot-arm-type level switch. An arm or a wheel can be attached to the pivot to cause it to activate. An arm is shown in the picture of this switch, and a wheel is shown in the diagram. A cable is threaded around the wheel or arm. A float is attached to one end of the cable, and a weight is attached to the other. When the level of liquid in the tank changes, the cable will move up or down and cause the switch to turn on or off.

It is important to understand that because these switches are pilot switches, they can safely switch only 10 to 15 A. This means that they can be used to control small motors or they can control larger motors by switching power on and off to a motor-starter coil.

Another type of pilot switch is called a *pressure switch*. This type of switch is used for sensing air pressure or water pressure and turning a set of contacts on or off. Figure 12–28 shows a typical pressure switch. The pressure switch may have a single point where it is activated on or off, or it may provide a span between the point where it turns on and off. If the switch has a span adjustment and a low-pressure adjustment, you should set the low-pressure adjustment where you want the switch to turn on and adjust the span to determine where you want the switch to turn off. For example, if you want the switch to control the pressure in a reservoir of an air compressor, you would set the low pressure to turn on when the pressure reaches 20 psi and then set the span to 40 psi. The span of 40 psi would cause the switch to turn off when the air pressure reached 40 psi more than the low-level setting. This means that switch would turn on the compressor when the air pressure dropped to 20 psi, and it would turn the compressor off when the pressure

reached 60 psi. Even though the switch has two points that turn it on and off, it still has only one set of single-pole or double-pole contacts that are activated.

12.5
ENCLOSURES

Enclosures are required to house disconnects, motor starters, and other motor controls. The enclosures are available in a variety of designs that will protect the devices inside from all types of environmental conditions. Figure 12–29 shows a list of typical enclosure types as

NEMA Enclosures for Starters	
Type	*Enclosure*
1	General purpose—indoor
2	Drip-proof—indoor
3	Dust-tight, rain-tight, sleet-tight—outdoor
3R	Rainproof, sleet-resistant—outdoor
3S	Dust-tight, rain-tight, sleetproof—outdoor
4	Watertight, dust-tight, sleet resistant—indoor
4X	Watertight, dust-tight, corrosion-resistant—indoor/outdoor
5	Dust-tight—indoor
6	Submersible, watertight, dust-tight, sleet-resistant—indoor/outdoor
7	Class I, group A, B, C, or D hazardous locations, air-break—indoor
8	Class I, group A, B, C, or D hazardous locations, oil-immersed—indoor
9	Class II, group E, F, or G hazardous locations, air-break—indoor
10	Bureau of Mines
11	Corrosion-resistant and drip-proof, oil-immersed—indoor
12	Industrial use, dust-tight, and driptight—indoor
13	Oiltight and dust-tight—indoor

Figure 12–28   Example of a typical pressure switch that is used as a pilot device. (Courtesy of Square D/Schneider Electric)

Figure 12–29   List of enclosure types provided by the National Electric Manufacturer's Association (NEMA). (Courtesy of Rockwell Automation's Allen-Bradley Business)

classified by the National Electrical Manufacturers Association (NEMA), and Fig. 12–30 shows examples of each type of enclosure.

Type 1 enclosures are designed for general-purpose applications in indoor locations. Thus the enclosure should not be exposed to extreme conditions, such as excessive moisture. The type 2 enclosure is rated for drip-proof conditions that exist indoors. This means that some moisture may come in contact with the enclosure, but it is not approved where equipment must be washed down or steam-cleaned daily.

Type 3 enclosures are designed for dust-tight, rain-tight, sleet-tight conditions that exist outdoors; type 3R enclosures are rainproof and sleet-resistant, and type 3S enclosures are designed for dust-tight, rain-tight, and sleetproof conditions. These enclosures are intended for use where protection is needed against falling rain, sleet, or dust. They are not intended to prevent condensation from forming on the inside of the enclosure or on internal components in the enclosure.

If the application is located where windblown water or sleet may be encountered, a type 4 enclosure should be used. The type 4 enclosures are designed to protect against windblown dust, rain, or water from direct-hose-down conditions. This enclosure is specified by NEMA for watertight, dust-tight, sleet-resistant indoor and outdoor applications. The 4× enclosure is designed for indoor and outdoor use and also provides protection in environments where corrosion resistance is required. The covers of these enclosures have protective gaskets that provide a barrier against these conditions. All hardware for the conduit must match the requirements to maintain the integrity of the design throughout the installation.

NEMA type 5 enclosures are designed for dust-tight applications located indoors. This type of enclosure is designed to protect the switch and contacts from a buildup of dust that may prevent proper operation of the switches and starters that are enclosed.

Type 6 enclosures are designed to provide submersible, watertight, dust-tight, sleet-resistant indoor and outdoor applications. The submersible watertight specification allows the switch to be fully submersed in water to a limited depth and maintain watertight integrity. This type of enclosure also depends on gaskets to maintain the watertight condition. As with other enclosures, conduits and connectors must meet the same NEMA standard so that the entire installation maintains the stated protection.

Type 7 enclosures provide class I protection for group A, B, C, or D hazardous locations, with air-break protection for indoor locations. These enclosures are designed to prevent dangerous gases from penetrating the enclosure and coming into contact with the open parts of switches or other controls. This provides protection in explosive atmospheres. Class I locations are hazardous because flammable gases or vapors may be present in sufficient quantities to cause explosions. Type 8 enclosures provide class I, group A, B, C, or D hazardous locations, oil-immersed for indoor applications. This type of enclosure is very similar to the type 7 enclosure except that it provides protection against oil immersion instead of air-break. This allows the enclosure to be used where oil and machine coolants are used extensively.

Figure 12–30 Examples of NEMA-rated enclosures. (Courtesy of Rockwell Automation's Allen-Bradley Business)

Table 310.16 Allowable Ampacities of Insulated Conductors Rated 0 Through 2000 Volts, 60…C Through 90…C (140…F Through 194…F), Not More Than Three Current-Carrying Conductors in Raceway, Cable, or Earth (Directly Buried), Based on Ambient Temperature of 30…C (86…F)

	Temperature Rating of Conductor (See Table 310.13.)						
	60…C (140…F)	75…C (167…F)	90…C (194…F)	60…C (140…F)	75…C (167…F)	90…C (194…F)	
Size AWG or kcmil	Types TW, UF	Types RHW, THHW, THW, THWN, XHHW, USE, ZW	Types TBS, SA, SIS, FEP, FEPB, MI, RHH, RHW-2, THHN, THHW, THW-2 THWN-2, USE-2, XHH, XHHW, XHHW-2, ZW-2	Types TW, UF	Types RHW, THHW, THW, THWN, XHHW, USE	Types TBS, SA, SIS, THHN, THHW, THW-2, THWN-2 RHH, RHW-2, USE-2, XHH, XHHW, XHHW-2, ZW-2	Size AWG or kcmil
	COPPER			ALUMINUM OR COPPER-CLAD ALUMINUM			
18	—	—	14	—	—	—	—
16	—	—	18	—	—	—	—
14*	20	20	25	—	—	—	—
12*	25	25	30	20	20	25	12*
10*	30	35	40	25	30	35	10*
8	40	50	55	30	40	45	8
6	55	65	75	40	50	60	6
4	70	85	95	55	65	75	4
3	85	100	110	65	75	85	3
2	95	115	130	75	90	100	2
1	110	130	150	85	100	115	1
1/0	125	150	170	100	120	135	1/0
2/0	145	175	195	115	135	150	2/0
3/0	165	200	225	130	155	175	3/0
4/0	195	230	260	150	180	205	4/0
250	215	255	290	170	205	230	250
300	240	285	320	190	230	255	300
350	260	310	350	210	250	280	350
400	280	335	380	225	270	305	400
500	320	380	430	260	310	350	500
600	355	420	475	285	340	385	600
700	385	460	520	310	375	420	700
750	400	475	535	320	385	435	750
800	410	490	555	330	395	450	800
900	435	520	585	355	425	480	900
1000	455	545	615	375	445	500	1000
1250	495	590	665	405	485	545	1250
1500	520	625	705	435	520	585	1500
1750	545	650	735	455	545	615	1750
2000	560	665	750	470	560	630	2000

CORRECTION FACTORS

Ambient Temp. (…C)	For ambient temperatures other than 30…C (86…F), multiply the allowable ampacities shown above by the appropriate factor shown below.						Ambient Temp. (…F)
21-25	1.08	1.05	1.04	1.08	1.05	1.04	70-77
26-30	1.00	1.00	1.00	1.00	1.00	1.00	78-86
31-35	0.91	0.94	0.96	0.91	0.94	0.96	87-95
36-40	0.82	0.88	0.91	0.82	0.88	0.91	96-104
41-45	0.71	0.82	0.87	0.71	0.82	0.87	105-113
46-50	0.58	0.75	0.82	0.58	0.75	0.82	114-122
51-55	0.41	0.67	0.76	0.41	0.67	0.76	123-131
56-60	—	0.58	0.71	—	0.58	0.71	132-140
61-70	—	0.33	0.58	—	0.33	0.58	141-158
71-80	—	—	0.41	—	—	0.41	159-176

* See 240.4(D).

Figure 12–31 American wire gauge (AWG) wires sizing table provided by the National Electrical Code (NEC). (Reprinted with permission from NFPA 70-2002, *National Electrical Code*®, Copyright ©1998, National Fire Protection Association, Quincy, MA 02269. This reprinted material is not the complete and official position of the NFPA on the referenced subject, which is represented only by the standard in its entirety.)

Type 9 enclosures provide protection against class II, group E, F, or G hazardous locations and air-break for indoor applications. Class II locations are hazardous because of the presence of combustible dust in quantities sufficient to explode or ignite.

Type 10 enclosures are rated for all applications within mines. This type of enclosure must be able to protect switch gear against explosive conditions. Type 11 enclosures provide corrosion resistance and drip-proof, oil-immersion protection for indoor applications. This type of enclosure is used where the vapors and fumes may be corrosive to switch gear or other motor controls that are mounted inside. The exterior of the enclosure is also resistant to corrosion from these fumes. This type of enclosure is also used for applications where machining operations are performed.

Type 12 enclosures are intended for indoor applications and provide protection against dust, falling dirt, and dripping noncorrosive materials. They are not intended for use against direct spraying of these materials. The last classification of enclosures is type 13.

12.6
CONDUCTORS

One part of the power-distribution system that is involved at all points in the system is the conductor. Conductors can be made of aluminum or copper wire.

Aluminum is generally used for long-distance distribution because of its lighter weight. Once the power is inside the plant, copper conductors are generally used for distribution. Copper conductors are solid in bus bar and busway applications, and they are stranded conductors for all other applications.

The conductors are sized by the amount of amperage they can carry. A table of typical conductor sizes and their ampacities is provided in Fig. 12–31. The standard used to determine the size of each conductor is called American wire gauge (AWG). These tables have been established by the National Electrical Code (NEC), and they are used to select the proper size of conductor to carry the load.

Conductors are also classified by the type of insulation used as a cover. The type of cover also determines the voltage rating of the wire. Typical voltage ratings for conductors used in motor-control applications include 300, 600, and 1,000 V. The voltage and current ratings of the conductor should not be exceeded under any circumstances.

Types of wire covering and abbreviations for each of these coverings are listed in the NEC tables, and they help determine the type of wire that should be selected for each application. The outer covering of the wire serves several purposes, including protecting the conductors from coming in contact with metal in the cabinet or other conductors. The cover also provides a location to stamp all specification data regarding the wire, including the voltage rating, AWG size, temperature specification, and type of covering.

QUESTIONS

Short Answer

1. Identify five types of limit switches and explain where each would be used.

2. Use the diagrams in Fig. 12–17 and explain how the drum switch can reverse the direction of rotation for a single-phase motor, a three-phase motor, and a DC motor.

3. Use the diagram in Fig. 12–24 to explain how the jog circuit must modify the traditional three-wire control circuit so it will not latch in.

4. Explain the operation of each of the three-level pilot devices shown in Fig. 12–27.

5. Use Fig. 12–29 to select the correct enclosure for the following applications: dust-tight indoors, rainproof and sleet-resistant outdoors, general-purpose indoors, oil-tight and dust-tight indoors.

True or False

1. A jogging circuit is used to interlock a motor starter during a sequence move.

2. A three-wire control circuit uses normally open auxiliary contacts of the motor starter to seal in the start push button.

3. In the diagram in Fig. 12–8, the float switch will close when the level that is being sensed increases.

4. A drum switch can be used to reverse the direction of a single-phase AC motor, a three-phase AC motor, or a DC motor.

5. The auxiliary contacts found between terminal 2 and terminal 3 on a motor starter are used to seal in the start push button in a three-wire control circuit.

Multiple Choice

1. In the diagram in Fig. 12–25, motor starter M1 will always be running before motor starter M2 can become energized because _____.
   a. the M1 coil requires a larger voltage than the M2 coil
   b. the M1 coil is in series with the M2 coil
   c. the M1 contacts are in series with the M2 coil

2. When two extra stop push buttons are added to a start-stop circuit, the stop buttons must be wired _____.
   a. in series with the other stop push button
   b. in parallel with the other stop push button
   c. in parallel with the start push button

3. In Fig. 12–9, which pressure switch will protect the circuit when it is in the auto position?
   a. Only pressure switch B
   b. Only pressure switch A
   c. Both pressure switches A and B

4. When two extra start push buttons are added to a start-stop circuit, the start push buttons must be wired _____.
   a. in series with the other stop push button
   b. in parallel with the other stop push button
   c. in parallel with the start push button

5. A 12-gauge wire with THWN insulation (according to the table in Fig. 12–31) can safely carry _____ A.
   a. 30
   b. 25
   c. 20

PROBLEMS

1. Draw the symbols for a normally open limit switch, a normally open, held-closed limit switch, a normally closed limit switch, and a normally closed, held-open limit switch.

2. Draw the symbols for normally open and normally closed pressure switches that are used for a high-pressure application and normally open and normally closed pressure switches that are used for a low-pressure application.

3. Draw a two-wire control circuit of a limit switch and a motor as a ladder diagram.

4. Use the table in Fig. 12–31 to select a wire that will carry 30 A and has THHN insulation that can withstand 90°C.

5. Draw a two-wire control circuit of a limit switch and a motor as a wiring diagram.

CHAPTER 13

Single-Phase AC Motors

OBJECTIVES

After reading this chapter you will be able to:

1. Explain the theory of operation of a simple induction AC motor.

2. Identify the main parts of an AC motor and explain their function.

3. Identify the parts of the centrifugal switch and explain their operation and function.

4. Explain the data found on a typical motor data plate.

Figure 13–1   Typical AC single-phase motor.

13.0
OVERVIEW OF SINGLE-PHASE AC INDUCTION MOTORS

The main loads on which you will work in most systems are motors of all types. You will work with both single-phase and three-phase motors while you are on the job. Because these motors are slightly different, you will learn about single-phase motors in this chapter and three-phase motors in the next chapter. Single-phase motors are somewhat smaller than three-phase motors; typically they are less than 2 hp. You typically find single-phase motors in shop power tools such as bench grinders, drill presses, small fans, and smaller pumps and conveyor motors. The motors used in these applications are called *open-type motors* because they have a shaft on one or both ends. Figure 13–1 shows a picture of a single-phase motor. The theory of operation, installation, and troubleshooting for open-type motors that are used to power fans, pumps, and belt-driven loads is also explained in this chapter. The material in these sections shows that some motors require special switches

and components to start and operate correctly. These devices are explained so that when you must troubleshoot a motor, you will fully understand their function. A diagram and picture are provided for each type of motor so that you will learn to identify each type of motor from its physical appearance and from its electrical diagram. As a technician you will be expected to identify, install, troubleshoot, and repair or replace any motors you encounter. This chapter provides a comprehensive overview for all single-phase motors found in electrical systems.

13.1
AC SPLIT-PHASE MOTOR THEORY

The AC split-phase motor is widely used in a variety of applications, and it has many of the basic parts of the other types of open motors. This motor has three basic parts: the stationary coils called the *stator*, the rotating shaft, called the *rotor*, and the *end plates*, which house the bearings that allow the rotor shaft to turn easily. Figure 13–2 shows an exploded view of a split-phase motor so that you can see where these parts are located inside the motor. From this photo you can see that the rotor is supported by the bearings in the end plates of the motor. The rotor is also mounted directly inside the stationary windings so that when voltage is applied to them, their magnetic field can be induced into the rotor.

If the AC split-phase motor requires a *centrifugal switch* to help it get started, it will be mounted inside the end plate. The actuator for the centrifugal switch is mounted on the end of the rotor's shaft, and it has fly-weights that swing out when the motor reaches 75 to 85% full rpm to open the centrifugal switch. The next

sections of this chapter show how these parts operate with each other to provide a rotating force to the motor's shaft to turn a fan, conveyor, or pump.

13.2
THE ROTOR

The rotating part of the motor is called the *rotor*. When voltage is applied to the coils of wire in the stator, they produce a very strong magnetic field. This magnetic field is passed to the rotor by induction, in much the same way as voltage is passed from the primary winding to the secondary winding of a transformer. After the rotor becomes magnetized, it begins to rotate. The speed of the rotor is determined by the number of poles and the frequency of the AC voltage that is applied to the motor. The rotor has a shaft that rotates to do the work of the motor, such as to turn a fan blade or to move the pulley for a conveyor. Figure 13–3 shows a diagram and a picture of a rotor that is used in most single-phase AC motors. As a rule, the rotor of an AC motor does not have any wire in it (only repulsion start or some synchronous AC motors have wire in their rotors, but these motors are not used too often in applications). The rotor is made from pressing laminated-steel plates on the frame. The frame of the rotor looks like the wire-frame exercise wheel that is used by hamsters. For this reason it has been named a *squirrel cage;* and when it is used in the rotor, it is called a squirrel-cage rotor. The squirrel cage is actually the frame for the rotor, and the parts that look like the squirrel cage are actually

Figure 13–2   Exploded view of a split phase motor.

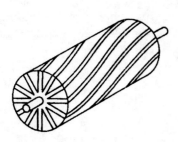

(a)

(b)

Figure 13–3   (a) Diagram of a squirrel-cage rotor for an AC motor. The rotor bars will become multiple-bar magnets when the rotor is magnetized. (b) Picture of a squirrel-cage rotor for an AC motor.

rotor bars, which act exactly like multiple-bar magnets. The laminated steel plates that are pressed onto the frame of the squirrel-cage rotor allow the rotor bars to be magnetized by induction when current is passed through the stator. This means that the permanent field in the stator acts like the primary of a transformer, and the rotor acts like the secondary windings of a transformer. The result of this is that the induction motor does not need brushes to pass current to the rotor to get its magnetic field to build. This is also why these types of motors are called *induction motors*. It is also important to understand that the laminated-steel sections in the rotor allow it to become magnetized very easily by the AC voltage and then quickly change the polarity of its magnetic field while the AC voltage oscillates from positive to negative as the AC sine wave changes. The polarity of the magnetic field in the rotor of the AC motor changes because the sine wave of the AC voltage changes from positive to negative naturally during each cycle of the sine wave. This causes the magnetic fields in the rotor to continually change polarity and to spin as long as AC voltage is applied to the motor.

13.3

LOCKED-ROTOR AMPERAGE AND FULL-LOAD AMPERAGE

When voltage is first applied to the motor, the run and start windings will draw a large amount of current, called *locked-rotor amperage* (LRA). This current is called locked-rotor amperage because the rotor has not started to turn at this time. The amount of current is large because the amount of resistance in the windings is very small and because the rotor has not yet started to rotate and create *counterelectromotive force* (CEMF) voltage. When the rotor starts to spin and comes up to full rpm, the current will return to normal, and it is called *full-load amperage* (FLA). Figure 13–4 shows a graph of the large LRA current that occurs when the motor is first started,

and it also shows the current returning to lower levels (FLA) when the motor is running at full-load speed.

The reason the current returns to its normal value after the rotor is spinning at full speed is because the rotor produces CEMF. When the rotor starts to spin, its magnetic field begins to pass the coils of wire in the run winding, and a voltage is generated when the magnetic field passes the coils. This generated voltage is out of phase with the applied voltage, which is why it is called counterelectromotive force, or CEMF. CEMF is also sometimes called *back EMF* or *counter-EMF*.

The amount of CEMF is equivalent to the speed at which the rotor is turning. If the rotor is turning at 90% of full speed, the CEMF will be approximately 90% of the applied EMF. For example, if the applied voltage is 230 VAC and the rotor is turning at approximately 90% rpm, the CEMF may be 200 VAC. If the impedance of the run winding is 10 ω, the full-load current is determined by dividing the difference in voltage (230 − 200), which is 30 VAC, by the impedance. This means the FLA current is approximately 3 A.

13.4

ROTOR SLIP AND TORQUE IN AN AC INDUCTION MOTOR

The amount of LRA and FLA that an AC induction motor draws also determines the amount of torque the motor's shaft will have. *Torque* is defined as the amount of *rotating force* that the rotor shaft has. You should remember that this force is needed to pump the piston of the compressor and turn the fan blades for a condenser fan. If the motor has a large amount of LRA during startup, it will also have a lot of starting torque. If the motor draws a large FLA when the motor is running, it will have a lot of running torque.

The amount of FLA is determined by the amount of slip a motor has. *Slip* is defined as the *difference* in the

Figure 13–4 Graph of the locked-rotor amperage (LRA) that occurs when a motor is first started and the full-load amperage (FLA) that occurs when the motor is running at full speed.

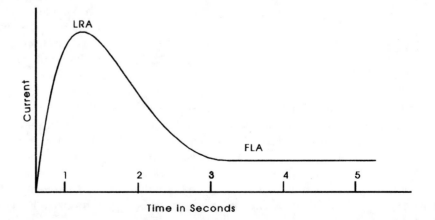

rated speed of a motor and its *actual running speed.* For example, a four-pole motor is rated for 1,800 rpm, but its actual running speed is approximately 1,725 rpm. The difference of 75 rpm is called slip. At first glance you would think that slip is not good for a motor because the motor is not running as fast as it should be, but you will see that the induction motor relies on slip to create the amount of current necessary to turn the load at the end of the motor shaft. If the amount of slip is minimum and the rotor is spinning close to the rated speed, the amount of CEMF is large, and the difference between the CEMF and the applied EMF may be as small as 5 VAC. When this occurs, the amount of current in the motor is reduced to less than 0.5 A, and consequently the amount of torque the rotor has is also reduced. It is also important to remember that the rotor gets the current required to build its magnetic fields through induction from the stator windings. The induction can occur only if the rotor is not turning at the same speed as the magnetic field is rotating in the stator.

When the rotor's torque is minimal, the shaft will not be able to turn its load, and it will begin to slow down, which decreases the amount of CEMF. This makes the difference between the CEMF and applied voltage larger, which makes the motor draw more current and produce more torque. All induction motors rely on slip to operate correctly and produce sufficient torque. If it is important that the motor run at its rated speed, a special type of motor called a *synchronous motor* could be used, because it runs at its rated speed with no slip. The synchronous motor is not used very often in industrial applications because it requires DC voltage to be supplied to its field.

It is also possible to use a variable-frequency drive to adjust the frequency of the voltage above 60 Hz to get the motor to run at its rated speed even if it has slip. Variable-frequency drives were introduced into industrial applications in the 1980s. You will learn more about variable-frequency drives in Chapter 19 that covers electronic devices. As a technician you will encounter a variety of variable-frequency drives in newer systems, where they are used to control the speed of conveyors, pumps, and fan motors.

13.5
END PLATES

The end plates play a very important part in the operation of the motor. They house the bearings that support the rotor to allow it to spin with a minimum amount of friction, and one end plate houses the end switch that is used to help the single-phase motor to get started. The bearings in a single-phase motor may be the ball-bearing type or a bushing type. The ball-bearing type is

Figure 13–5   End plates are shown with a rotor. The bearings are mounted in the end plates, and they support the rotor shaft so that it can spin freely.

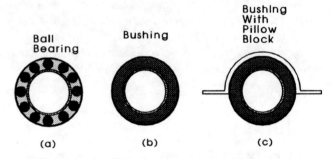

Figure 13–6   (a) An oil-type bearing (bushing) with insulator; (b) sealed-type bronze bushing and sealed ball bearing; (c) pillow block-type bearing with bronze bushing.

more expensive and is generally used in larger single-phase motors. The bushing is used in smaller and less expensive single-phase motors. Figure 13–5 shows end plates, and Fig. 13–6 shows an example of a ball bearing and a bushing. The ball bearing shown in Fig. 13–6a uses a number of small steel balls, which create a rolling surface between the inner hub and outer hub. The shaft of the motor is pressed securely into the inner hub of the bearing so that the shaft and inner hub of the bearing move together as though they were welded. As the rotor shaft spins, the inner hub rolls on the balls, and the balls make contact with the outer hub. This means that only a small portion of each ball is touching the inner and outer hub at any time, which keeps friction to a minimum. The ball bearing for smaller motors is lubricated with grease when it is manufactured, and it usually is sealed for the life of the bearing, which means that motors with these types of ball bearings will not need to be lubricated. Some larger motors use bearings that must be greased periodically. A grease fitting is provided so that the ball bearing can be greased once or twice a year, depending on the number of hours the motor runs.

The bushing shown in Fig. 13–6b is made of a brass or bronze alloy sleeve with a lubrication hole in it. An *oiler pad,* which is also called an *oil wick,* is mounted on the outer side of the bushing, so oil from the pad can find its way through the hole to the inside of the bushing. The oil creates a "bearing" surface between the shaft of the rotor, which spins while the bushing remains stationary. As long as a sufficient amount of oil remains on the inside of the bushing, between it and the rotor shaft, the amount of friction is held to a minimum. If the bushing ever runs dry, the rotor shaft will be allowed to touch the bushing surface and both surfaces will begin to wear severely. It is vitally important to ensure that the lubrication ports for the bushing-type motor are always located on the top side of the motor so that gravity can draw the lubrication through the lubrication pad into the bushing. It is also important to remember that the motor that uses a bushing must be lubricated at least once a year.

13.6
THE STATOR

The stator is actually made of two separate windings, called the *run winding* and the *start winding.* Each of these windings is subdivided into poles, which represent the poles of a magnet. We use a four-pole motor for our examples in this chapter. This means that the run winding has four poles and the start winding has four poles. Figure 13–7 shows a picture stator so that you can more easily see the coils of wire that are mounted in it. From this picture you can see the four poles of the run winding and the four poles of the start winding. The windings are pressed over pole pieces that also become magnetized when the coils have current flowing in them. The pole pieces are made from laminated steel like the rotor, so the poles can quickly change polarity when AC voltage is applied.

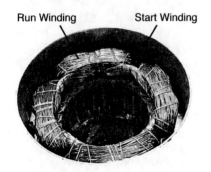

Run Winding   Start Winding

Figure 13–7 Close-up picture of start and run windings in the stator. The run winding has larger wire than the start winding.

The AC sine wave starts at 0 V and peaks positive, returns to 0 V and peaks negative, and returns to 0 V again. It continues to oscillate from positive to negative and back as long as voltage is applied. This action causes the magnetic field in each winding in the stator to be positive and then negative and continually change. The changing magnetic field actually results in the magnetic field rotating around the poles inside the stator. Because the rotor also has a magnetic field, it will begin to spin and follow the rotating magnetic field in the stator. It will be easier to understand the function of the stator if you study the operation of the start winding and run winding separately.

13.7
THE START WINDING

If the AC induction motor only had one winding, the rotor would not tend to rotate when single-phase AC voltage was first applied to the stator windings in the AC induction motor. The rotor needs a small amount of phase shift between the applied voltage and current in the magnetic field created in the stator to begin to spin. There are several ways to provide the phase shift required to start the rotor spinning. One way to provide the phase shift is to use two windings in the stator and physically offset them. A second way to cause a phase shift in the stator is to put more wire in one set of windings than in the other. For this reason, the stator has a start and a run winding, and the start winding is made from smaller wire than the run winding so it can have more turns of wire and thus have more wire.

Figure 13–7 shows a close-up picture of a stator that clearly shows the run and start windings. From this picture you can see that the run winding is made of larger wire and is physically offset from the start winding. Figure 13–8 shows how the run winding and start winding are placed in the stator offset from each other. Figure 13–8 also shows an electrical diagram of the start winding and run winding for an eight-lead single-phase AC motor and for a four-lead single-phase motor.

The electrical diagram in Fig. 13–8a shows that the start winding is made of smaller wire and has more turns than the run winding. The start-winding coil has more turns of wire than the run-winding coil, so more feet of wire are used in the start winding. Because the start winding has more wire, it takes the current longer to pass through it than through the run winding. It is important to understand that when voltage is applied to the AC induction motor, voltage starts into the run winding and the start winding at the same time. Because

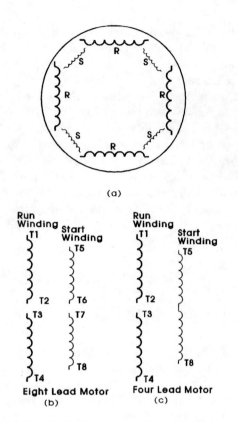

(a)

Run
Winding Start
T1   Winding
       T5

T2   T6
T3   T7

T4   T8

Eight Lead Motor
(b)

Run
Winding Start
T1   Winding
       T5

T2
T3

T8

T4

Four Lead Motor
(c)

Figure 13–8   (a) A diagram that shows the four-pole run winding and four-pole start winding in the stator of an AC induction motor. Notice that the start winding is physically offset in the stator from the run winding to help provide more of a phase shift and that the run winding is made of larger-gauge wire and has fewer turns than the start winding. (b) A diagram of the run winding and start winding of an eight-lead induction motor. Notice that the two sections of the run windings are identified as T1–T2 and T3–T4, and the two sections of the start winding are identified as T5–T6 and T7–T8. (c) A diagram that shows a four-lead induction motor. The run winding is identified as T1–T4, and the start winding is identified as T5–T8.

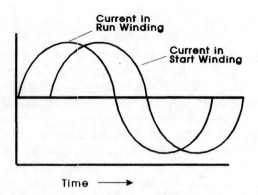

Current in
Run Winding

Current in
Start Winding

Time

Figure 13–9   A diagram of the phase shift that occurs when voltage is applied to the start and run windings of an AC motor. Because the start winding has more wire than the run winding, a phase shift is created between the magnetic fields in the two windings.

Because the start winding is made from very small wire so that its coils can have many more turns of wire than the run winding, the start winding will draw a large amount of current when voltage is applied to it. If the start winding were allowed to remain in the circuit continually, this large current would damage it and eventually burn a hole in the windings. For this reason a centrifugal switch is used to disconnect the start winding after the motor's rotor is spinning. The centrifugal switch can determine when the speed of the rotor has reached 75 to 85% rpm and disconnect the start winding from the circuit until it is needed the next time the motor is started.

13.8
THE OPERATION OF THE CENTRIFUGAL SWITCH

The centrifugal switch is also called the *end switch* in a motor because it is mounted in the end plate. Even though the centrifugal switch is physically mounted in the end plates, it is important to understand that it is electrically connected to the start winding that is located in the stator. The centrifugal switch is actually part of a larger assembly consisting of an activation mechanism that has flyweights that begin to swing out when the rotor reaches 75% full rpm. The activation mechanism is mounted on the end of the rotor shaft so it can turn when the rotor shaft is turning. The activation mechanism is mounted on the shaft in such a way that it makes contact with the centrifugal switch and holds the switch contacts in the closed position when the shaft is not rotating.

Figure 13–10 shows a centrifugal switch and the activation mechanism. Figure 13–10a shows a centrifugal

the start winding has more wire and is offset physically from the run winding, a phase shift is created.

This phase shift gives the rotor a sufficient amount of starting torque (rotational force) to get the rotor to start turning and to keep it turning until the motor comes up to full speed. The phase shift occurs between the run winding and the start winding, but it also can be described as occurring between the voltage waveform and the current waveform because inductors (motor windings) cause the current waveform to lag the waveform of voltage. In later sections you will see how capacitors can be added to the start windings to enhance the amount of phase shift so the motor will have more starting torque. Figure 13–9 shows the phase shift between the current in the start winding and that in the run winding.

Figure 13–10   (a) Centrifugal switch removed from an end plate and the activation mechanism removed from the armature shaft; (b) the centrifugal switch is mounted in the end plate and the end plate is mounted on the end of the motor shaft so that the activation mechanism is touching the centrifugal switch.

switch that is removed from an end plate. Figure 13–10b shows a centrifugal switch mounted in the end plate, with its flyweights at rest as you would find them when the rotor is not turning.

The centrifugal switch is a set of electrical contacts mounted on spring steel. Spring steel is used because it can be flexed so that the spring tension in the steel keeps the contacts in the open position. The activation mechanism that is mounted on the rotor shaft comes into contact with the spring steel of the switch when the end plate is attached to the stator. You should remember from the

cutaway diagram in Fig. 13–11b that the rotor shaft goes through the bearing in the end plate, and when the end plate is drawn tight against the stator with screws, the activation mechanism touches the spring steel and causes the switch contacts to press against each other to complete the electrical circuit through the switch.

When voltage is applied to the windings in the motor, current flows through the run winding and through the start winding because the activation mechanism is holding the contacts of the centrifugal switch closed. As the rotor begins to spin, centrifugal force begins to pull the flyweights outward. Because the flyweights are held together with a spring on each side, they will not swing out all the way until the rotor is spinning approximately 75 to 85% of full rpm. The spring tension that holds the flyweights inward will determine the speed at which the flyweights will swing out completely.

When the flyweights swing out completely, they allow the spring tension on the activation mechanism to move the mechanism to a retracted position, which allows the spring tension in the contacts of the centrifugal switch to move the contacts to the open position. When the contacts of the centrifugal switch open, current to the start winding is interrupted, and the motor continues to run with current flowing only through the run winding of the motor.

Figure 13–11a shows a diagram of the activation mechanism when the rotor is not turning. In this part of the diagram you can see that the flyweights are held in position by spring tension. When the flyweights are in this position, the activation mechanism is pressed against the switch contacts, holding them in the closed position so current can flow through them. In Fig. 13–11b the rotor is turning at nearly full speed. The centrifugal force that is created when the rotor shaft is spinning at full speed causes the flyweights to swing out away from the rotor shaft. When the flyweights swing out completely, the activation mechanism is pulled away from

Figure 13–11   (a) Flyweights and activation device for a centrifugal switch. The activation device is mounted on the end of the rotor shaft so that it can come into contact with the centrifugal switch. A spring holds the flyweights against the shaft when the rotor is not turning. (b) When the rotor is spinning at 75 to 85%, the flyweights move outward from the shaft because of centrifugal force. The action of the flyweights causes the activation device to drop away from the switch contacts so that they can move to their open position.

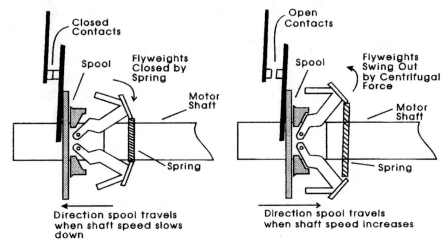

the contacts, which allows the tension of the spring steel to open the contacts.

13.9
THE RUN WINDING

The run winding is the part of the stator that remains in the motor circuit at all times. Figure 13–12 shows an electrical diagram for a single-phase motor, and you can see that the run winding is easily identified because it uses larger wire and fewer coils in its windings. In most single-phase motors, the run winding is numbered, as shown in the diagram. If the motor is a dual-voltage motor, its run winding is made in two sections, one with terminal ends

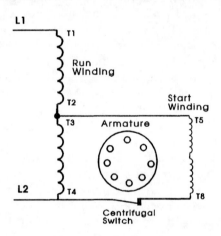

Split-phase open motor
wired for 230 volts

(a)

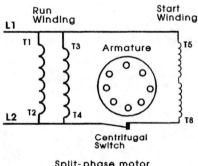

Split-phase motor
wired for 115 volts

(b)

Figure 13–12   (a) Electrical diagram of an AC induction motor wired for high voltage (230 or 208 VAC). The two sections of the run winding are numbered T1 and T2, and T3 and T4 are connected in series. (b) Electrical diagram for an AC induction motor wired for low voltage (115 VAC). The run winding in this diagram is shown in two sections that are connected in parallel.

T1 and T2 and the other with terminal ends T3 and T4. If the motor is wired for high voltage (220 or 208 VAC), the windings will be wired in series with each other.

Figure 13–12b shows the two sections of the run winding connected for low voltage (115 VAC). If the motor is wired for low voltage, the two sections of the run windings are connected in parallel with each other. These two diagrams are used in all electrical system diagrams for installing and troubleshooting motors in industrial applications.

You can test the run winding and start winding with an ohmmeter to determine which one is which. For example, in the diagram each part of the run winding has 4 ω of resistance and the start winding has 12 ω.

13.10
MULTIPLE-SPEED MOTORS

The coils of wire in the run winding can be separated into sections called *poles*. The motor has an even number of poles. The speed of an induction motor is determined by the number of poles and the frequency. The speed can be calculated from the following formula:

$$rpm = \frac{frequency \times 120}{poles}$$

A typical motor has two, four, six, or eight poles. The speed of each type of motor is listed in the following table.

No. of poles	rpm
2	3,600
4	1,800
6	1,200
8	900

Motors for typical industrial applications need to be able to operate at a variety of speeds. This means that the motor may be manufactured with a set number of windings to provide a set speed, or the motor may be reconnected in the field to give the motor a different speed.

Another way to change the speed of a motor is to tap the run winding so that more or less inductive reactance is used. This means that if you have a very long run winding, you can place taps on it just like the multiple-tapped transformer. Each tap uses less of the total run winding, and consequently the change in inductive reactance allows the speed of the motor to change. The amount of change in rpm for each tap is approximately 15 to 20%. Figure 13–13 shows an example of a motor with a multitapped run winding. From the diagram you can see the taps are in the run winding, and

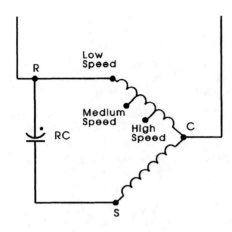

Low Speed = Red Wire
Medium Speed = Blue Wire
High Speed = Black Wire

Figure 13–13 The electrical diagram of a multitapped fan motor. The run winding is tapped so that it can provide three different choices for the number of poles that are used. The black wire is used for the high speed, the blue wire is used for the medium speed, and the red wire is used for the low speed.

you can see the color code for the wires. If you want the motor to run at high speed, you use the black wire for the run winding, and if you want the motor to run at medium speed, you use the blue wire; the red wire allows the motor to run at low speed. Generally the multiple speeds are used for fan motors, where the high speed is used for cooling and the low speed is used for heating. The medium speed is used where a slightly lower speed for air conditioning or a higher speed than the lowest speed for heating is needed. It is important to remember that as the tap is changed for this motor, it will use less of the winding to run faster, but it will also lose power. In some motors the common terminal is connected to the black, red, or blue wire instead of the run terminal.

13.11
MOTOR DATA PLATES

The motor's *data plate* lists all the pertinent data concerning the motor's operational characteristics. It is sometimes called the *name plate*. Figure 13–14 shows an example of a data plate for a typical AC motor. The data plate contains information about the ID (identification number), FR (frame and motor design), motor type, phase, horsepower rating, rpm, volts, amps, frequency, service factor (SF), duty cycle (time), insulation class, ambient temperature rise, NEMA design, and

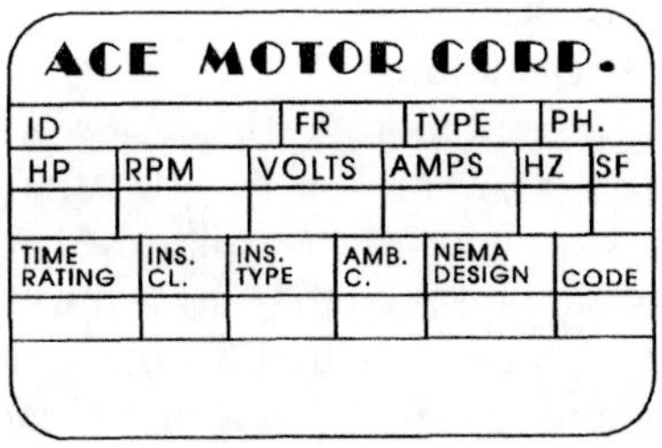

ACE MOTOR CORP.					
ID		FR	TYPE	PH.	
HP	RPM	VOLTS	AMPS	HZ	SF
TIME RATING	INS. CL.	INS. TYPE	AMB. C.	NEMA DESIGN	CODE

Figure 13–14 Data plate for a typical AC motor.

code. Each of these features is discussed in detail in the next sections.

13.11.1
Identification Number (ID)

The identification number (ID) for a motor is basically a model and serial number. The model number includes information about the type of motor, and the serial number is a unique number that indicates where and when the motor was manufactured. These numbers are important if a motor is returned for warranty repairs or if an exact replacement is specified as a model-for-model exchange.

13.11.2
Frame Type

Every motor has been manufactured to specifications that are identified as the *frame size*. These data include the distance between mounting holes in the base of the motor, the height of the shaft, and other critical data about physical dimensions. These data are then given a number such as 56. The frame number indicates that any motor that has the same frame number will have the same dimensions, even though it may be made by a different company. This allows users to stock motors from more than one manufacturer or replace a motor with any other with the same frame information and be sure that it will be an exact replacement.

13.11.3
Motor Type

The motor-type category on the data plate refers to the type of ventilation the motor uses. One type is the *open* type, which provides flow-through ventilation from the fan mounted on the end of the rotor. In some motors that are rated for variable-speed duty (used with variable-frequency drives), the fan is a separate motor that is built into the end of the rotor. The fan motor is connected directly across the supply voltage, so it will maintain a constant speed to provide constant cooling regardless of the motor speed.

Another type of motor is the *enclosed* type. The enclosed motor is not air-cooled with a fan; instead it is manufactured to allow heat to dissipate quickly to and from the inside of the motor outward to the frame. In most cases the frame has fins built into it on the outside to provide more area for cooling air to reach.

13.11.4
Phase

The phase of the motor is indicated as single-phase or three-phase. The number may be given as 1 or 3, or it may be written out as one or three.

13.11.5
Horsepower (HP) Rating

The horsepower (hp) rating is given as a fractional horsepower (a number less than 1.0) or a horsepower greater than 1. Fractional horsepower values may be given as fractions ($\frac{1}{2}$) or decimals (0.5).

13.11.6
Speed (rpm)

The motor speed is indicated as in terms of rpm. This is the rated speed for the motor and does not account for slip. The actual speed of the motor will be less because of slippage. As you know, the speed of the motor is determined by the number of poles and the frequency of the AC voltage. Typical speed for a two-pole motor is 3,600 rpm, for a four-pole motor it is 1,800 rpm, and for an eight-pole motor it is 1,200 rpm. The actual speed of a motor rated for a 3,600-rpm motor is approximately 3,450; for the 1,800-rpm motor it is approximately 1,750, and for the 1,200-rpm motor, it is approximately 1,150. The actual amount of slip is indicated by the *motor-design letter.* (The motor-design letter is explained in Section 13.11.14.)

13.11.7
Voltage Rating

Voltage ratings for single-phase motors are typically 120, 208, or 240 V. Three-phase motors have typical voltage ratings of 208, 240, 440, 460, 480, and 550 V. Other voltages may be specified for special types of motors. You should always ensure that the power supply voltage rating matches the voltage rating of the motor. If the rating and the power supply voltage do not match, the motor will overheat and be damaged.

13.11.8
Amps Rating

The amps rating is the amount of full-load current (FLA) the motor should draw when it is under load. This rating helps the designer calculate the proper wire size, fuse size, and heater size in motor starters. The supply wiring for the motor circuit should always be larger than the amps rating of the motor. The NEC (National Electric Code) provides information to help you determine the exact fuse size and heat sizes for each motor application.

13.11.9
Frequency

The frequency of a motor is listed in hertz (Hz). A typical frequency rating for motors in the United States is 60 Hz. Motors manufactured for use in some parts of Canada and all of Europe and Asia are rated for 50 Hz. You must be sure that the frequency of the motor matches the frequency of the power supplied to the motor.

13.11.10
Service Factor (SF)

The *service factor* (SF) is a rating that indicates how much a motor can be safely overloaded. For example a motor that has an SF of 1.15 can be safely overloaded by 15%. Thus, if the motor is rated for 1 hp, it can actually carry 1.15 hp safely. To determine the overload capability of a motor, you multiply the rated hp by the service factor. The motor is capable of being overloaded because it is designed to be able to dissipate large amounts of heat.

13.11.11
Duty Cycle

The *duty cycle* of a motor is the amount of time the motor can be operated out of every hour. If the motor's duty cycle is listed as continuous, it means the motor can be run 24 h a day and does not need to be turned off to cool down. If the duty cycle is rated for 20 min, it means the motor can be safely operated for 20 min before it must be shut down to be allowed to cool. A motor with this rating should be shut down for 40 min of every hour of operation to be allowed to cool.

Another way to specify the duty cycle of a motor is to use the *motor rating.* The motor rating on a data plate refers to the type of duty the motor is rated for. The types of duty include continuous duty, intermittent duty, and heavy duty, which includes jogging and plugging duty. Continuous duty includes applications where the motor is started and allowed to operate for hours at a time. Intermittent duty includes operations where the motor is started and stopped frequently. This type of application allows the motor to heat up because it will draw LRA more often than will a motor rated for continuous duty.

Motors that are rated for jogging and plugging are built to withstand very large amounts of heat that build up when the motor draws large LRA during starting and stopping. Because the motor can be reversed when it is running in the forward direction for plugging applications, it builds up excessive amounts of heat. Motors with this rating must be able to get rid of heat as much as possible to withstand the heavy-duty applications.

13.11.12
Insulation Class

The *insulation class* of a motor is a letter rating that indicates the amount of temperature rise the insulation of the motor wire can withstand. The numbers in the insulation class are listed in degrees Celsius (°C). The table in Fig. 13–15 shows typical insulation classes for motors. The insulation class and other temperature-related features of a motor will help determine the temperature rise the motor can withstand.

13.11.13
Ambient Temperature Rise

The ambient temperature rise is also called the Celsius rise. It is the amount of temperature rise the motor can withstand during normal operation. This value is listed in degrees Celsius. A typical open motor can withstand a rise of 40°C (104°F), and an enclosed motor can withstand a 50°C (122°F) rise. This means the motor should not be exposed to environments where the temperature is 104°F above the ambient. If the ambient is considered to be 72°F, the motor is limited to a temperature of 176°F. Another classification that helps determine the amount of temperature a motor can withstand is the type of insulation the motor winding has. The classes of

Class	Temperature Rise, °C
A	105
B	130
F	155
H	180

Figure 13–15 Insulation class for motors. This table indicates the amount of temperature for which the wire's insulation is rated.

insulation used with motors and the amount of temperature these classes can handle are listed in Section 13.11.14.

13.11.14
NEMA Design

The National Electric Manufacturers Association (NEMA) provides design ratings that may also be listed as motor design. The motor design is identified on the data plate by the letter A, B, C, or D. This designation is determined by the type of wire, insulation, and rotor used in the motor and is not affected by the way the motor might be connected in the field.

Type A motors have low rotor-circuit resistance and have approximate slip of 5 to 10% at full load. These motors have low starting torque with a very high locked-rotor amperage (LRA). This type of motor tends to reach full speed rather rapidly.

Type B motors have low-to-medium starting torque and usually have slip of less than 5% at full load. These motors are generally used in fans, blowers, and centrifugal pump applications.

Type C motors have a very high starting torque per ampere rating. This means that they are capable of starting when the full load is applied for applications such as conveyors and crushers and reciprocating compressors such as air-conditioning and refrigeration compressors. These motors are rated to have slip of less than 5%.

Type D motors have a high starting torque with a low LRA rating. This type of motor has a rotor made of brass rather than copper segments. It is rated for slip of 10% at full load. Normally, this type of motor will require a larger frame to produce the same amount of horsepower as a type A, B, or C motor. These motors are generally used for applications with a rapid decrease of shaft acceleration, such as a punch press that has a large flywheel.

These standards are set by NEMA, and a motor must meet all the requirements of the standard to be marked as a type A, B, C, or D motor. This allows motors made by several manufacturers to be compared on an equal basis according to application.

13.11.15
NEMA Code Letters

NEMA code letters use letters of the alphabet to represent the amount of locked-rotor amperage (LRA) in kVA per horsepower that a motor will draw when it is started. These letters are listed in Fig. 13–16. From the table you can see that letters at the beginning of the alphabet indicate low LRA ratings, and letters toward the end of the alphabet indicate higher LRA ratings. It is important to remember that the number in the table is not

NEMA Code Letter	Locked-Rotor kVA per hp
A	0–3.15
B	3.15–3.55
C	3.55–4.00
D	4.00–4.50
E	4.50–5.00
F	5.00–5.60
G	5.60–6.30
H	6.30–7.10
J	7.10–8.00
K	8.00–9.00
L	9.00–10.00
M	10.0–11.2
N	11.2–12.5
P	12.5–14.0
R	14.0–16.0
S	16.0–18.0
T	18.0–20.0
U	20.0–22.4
V	22.4 and up

Figure 13–16 Table with locked-rotor amperage (LRA) ratings. The ratings are listed as the amount of kVA per horsepower. (Courtesy of National Electrical Manufacturers Association (NEMA))

Figure 13–17 Useful electrical formulas for determining amperes, horsepower, kilowatts, and KVA.

the amount of LRA the motor will draw, but rather it is the number that must be multiplied by the horsepower rating of the motor.

13.11.16
Useful Electrical Formulas for AC Motors

From time to time when you are working with a motor, you will need to calculate horsepower, current, kilowatts, VA, or other values. Figure 13–17 shows a table of the most common formulas used with AC single-phase and three-phase motors. The reason you will need the wide variety of formulas shown in this table is that some motor data plates will list information in horsepower, while others will list volts and amps or watts. These formulas will make it easy to convert motor data between all the different variables that you may be able to determine from the data plate or from measurements you might take while the motor is running. Some manufacturers supply the formulas in this table in a chart or in a slide chart that will provide the calculation and answer for each problem you are trying to solve. For example, if you know the voltage and current for your motor, you can move the slide chart to these numbers and all other horsepower data about this motor will be provided in a window on the slide chart. This information will be very useful when you are trying to replace a motor or if you are trying to troubleshoot a problem.

To Find	Formulas for Single Phase AC Motors	Formulas for Three Phase AC Motors
How to find current (amps) when horsepower (HP), volts, efficiency and power factor (PF) are known	$I = \dfrac{HP \times 746}{volts \times Eff \times PF}$	$I = \dfrac{HP \times 746}{1.73 \times volts \times Eff \times PF}$
How to find current (amps) when Kilowatts (kW), volts, and power factor (PF) are known.	$I = \dfrac{kW \times 1000}{volts \times PF}$	$I = \dfrac{kW \times 1000}{1.73 \times volts \times PF}$
How to find current (amps) when kilovolt-amps (kVA), and volts are known.	$I = \dfrac{kVA \times 1000}{volts}$	$I = \dfrac{kVA \times 1000}{1.73 \times volts}$
How to find output horsepower (HP) when current volts, efficiency and power factor (PF) are known.	$HP = \dfrac{amps \times volts \times Eff \times PF}{746}$	$HP = \dfrac{amps \times volts \times 1.73 \times Eff \times PF}{746}$
How to find kilovolt-amps (kVA) when current and volts are known.	$kVA = \dfrac{amps \times volts}{1000}$	$kVA = \dfrac{amps \times volts \times 1.73}{1000}$
How to find kilowatts (kW) when current, volts and power factor (PF) are known.	$kW = \dfrac{amps \times volts \times PF}{1000}$	$kW = \dfrac{amps \times volts \times 1.73 \times PF}{1000}$

Formulas for Single Phase and Three Phase AC Motors

13.12

WIRING SPLIT-PHASE MOTORS FOR A CHANGE OF ROTATION AND CHANGE OF VOLTAGE

The wiring diagram for the run and start windings of the split-phase motor were shown in Fig. 13–12. The terminals in the run and start windings can be reconnected in the field to change the rotation of the motor or to allow the motor to operate at a different voltage. Because the split-phase motor and the capacitor-start motor are very similar, these connections are presented in Section 13.16 after you learn more about capacitor-start, induction-run motors.

13.13

OVERVIEW OF CAPACITOR-START, INDUCTION-RUN MOTORS

The *capacitor-start, induction-run* (CSIR) motor is basically a split-phase induction motor that adds a capacitor with the start winding to create a larger phase shift between the start and run windings to start the motor. The capacitor-start, induction-run motor is generally used in applications where the load is connected directly to the shaft of the motor. For example, a capacitor-start, induction-run motor can be used to turn direct-drive fans and larger pump loads. Some capacitor-start, induction-run motors are used to drive belt-driven compressors. The torque for a capacitor-start, induction-run motor is larger than for the split-phase-type motor. A typical capacitor-start, induction-run motor is shown in Fig. 13–18.

13.14

THEORY OF OPERATION FOR A CAPACITOR-START, INDUCTION-RUN MOTOR

The capacitor-start, induction-run motor operates in a manner very similar to the AC split-phase motor described at the beginning of this chapter. It has a rotor and a stator with a run winding and a start winding. A start capacitor is connected in series with the start winding to create a larger phase shift than the split-phase motor has. The start capacitor is generally rated

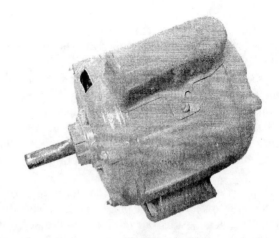

Figure 13–18   A capacitor-start, induction-run motor used for HVAC and refrigeration applications such as direct-drive fan motors and pump motors.

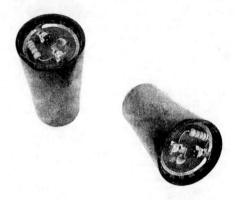

Figure 13–19   The start capacitor is mounted in a black plastic case that has a round, cylindrical shape.

from 50 to 90μ F (microfarads). The capacitance and voltage rating for the capacitor are listed on its case. Figure 13–19 shows a picture of a start capacitor. The start capacitor is so named because it is always connected in series with the start winding in the capacitor-start, induction-run motor to provide a larger phase shift, which will also create more starting torque. The start capacitor is also what gives the capacitor-start, induction-run motor its name.

The start capacitor has several physical features that make it easy to recognize when you are servicing equipment. One, which makes it especially easy to recognize, is that it is generally mounted in a black plastic case. Also, the black plastic case for the start capacitor is usually round in shape. When you look at the picture of the start capacitor in Fig. 13–19, you can see that it is mounted in a round, black plastic case.

The start capacitor is made of two sheets of conducting foil, called *plates* that are separated (sandwiched) with three sheets of insulating paper called the *dielectric.* Figure 13–20a shows an example of the foil sheets and dielectric sheets before they are rolled up. Figure 13–20b shows a diagram of the sheets rolled up so that they will fit into the round, plastic case. A lead wire is soldered onto each piece of foil and brought out to the top of the capacitor to be used as the capacitor's terminals. Figure 13–20c shows the electrical symbol for the capacitor. Notice that the symbol shows the capacitor's conducting plates separated by the dielectric so current cannot pass directly from one plate to the other.

When voltage is applied to the capacitor in a circuit, current will flow as far as the plate on the left side of the capacitor. Because the capacitor plates are separated by dielectric paper, current cannot flow directly through the capacitor to the other plate, so the voltage must reverse and move backward through the circuit until it reaches the other capacitor plate. This process is referred to as *charging the capacitor and discharging the capacitor.* If the voltage that causes the current to flow is AC voltage, the AC sine wave will continually charge and discharge the capacitor, which keeps current continually flowing through the components in the circuit.

Because it takes time for the voltage to charge up the first plate and then reverse itself and discharge the

plate, the capacitor causes a phase shift of up to 90° between the voltage waveform and the current waveform in the circuit. It is important to remember that the capacitor will always cause the voltage waveform to lag the current waveform.

13.15
APPLYING VOLTAGE TO THE START WINDING AND CAPACITOR

Figure 13–21 shows an electrical diagram of a capacitor-start, induction-run motor so that you can see that the start capacitor (SC) is connected in series with the centrifugal switch and the start winding. From this diagram you can see that when voltage is first applied to the leads of the motor, current will flow from L1 through the run windings and back to L2 or N. At the same time, voltage is applied to the first capacitor plate. Because current cannot flow through the capacitor to the other plate, the voltage must reverse itself and move back through the run winding; eventually it finds its way back to the start winding and reaches the other plate of the capacitor. Because the voltage must take the long way around the motor circuit to get to the other plate, it creates a phase shift between the current flowing in the run winding and the current flowing in the start winding. This process continues as long as the centrifugal switch remains closed and current flows through the start winding. The capacitor causes a larger phase shift than the split-phase motor can develop without a capacitor. Because the capacitor causes a larger phase shift than the split-phase motor can develop with just the start winding, the capacitor-start, induction-run

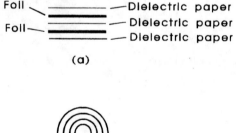

(a)

(b)

(c)

Figure 13–20 (a) Two sheets of conducting foil that are separated by sheets of dielectric paper. The dielectric paper acts as an insulator between the conducting foil. The two sheets of foil are called the plates. (b) The layers of conducting foil and dielectric paper are rolled up so that they will fit into the round, black plastic case. (c) The electrical symbol for the capacitor shows the two conducting plates separated by the insulator.

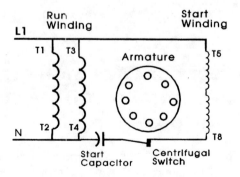

Figure 13–21 Electrical diagram of a capacitor-start, induction-run (CSIR) motor. Notice the start capacitor (SC) is connected in series with the centrifugal switch and start winding.

motor will have more starting torque than the split-phase motor.

When the motor's rpm increases to 75 to 85% of full rpm, the centrifugal switch opens and causes the current to stop flowing through the start winding. After the centrifugal switch opens the start-winding circuit, current will continue to flow through the run winding so the motor will act as an induction motor when it is running. This is why the motor is called a *capacitor-start, induction-run motor.*

13.16

WIRING SPLIT-PHASE AND CAPACITOR-START, INDUCTION-RUN MOTORS FOR 115 V

The diagram shown in Fig. 13–21 shows the electrical diagram for a capacitor-start, induction-run motor that is connected for 120 V. This connection is also called the *low-voltage connection.* The diagram for a split-phase motor is identical except it does not have a start capacitor connected to its start winding. When you must connect a split-phase motor or capacitor-start, induction-run motor for 120 V, you should connect terminals T1, T3, and T5 together with the supply voltage wire L1, and you should connect terminals T2, T4, and T8 together with the neutral supply voltage wire. You should notice from the diagram that these connections place the two sections of the run winding in parallel with each other and with the start winding. You should also notice that the start capacitor is connected in series with the centrifugal switch. This means that each section of the run winding will have 120 V applied to it, and the start winding will have 120 V.

13.17

WIRING A CAPACITOR-START, INDUCTION-RUN MOTOR FOR 230 V

At times you may need to reconnect the capacitor-start, induction-run motor or which you are working for a higher voltage. Figure 13–22 shows an electrical diagram of a capacitor-start, induction-run motor that is connected for 240 V. This connection is also called the *high-voltage connection.* Again, the diagram for the split-phase motor is similar except it does not have

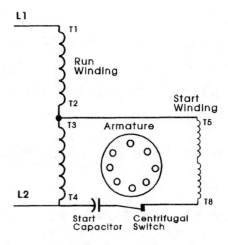

Figure 13–22   Electrical diagram for a capacitor-start, induction-run motor connected for 240 V.

the start capacitor. From this diagram you can see that the two sections of the run winding are connected in series. This connection ensures that 120 V is applied to each winding as a voltage drop. You should understand that the insulation on the wire that is used for the run and start windings is rated for 120 V. The two run windings are connected in series, so you are ensuring that each winding will receive only 120 V, because the 240 V will be dropped equally across each run winding. You should also notice that the capacitor is always connected to the centrifugal switch, so it will always be connected in series with the start winding.

It is important at this time to remember that the motor may be rated for 208 V for its higher voltage rather than 240 V. The diagram for the higher voltage is the same regardless of whether the motor data plate indicates the motor is rated for 240 V or 208 V. The important point to remember is that the motor must match the exact amount of incoming voltage used to supply the motor. This means that you should always measure the exact amount of voltage before you connect a motor.

You should also notice that the start winding is connected so that it receives only 120 V. This means that the start winding is connected so that it is actually in parallel with the lower section of the run winding (T3—T4). One end of the centrifugal switch is connected permanently to one terminal of the start capacitor. The other terminal of the start capacitor is connected to terminal T4. This ensures that it will always be connected across the lower section of the start winding. You can see that if the motor is connected for 240 V or 120 V, one end of the centrifugal switch is always connected through the capacitor to terminal T4.

13.18

WIRING A SPLIT-PHASE AND CAPACITOR-START, INDUCTION-RUN MOTOR FOR A CHANGE OF ROTATION

When you are working in the field, you will find times when you must change the direction of rotation of the motor's shaft. Identify the rotation of the shaft by standing behind the motor (at the opposite end from where the shaft comes out of the end plate) and observe the direction of rotation. To change the rotation of the motor shaft, you must change the direction current flows through the start winding. The easiest way to do this is to exchange the T5 and T8 wires.

The capacitor-start, induction-run motor that is connected for high voltage can also be wired so that the direction of its shaft's rotation can be reversed. When the motor is wired for clockwise rotation, terminal T5 of the start winding is connected to terminals T2 and T3 of the run winding, and terminal T8 of the start winding is connected to terminal T4 of the run winding. Figure 13–22b shows the electrical diagram of a high-voltage (240-V) capacitor-start, induction-run motor connected for counterclockwise rotation. In this diagram you can see that terminal T8 of the start winding is connected to terminals T2 and T3 of the run winding, and terminal T5 of the start winding is connected to terminal T4 of the run winding.

13.19

CHANGING SPEEDS WITH A CAPACITOR-START, INDUCTION-RUN MOTOR

The split-phase and capacitor-start, induction-run motors are generally not designed to have their speed changed. Some more expensive capacitor-start, induction-run motors have additional sets of poles provided so that their speeds can be changed by changing the number of poles used in each run winding, but typically this type of motor cannot be reconnected for a change of speed. Thus if you need to change the speed of the motor, you would need to replace the motor with one of a different speed. It is also important to understand that if the motor is used to drive a belt-driven fan, the speed of a fan can be altered slightly by changing the size of the pulley on the motor.

13.20

OVERVIEW OF PERMANENT SPLIT-CAPACITOR MOTORS

The *permanent split-capacitor* (PSC) motor is similar to the split-phase induction motor and capacitor-start, induction-run motor in that it has a start winding and a run winding, but it is different in that it does not use a centrifugal switch to remove power from the start winding when the motor reaches full speed. Instead of a centrifugal switch, the PSC motor uses a run capacitor that is connected in series with the start winding to limit the amount of current allowed to flow through the start winding when the motor reaches full speed. The run capacitor is mounted in a metal container so that it can dissipate the heat better, which allows it to remain in the circuit even after the motor is running at full speed. The permanent split-capacitor motor is generally used in applications where the load is connected directly to the shaft of the motor. For example, a permanent split-capacitor motor can be used to turn direct-drive fans, such as squirrel-cage-type fans or blade-type fans, and motors for conveyor applications and pump applications. The torque for a permanent split-capacitor motor is larger than that of a split-phase-type motor and about the same as or slightly less than a capacitor-start, induction-run motor. Two typical permanent split-capacitor motors are shown in Fig. 13–23. You should be able to see that the run capacitor is mounted on the top of each motor. Figure 13–23 shows one type of PSC motor that has a shaft extending out of both ends of its end plates. This type of PSC motor is used in window air

Figure 13–23   Typical permanent split-capacitor motors used for applications such as conveyors that are belt-driven or direct-drive blade-type fan motors and pump motors.

conditioners; a blade-type fan is mounted on one end to operate as the condenser fan, and a squirrel-cage fan is on the other end to operate as the evaporator fan.

13.21

BASIC PARTS AND THEORY OF OPERATION FOR A PSC MOTOR

The basic parts of the PSC motor include the start winding and the run winding, which are mounted in the stator, and the end plate, which supports the rotor that has the motor shaft on one or both ends. The permanent split-capacitor motor operates in a manner very similar to the AC split-phase motor. It has a rotor and a stator with a run winding and a start winding. A run capacitor is connected in series with the start winding to create a larger phase shift than the split-phase motor has. The run capacitor is generally rated from 5 to μ60 F. The capacitance and voltage rating for the capacitor are listed on its case. Figure 13–24 shows a picture of several examples of run capacitors. The run capacitor is so named because it remains in the motor circuit even after the motor is running at full speed.

The run capacitor has several physical features that make it easy to recognize when you are servicing equipment. The easiest way to recognize a run capacitor is that it is generally mounted in a metal container. A second characteristic is that the metal container for the run capacitor is usually oval or rectangular in shape.

The run capacitor is similar to the start capacitor in that two sheets of conducting foil, called *plates*, are separated (sandwiched) by three sheets of insulating paper, called the *dielectric*. The foil plates and dielectric are rolled up and placed in the metal container, and an electrical terminal is connected to each plate.

You should remember that when voltage is applied to the capacitor in a circuit, current will flow as far as the plate on the left side of the capacitor. Because the capacitor plates are separated by dielectric paper, current cannot flow directly through the capacitor to the other plate, so the voltage must reverse and move backward through the circuit until it reaches the other capacitor plate. Recall that this process is referred to as charging the capacitor and discharging the capacitor. If the voltage that causes the current to flow is AC voltage, the AC sine wave will continually charge and discharge the capacitor, which keeps current continually flowing through the components in the circuit.

Because it takes time for the voltage to charge up the first plate and then reverse itself and discharge the plate, the capacitor causes a phase shift of up to 90° between the voltage waveform and the current waveform in the circuit. It is important to remember that the capacitor always causes the voltage waveform to lag the current waveform.

The reason a centrifugal switch is not needed to remove the start winding from the circuit after the motor is running at full speed is that the CEMF that is generated by the rotor is present across the start winding. When this counter-EMF is present, it opposes the applied voltage, and the resulting current in the start winding is determined by the difference of the applied voltage and CEMF. If the motor is running at near full rpm, the CEMF is high, and the amount of current in the start winding is minimal. If the motor encounters a large load and its slip increases because the speed of its shaft slows down, the CEMF decreases and the difference between the applied and CEMF becomes larger, which allows the start winding to draw extra current. This extra current helps provide enough additional torque to get the motor shaft back to its original speed. In this manner, the run capacitor allows the start winding to stay in the circuit and add current when additional torque is required to get the shaft to return to its normal speed. Thus, the PSC motor is able to regulate its speed. This is an important feature for the PSC motor if it is used for moving air or pumping cooling water for a chiller.

13.22

APPLYING VOLTAGE TO THE START WINDING AND CAPACITOR OF A PSC MOTOR

Figure 13–25 shows an electrical diagram of a typical permanent split-capacitor motor. You can see that the run capacitor (RC) is connected in series with the start

Figure 13–24 Examples of typical run capacitors. Notice that run capacitors are mounted in metal containers that are oval or rectangular in shape.

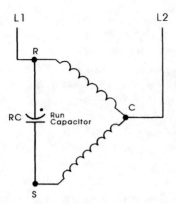

Figure 13–25   Electrical diagram of a permanent split-capacitor (PSC) motor. Notice the run capacitor (RC) is connected in series with the start winding, and this type of motor does not have a centrifugal switch.

winding. When voltage is first applied to the leads of the motor, current flows from L1 through the run windings and back to L2 or N. At the same time, voltage is also applied to the first capacitor plate. Because current cannot flow through the capacitor to the other plate, the voltage must reverse itself and move back through the run winding until it finds its way back to the start winding, where it reaches the other plate of the capacitor. Because the voltage must take the long way around the motor circuit to get to the other plate, it creates a phase shift between the current flowing in the run winding

and the current flowing in the start winding. In the PSC motor this process continues as long as voltage is applied to the motor windings. Because the run capacitor remains in the start-winding circuit at all times for this motor, it is called a *permanent split-capacitor motor.*

In the PSC motor the initial current flow into the motor windings is rather large because the rotor is not turning at first. This current is called *locked-rotor amperage* (LRA). Because a run capacitor is mounted in series with the start winding, the current cannot flow directly through the capacitor. Instead, the voltage moves as far as the first plate, when L1 voltage is applied, and then stops. At this point one plate of the capacitor has a large positive charge, which creates a potential differential across the plates of the capacitor. This potential differential causes the voltage to move from the first plate back through the run winding and start winding, until the voltage reaches the opposite plate of the capacitor. This causes a large phase shift between the current in the run winding and start winding, which is sufficient to cause the rotor to spin.

When the rotor begins to spin, it begins to generate counter EMF (CEMF) which causes the current flow to diminish to a point of normal current flow called full-load amperage (FLA). The actual amount of current the motor draws depends on the difference between the rated speed and the actual speed. This difference is called *slip.* The rotor continues to rotate as long as voltage is applied to the windings.

QUESTIONS

Short Answer

1. Identify the main parts of an AC split-phase motor.

2. Explain how you can identify the run winding and start winding in the stator of an AC split-phase motor.

3. Explain how a magnetic field is developed in the rotor of an AC motor.

4. Explain how the squirrel-cage rotor acts like the bar-magnet rotor in an AC motor.

5. Explain why a rotor rotates when a voltage is applied to the stator of an AC motor.

True or False

1. Torque is rotational force.

2. Slip is the difference between the actual speed of a motor and its rated speed.

3. The AC split-phase motor is called an induction motor because it uses a transformer with a capacitor to start.

4. The run winding is physically offset from the start winding in the stator of an AC motor to help it provide sufficient phase shift to start the rotor.

5. The start winding uses larger wire than the run winding.

Multiple Choice

1. The start capacitor is _____.
   a. connected in series with the run winding of a PSC motor
   b. connected in series with the start winding of a CSIR motor
   c. connected in series with the run winding of a CSIR motor

2. The run capacitor is _____.
   a. connected in series with the run winding of a PSC motor
   b. connected in series with the start winding of a PSC motor
   c. connected in series with the start winding of the CSIR motor

3. The start winding of a split-phase motor should have _____ the run winding.
   a. less resistance than
   b. more resistance than
   c. the same resistance as

4. The permanent split-capacitor (PSC) motor _____.
   a. uses a run capacitor and a centrifugal switch
   b. uses a start capacitor and a centrifugal switch
   c. uses a run capacitor but does not have a centrifugal switch

5. The capacitor-start, induction-run (CSIR) motor _____.
   a. uses a start capacitor that is connected in series with its start winding
   b. uses a start capacitor that is connected in series with its run winding
   c. uses a run capacitor that is connected in series with its start winding

PROBLEMS

1. Calculate the speed of a two-pole, four-pole, six-pole, and eight-pole motor.

2. Draw the sketch of a typical data plate (name plate) for an AC motor and identify each of the items listed in Section 13.11.

3. Draw the electrical diagram of a split-phase motor wired for 115 VAC.

4. Draw the electrical diagram of a capacitor-start, induction-run motor that is wired for 230 VAC.

5. Draw the electrical diagram of a PSC motor that is wired for 230 VAC.

CHAPTER 14

Three-Phase Motors

OBJECTIVES

After reading this chapter you should be able to:

1. Explain the theory of operation of a three-phase motor.

2. Identify the main parts of the three-phase motor.

3. Wire the three-phase motor for high or low voltage.

4. Change the rotation of a three-phase motor.

5. Wire the three-phase motor for a delta or wye configuration.

14.0
THREE-PHASE MOTOR THEORY

When industrial applications require larger motors (more than 1 hp), three-phase motors are usually used. The three-phase motor can produce extremely large starting and running torque. In this section you learn about how three-phase AC motors have some similarities to single-phase AC motors. Today all maintenance technicians must be able to work on jobs that utilize three-phase motors. You must be able make small changes on three-phase motors in the field so they can operate at a different speed, different voltage, or different rotation. This chapter makes these field changes, installation, and troubleshooting three-phase motors easy to perform.

The nature of three-phase AC voltage is that it has three independent sources of voltage that are 120° apart. Figure 14–1 shows a diagram of three-phase voltage. Notice that this diagram shows three separate sine waves that are 120° apart. This natural phase shift in the voltage provides the necessary shift in the magnetic field when the voltage is applied to the stationary fields (sta-

tor) of the AC motor. This means that three-phase voltage can provide the phase shift required to start a motor naturally, without any capacitors or start windings.

When the three-phase voltage is applied to the stator winding of a motor, the phase shift causes the magnetic field in the stator to actually rotate or move around the stator at the speed of the frequency of the AC voltage. The phase shift in the stator also causes the squirrel-cage rotor in the three-phase motor to become magnetized. When the rotor is magnetized, its field follows the rotating magnetic field in the rotor and causes the rotor to spin.

Because the three-phase voltage has a natural phase shift, it creates a very strong rotating magnetic field in the motor with the same natural phase shift. The strength of the magnetic field creates the strong torque at the shaft of the motor. This provides a means to start the motor under heavy loads such as the high pressure that a compressor must start against. The windings of a three-phase motor may be connected in a number of configurations to provide more or less starting torque and running torque, and the windings can be connected for high-voltage (480 VAC) applications and low-voltage (240 VAC) applications. The direction of rotation of the three-phase motor may be reversed by interchanging any two of the three supply-voltage wires so that the phase relationship is reversed.

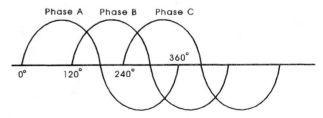

Figure 14–1   Example of three-phase voltage. Notice this voltage consists of three independent sine waves.

14.1
BASIC PARTS OF A THREE-PHASE MOTOR

Figure 14–2 shows a typical three-phase motor. The three-phase motor consists of a stator that has three separate windings mounted out of phase from each other. Figure 14–3 shows an exploded-view picture of the three-phase motor. The windings are equally spaced around the stator, and the wire in each winding is the same gauge size. This means that the three windings in the stator are equal, and if you measure their resistance, it should be the same for each winding. The rotor is also shown in this picture. From this picture you can see that the rotor is a squirrel-cage rotor. This squirrel-cage rotor is very similar to the rotors in other AC motors in that it is made of laminated-steel plates pressed on the squirrel-cage frame. The steel plates allow the rotor to become magnetized very easily and to give up its magnetic field very easily. This allows the magnetic bars in the rotor to become magnetized positively and negatively as the rotor spins inside the stator. Because a three-phase motor has three separate voltages, the three-phase motor has three separate and distinct magnetic fields in the rotor at the same time, which gives the rotor optimum torque.

14.2
THEORY OF OPERATION FOR A THREE-PHASE MOTOR

The theory of operation for a three-phase motor is very easy to understand. When three-phase voltage is applied to the three windings of the motor, three separate magnetic fields are developed, which immediately begin to rotate through the stator windings. Because the magnetic field rotates around the stator, the three magnetic fields in the rotor of the motor "chase" the magnetic fields and cause the rotor to begin to spin. The speed of rotation of the magnet field is fixed, and it is based on the number of windings and the frequency of the applied voltage. This means that the three-phase motor runs at a constant speed unless the frequency of the applied voltage is varied. In some industrial applications, a variable-frequency drive is used to control the speed of the motor by providing less than 60 Hz at times when the speed of the motor does not have to be maximum.

The natural phase shift in the three-phase voltage and the location of the three-phase windings in the stator of the motor allow the motor to develop the magnetic field in the rotor through induction so no extra components or circuits—such as a centrifugal switch, current relay, potential relay, or capacitors—are needed to provide extra torque from a start winding. As long as voltage is applied to the stator, a magnetic field is provided in the rotor by induction, and it always tries to cause the rotor to spin with the rotating magnetic field in the stator, which allows the rotor to provide torque at its shaft that can be used to turn the shaft to do work.

Figure 14–2 A typical three-phase motor.

Figure 14–3 Exploded view of a three-phase motor.

14.3
WIRING A SIMPLE THREE-PHASE MOTOR

The simplest three-phase motor you will encounter has only three wires brought out of its case for external connections. This makes all connections very simple. A diagram of this type of three-phase motor is shown in Figure 14–4. This diagram shows a motor-starter coil that is controlled by a thermostat. The motor terminals are identified as T1, T2, and T3. When you make the field wiring connections to these terminals, you can connect L1 to T1, L2 to T2, and L3 to T3. Because this type of motor provides only three wires for external connection, you cannot change the connections of the windings to change the speed of the motor or its torque. You can change its direction of rotation by switching L1 to T2 and L2 to T1. At times you may need to reverse the rotation of the three-phase motor for different applications.

14.4
CHANGING CONNECTIONS IN THREE-PHASE MOTORS TO CHANGE TORQUE, SPEED, OR VOLTAGE REQUIREMENTS

At times as a technician, you will need to make minor changes to three-phase motors that are used to move conveyor belts, turn fans, pump water and hydraulic fluid, or compress air. These changes include providing more starting torque, operating the motor at different speeds, or changing connections so that the motor can run at a different voltage. Because these motors may cost hundreds of dollars, it is important that you be able to make these changes while the motor is located on its machine right on the factory floor. These changes will

also allow you to use the existing motor rather than order an expensive new motor. After you learn about these field wiring changes, you will become confident in making these changes in the field.

14.5
WIRING A THREE-PHASE MOTOR IN A WYE CONFIGURATION

As you know, the three-phase motor has three equal windings, and it does not need any starting switches or capacitors connected to any of its windings. Because the motor has three equal windings, they can be connected to a wye configuration or a delta configuration. Figure 14–5 shows the three-phase motor windings for a six-lead motor connected in a typical wye configuration. The ends of each lead are numbered so that they can be changed or reconnected for different wiring configurations. The ends of the leads for the first winding are identified as T1 and T4, the second windings are T2 and T5,

Figure 14–5   A three-phase motor connected in a wye configuration.

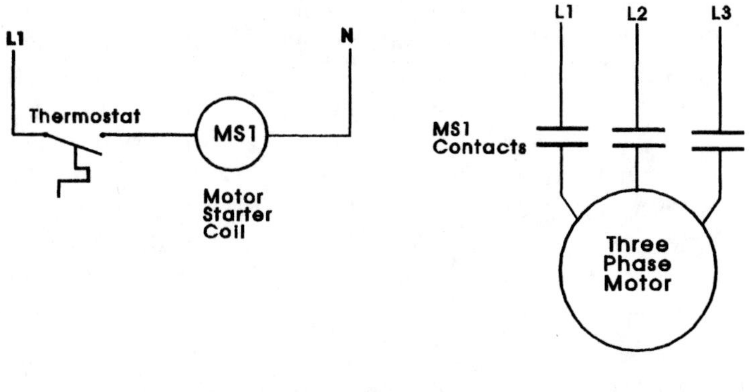

Figure 14–4   A three-phase motor is controlled by a motor starter. The motor-starter coil is controlled by a thermostat.

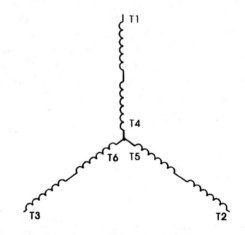

and the third windings are T3 and T6. You can see that terminals T4, T5, and T6 are all connected together at one point in the center of the motor, which is called the *wye point*. The most important thing to remember about connecting a three-phase motor in a wye configuration is that it will have less starting torque than if the same motor leads are connected in a delta configuration. This means that the wye-connected motor draws less current than the delta-connected motor. These are called "wye" windings because they look like the letter Y upside down.

14.6

WIRING A THREE-PHASE MOTOR IN A DELTA CONFIGURATION

The six-lead motor shown in the previous figure can also be connected in a delta configuration. Figure 14–6 shows a six-lead motor connected in a delta configuration. In this diagram you can see that the windings are connected in such a way that the three corners of the delta configuration are identified as T1, T2, and T3. In this configuration the incoming power supply is connected to T1, T2, and T3. You should remember that the delta-connected motor has more starting torque and draws more current than the wye-connected motor. The configuration is called "delta" since it looks like the Greek letter delta, or Δ.

14.6.1
Wiring a Three-Phase, Six-Wire Open Motor for High Voltage or Low Voltage

All the terminal leads of a three-phase open motor are brought outside the stator so the motor can be reconnected to operate with a high-voltage supply (480 VAC) or a low-voltage supply (230 VAC). The number of leads that are brought out of the stator for a three-phase motor may be six, nine, or twelve. If six leads are brought out of the stator, you can use the diagram in Fig. 14–7a to connect the leads for low voltage (240 or 208 V) and Fig. 14–7b to connect the leads for high voltage (480 V). It is important to understand that if the motor is a six-lead, three-phase motor, it is wired so it can be connected for high or low voltage. You should notice that the same connections are used for both the delta- and the wye-connected six-lead motor.

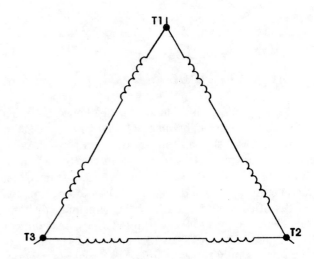

Figure 14–6   A three-phase motor connected in a delta configuration.

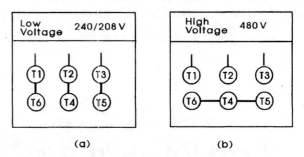

Figure 14–7   (a) Connections for a low-voltage, wye-connected and low-voltage, delta-connected motor; (b) connections for a high-voltage, wye-connected motor and a high-voltage, delta-connected motor.

14.7

REWIRING A NINE-LEAD, THREE-PHASE, WYE-CONNECTED MOTOR FOR A CHANGE OF VOLTAGE

A number of three-phase motors have nine leads brought out of the stator. The reason these motors have nine leads is that each of the three windings is broken into two pieces. The two sections that make up the three windings can be connected in a wye configuration or a delta configuration. Figure 14–8 shows the nine leads connected in a wye configuration. From this diagram you can see that the two ends of the first half of the first winding are identified as T1 and T4, and the ter-

Figure 14–8   (a) Connections for a nine-lead, three-phase, wye-configured motor wired for high voltage (480 V); (b) connections for a nine-lead, three-phase, wye-configured motor wired for low voltage (208 or 240 V).

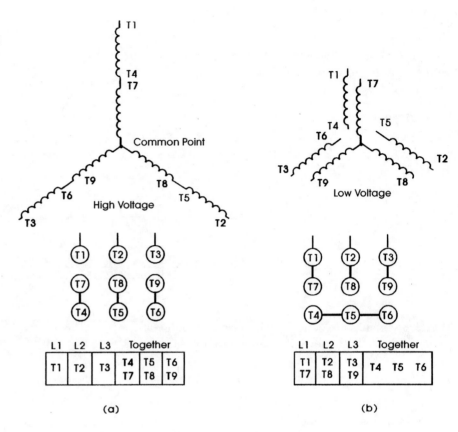

(a)   (b)

minals for the second half of the first winding are identified as T7 and common. The common point is where the three windings are connected together to form the wye point.

The terminal ends for the first half of the second winding are identified as T2 and T5, and the terminals for the second half of the second winding are identified as T8 and common. The terminal ends for the first half of the third winding are identified as T3 and T6, and the terminals for the second half of the second winding are identified as T9 and common. The diagram in Figure 14–8 shows the correct way to connect the terminals of the wye-connected motor for high voltage (480 V) and low voltage (208 or 240 V). It is also important to note that if the three-phase motor brings out nine leads, it is internally connected as a wye or a delta motor. If the motor is connected as a nine-lead, wye-connected motor and you want to change it to delta because you need more starting torque, you will not be able to make this conversion. The reason is that the motor is internally connected so that it can be wired only as a wye-connected motor. Thus if you need a delta-connected motor, you will need to purchase a new nine-lead motor. This is a minor drawback of the three-phase motor, because you very seldom change the motor from delta to wye or wye to delta once the motor is installed.

14.8
REWIRING A NINE-LEAD, THREE-PHASE, DELTA-CONNECTED MOTOR FOR A CHANGE OF VOLTAGE

A nine-lead, three-phase, delta-connected motor can also be rewired for high and low voltage in the field. The nine-lead motor has each winding broken into two sections. Figure 14–9 shows the nine leads connected in a delta configuration. From this diagram you can see that the two ends of the first half of the first winding are identified as T1 and T4, and the terminals for the second half of the first winding are identified as T1 and T9.

The terminal ends for the first half of the second winding are identified as T2 and T5, and the terminals for the second half of the second winding are identified as T2 and T7. The terminal ends for the first half of the third winding are identified as T3 and T6, and the terminals for the second half of the third winding are identified as T3 and T8. The diagram in Fig. 14–9 shows the correct way to connect the terminals of the delta-connected motor for high voltage (480 V) and low voltage (208 or 240 V). The important point to remember is

Figure 14–9   (a) Connections for a nine-lead, three-phase, delta-configured motor wired for high voltage (480 V); (b) connections for a nine lead, three-phase, delta-configured motor wired for low voltage (208 or 240 V).

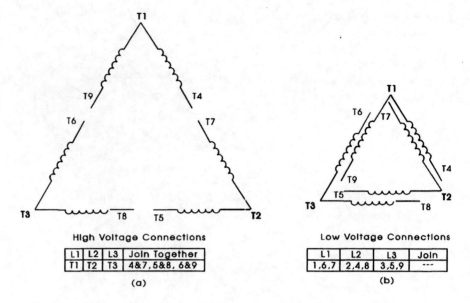

High Voltage Connections

L1	L2	L3	Join Together
T1	T2	T3	4&7, 5&8, 6&9

(a)

Low Voltage Connections

L1	L2	L3	Join
1,6,7	2,4,8	3,5,9	---

(b)

that the windings are connected in series for high voltage, and they are connected in parallel for low voltage.

14.9

THE 12-LEAD, THREE-PHASE MOTOR

Some larger three-phase motors can be connected as either a wye motor or a delta motor. In order for this change to occur, the motor must have either 6 or 12 leads brought out of the stator. If the motor has 9 leads brought out, the motor cannot be changed from wye to delta in the field. For example, in some applications you must start the motor as a wye-connected motor so it does not draw too much LRA and then you can redo it as a delta-connected motor after it is running so that it will provide sufficient running torque. This type of motor is called a *wye-delta motor application,* and the configuration is changed, while the motor is running, by a wye-delta motor starter. The wye-delta motor starter is a special motor starter that has two independent sets of three-phase contacts connected to it. The motor is connected in a wye configuration on the first set of contacts, and it is connected as a delta-configured motor on the second set of contacts. When the motor is started, the coil for the first set of contacts is energized, and the motor is connected as a wye-connected motor when voltage is first applied. When the motor reaches full speed, the relay of the second set of contacts is energized, and the motor is rewired as a delta-configured motor. The motor for the wye-delta starting must be a 6- or 12-lead, three-phase motor. Figure 14–10 shows a 12-lead motor connected as a wye motor, and Figure 14–11 shows a 12-lead motor connected as a delta-connected motor.

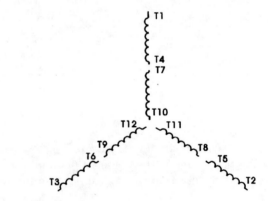

Figure 14–10

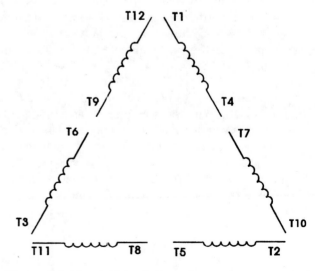

Figure 14–11   A 12-lead, three-phase motor connected in a delta configuration.

14.10

WIRING A THREE-PHASE MOTOR FOR A CHANGE OF ROTATION

At times you will need to change the direction of rotation for the three-phase motor while it is connected to its application. This is perhaps the most common change you will make to the three-phase motor. As a technician, you will need to change the direction that a pump motor runs when it is turning the wrong way, or you may be requested to change the direction a fan mo-

tor or conveyor motor is turning to ensure it is moving in the proper direction. In these cases all that you need to do is exchange any two of the supply leads. For example, you can change L1 and L2 so that L2 is connected to T1 and L1 is connected to T2. Figure 14–12a shows the three-phase wye motor connected for clockwise rotation. Notice that L1 is connected to T3 and L2 is connected to T1 In Fig. 14–12b you can see that the motor is connected to operate in a counterclockwise direction. L1 and L2 are exchanged so that L1 is connected to T1 and L2 is connected to T3. Figure 14–13a and b show similar diagrams for a delta-wired motor.

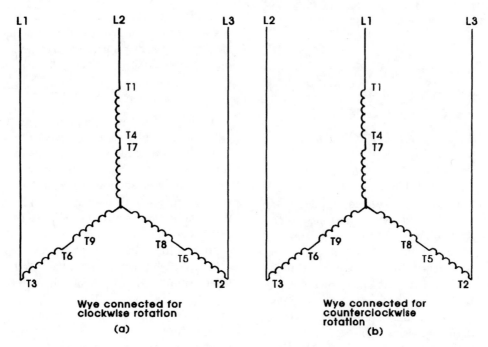

Figure 14–12 (a) A three-phase, wye-connected motor connected for clockwise rotation; (b) a three-phase, wye-connected motor connected for counterclockwise rotation. Notice T3 is now connected to L2 and T2 is now connected to L3.

Figure 14–13 (a) A three-phase, delta-connected motor connected for clockwise rotation; (b) a three-phase delta-connected motor connected for counterclockwise rotation. Notice in part b that T1 and T12 are now connected to L3, that T2 and T10 are now connected to L2, and that T3 and T11 are connected to L1.

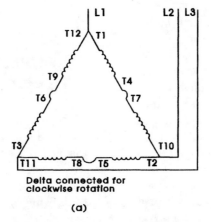
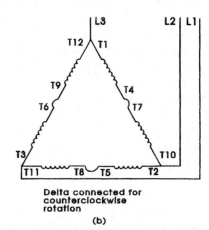

14.11

CONTROLLING A
THREE-PHASE MOTOR

The three-phase motor is generally turned on or off with a motor starter. The voltage to the coil of the motor starter is controlled by a stop switch and start switch. When the contacts in the switches close, voltage is applied to the coil of the motor starter, and it draws current, becomes magnetized, and pulls the three sets of normally open contacts to their closed position. When the contacts close, three-phase voltage is applied to the motor, and it begins to run. When the stop button is depressed, it de-energizes the relay coil, and the three sets of contacts open and turn the motor off.

14.12

CONTROLLING THE SPEED
OF A THREE-PHASE MOTOR

The speed of a three-phase motor can be changed by changing the number of poles in the motor windings or by changing the frequency of the voltage supplied to the motor. Prior to the 1980s the only way that the frequency of the supply voltage could be easily changed was by changing the speed of the generator that produced the electricity. Because this was not practical, the preferred method of changing the speed of the three-phase motor was to reconnect the motor in the field so it would have a different number of poles. This basically allowed the motors to have two speeds.

In the 1980s more emphasis was placed on efficiency, so it became important to change the speeds of three-phase fans and three-phase pump motors so that they would run only at the speed necessary to do the job. In the late 1970s and 1980s electronic technology began to produce products that could control larger amounts of voltage and current so that the frequency of the voltage sent to a motor could be changed. Chapter 16 goes into detail about these devices. In this chapter all you need to know is that the variable-frequency drive is specifically designed to change the frequency of voltage that is used to power a three-phase or single-phase motor. Because the frequency is changed, the speed of the motor can be adjusted. For example, if the motor needs to operate at a higher speed, the frequency would be increased above 60 Hz. If the speed needs to be decreased, the frequency would be lowered below 60 Hz. Figure 14–14 shows a picture of a variable-frequency drive, and Figure 14–15 shows a block diagram of a variable-frequency drive. From the block diagram you

Figure 14–14   An Allen-Bradley three-phase, variable-frequency drive that is used to control the speed of three-phase motors. (Courtesy of Rockwell Automation's Allen-Bradley Business)

can see that the drive is supplied with three-phase voltage. The first section of the drive is the rectifier section. In this section three sets of diodes are used to rectify the AC voltage to DC voltage. The second section of the drive is a filter section. You can see that the voltage waveform is half a sine wave as it enters the filter section, and it changes to filtered pure DC as it leaves the filter section. The final section of the drive is the transistor section, and the transistors can be turned on and off by a triggering circuit so that the output waveform looks similar to the waveform of the original three-phase voltage that supplies this circuit. The main difference between the input voltage and the output voltage is that the frequency of the output voltage can be changed to any value between 1 and 120 Hz. Thus the variable-frequency drive can create frequencies from 1 to 200% of 60 Hz, but the motor speed in practice is generally controlled at between 75 and 125% of the rated motor rpm.

14.13

TROUBLESHOOTING THE
THREE-PHASE MOTOR

At times in the field you will be expected to troubleshoot a three-phase motor that will not start or a motor that is malfunctioning. When you are in the field, you can test a three-phase motor to check voltage and current and to see if each of the windings has continuity. The following sections explain how to make each of these tests. The most frequent test for a three-phase mo-

Figure 14–15 A block diagram of a variable-frequency drive. AC three-phase voltage enters the drive and is converted by the rectifiers to half-wave DC. The filter section converts the half-wave DC to pure DC. The transistors convert the pure DC back to variable-frequency AC.

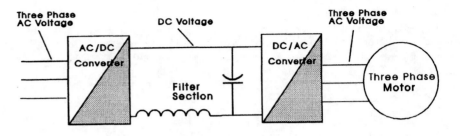

tor is the test of supply voltage at the disconnect or fuse box and for voltage at each of the three motor terminals.

If the three-phase motor fails to run when it is energized, you will need to make two basic tests. The first test involves testing each of the three lines that supply voltage to the motor to ensure that three-phase voltage is flowing through the wires to the terminals of the motor. You can make this test by testing for voltage right at the terminals of the motor. If the voltage supply for this motor is 240-V three-phase, you should measure 240 V between T1 and T2, T2 and T3, and T3 and T1. If you have voltage at two of the three terminals, it indicates that you have lost one phase of the three-phase supply voltage; the most likely problem is a blown fuse. In some cases the motor can continue to run when one of the phases is lost, but the motor will not have sufficient torque to restart, and the overload in the motor starter will be tripped when you find the motor during a troubleshooting call. Be sure that the voltage test indicates you have the correct amount of voltage at each of the three tests you make at the terminals on the motors. The voltage tests are similar to the test you learned about to determine if the voltage is present in a three-phase disconnect switch. If the motor is a three-phase hermetically sealed compressor, it may have internal overloads and you may need to allow the compressor to cool down for several hours before the overloads will automatically reset. If the compressor continues to trip its overloads, you will need to have a technician test the system for loss of refrigerant or other problems that cause compressors to overload and overheat.

If the correct amount of voltage is present at all three terminals and the motor will not run, you must suspect that one of the windings in the motor is open and has infinite (∞) resistance. (Remember that the motor may also have an internal thermal overload that would give a similar symptom. Be sure to allow several hours for the overload to cool down before you make a decision about changing a compressor.) If you suspect the winding is open, you will need to turn the power off to the system, disconnect all the wires from the motor terminals, and test each winding for resistance. It is important to understand that if you leave the motor connected to the circuit, you may read back-feed resistance through a transformer or other motor winding on the system. Figure 14–16 shows the proper location for placing a voltmeter to make these measurements.

14.14
TESTING THE THREE-PHASE MOTOR FOR CONTINUITY

If the voltage test indicates that the proper amount of voltage is present at the terminals, but the motor does not start and hums or does not make any noise when it does not start, you should suspect one or more motor windings have an open, and you must test the motor windings for resistance. Always remember that the voltage to the motor should be turned off for the continuity test, and the windings should be isolated if possible. The test for resistance should be between T1 and T2, T2 and T3, and T3 and T1. If any of these windings are open, the motor must be replaced. As stated before, some compressors have an internal thermal overload built into the windings, and it may open if the internal temperature gets too hot. The motor windings may get too hot if the amount of refrigerant in the system is too low to provide cooling for the motor or if the motor is overloaded. Remember the hermetic motor is cooled only by the extra refrigerant brought back to the motor specifically to cool the windings. If the overload is open, it will not allow current to flow through the windings, and hence the motor will not run. Because the overload may be the problem, it is always a good practice to wait 1 to 2 hours for the compressor to cool down and to retest the windings for continuity before you change it out.

It is also important to understand that some refrigeration compressors use the motor windings as a heater when the motor is not running. This is accomplished by connecting capacitors and resistors to allow a small amount of bleed current to flow through the windings to act as a small heater. This small amount of heat keeps the windings warm enough so that the refrigerant oil does not migrate from the compressor to part of the system that is warmer. A problem may occur when you test the compressor if you do not make sure that all this circuit is isolated from the motor windings when you are testing for an open circuit. Figure 14–17 shows the proper locations at which to place the ohmmeter to make the continuity test.

Figure 14–16   The proper location for placing voltmeter terminals to test for three-phase voltage on the three-phase motor.

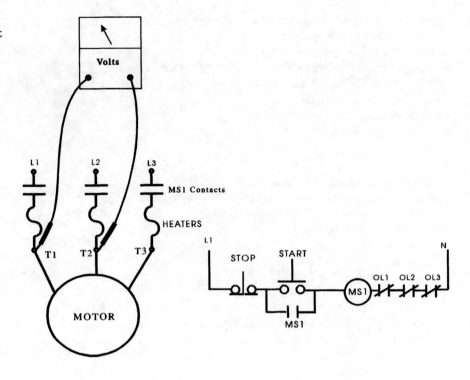

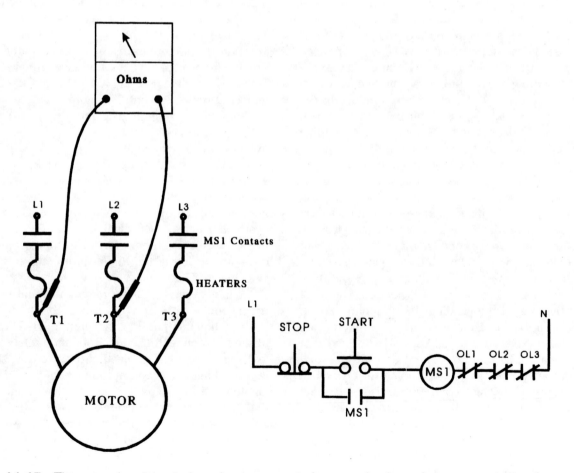

Figure 14–17   The proper locations to place ohmmeter terminals to test the three-phase motor windings for continuity to check for an open circuit.

14.15

TESTING A THREE-PHASE MOTOR FOR CURRENT

If the three-phase motor will start, you must check it for the proper amount of current to ensure that it is not overloaded. Figure 14–18 shows the proper location at which to place the clamp on the ammeter around each of the three-phase wires that supply voltage to the motor. Because the supply wires are connected in series,

the current measurement indicates how much current the motor is drawing through each winding. It is important to understand that the amount of current measured in each of the three wires should be nearly the same. If one of the leads is drawing excess current, it indicates the motor is beginning to fail. If all three measurements are larger than the data plate rating, the motor is being overloaded, and you need to determine the cause of the overload. Sometimes three-phase pump motors draw excess current when the pump is overloaded.

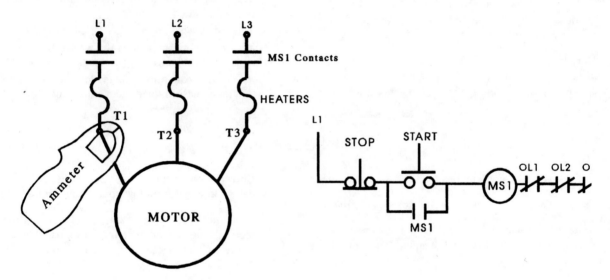

Figure 14–18   The proper location to place a clamp on an ammeter to test the three-phase motor for the proper amount of current.

QUESTIONS

Short Answer

1. Explain why a three-phase motor can start without using any start relays or capacitors.

2. Identify the main parts of a three-phase open motor.

3. Explain why you would need to reconnect the terminals of a three-phase motor to change its direction of rotation.

4. Explain why you would need to reconnect the terminals of a three-phase motor for a change of voltage.

5. The three-phase motor you are testing for continuity is a sealed compressor, and your test indicates an open circuit. Explain why you should wait 2 h or more before you make the decision to change the compressor.

True or False

1. The direction of rotation of a three-phase motor can be reversed by exchanging any two of its three terminals.

2. One of the windings in a three-phase motor is smaller than the other two, so a phase shift can be created to help the motor get started.

3. The three-phase motor has more starting torque than an equal-size single-phase motor.

4. It is possible to change the internal winding connections on a nine-lead, three-phase motor so that it can operate on 480 VAC or 240 VAC.

5. It is possible to change the connections on a three-phase motor so it will operate as a wye or a delta motor.

Multiple Choice

1. If you change motor terminal T1 to L2 of the supply voltage and terminal T2 to L1 of the supply voltage, a three-phase motor will _____.
   a. not run correctly because it must be wired L1 to T1, L2 to T2, and L3 to T3
   b. change the direction of its rotation
   c. be able to run on 240 VAC instead of 480 VAC

2. A three-phase motor does not require any current relays, potential relays, or centrifugal switches because _____.
   a. it has three equal windings instead of a start winding
   b. it is a low-torque motor, and these relays would cause its torque to become too large
   c. these switches are mainly used to control for over-current conditions.

3. The phase shift that is required to create torque for a three-phase motor is _____.
   a. created by adding capacitors to the motor windings
   b. found naturally in the three-phase voltage used to provide power for the motor
   c. created by making each of the windings in the three-phase motor slightly different

4. The reason you need to know how to reconnect a three-phase motor for a change of voltage is _____.
   a. because the voltage rating of the motor that you have in stock is rated 480 VAC and the motor that it is replacing is rated for 240 VAC
   b. because the motor may need more speed
   c. because some three-phase motors need DC voltage instead of AC voltage

5. If the three-phase motor runs but draws excessive current, you should suspect _____.
   a. a faulty centrifugal switch or current relay
   b. the motor is running in the wrong direction and you should exchange any two leads
   c. one of the motor windings is open or one phase of the three-phase voltage is not supplying voltage

PROBLEMS

1. Draw a sketch of a wye-connected, three-phase motor.

2. Draw a sketch of a delta-connected, three-phase motor.

3. Draw the connections that you would use to connect a wye-wired motor for low voltage and for high voltage.

4. Draw the connections that you would use to connect a delta-wired motor for low voltage and for high voltage.

5. Draw the electrical diagram of a nine-lead, three-phase motor connected for wye configuration and for delta configuration and explain why each is used.

CHAPTER 15

DC Motors

OBJECTIVES

After reading this chapter you will be able to:

1. Identify the basic parts of a DC motor and explain their operation.

2. Explain how to change the speed of a DC motor.

3. Explain how to change the rotation of a DC motor.

4. Explain the difference between series, shunt, and compound DC motors.

15.0
INTRODUCTION

DC motors are commonly used to operate machinery in a variety of applications on the factory floor. DC motors are one of the most versatile types of energy converters used in industry today. You will find DC motors that provide speed control and directional control. DC motors can easily have their speed changed by varying the voltage sent to them. The electronic motor drives today allow voltage and current to be varied to control the speed and torque of the DC machines. You will find a large variety of DC motors on everything from robots to other automation system and also on very large variable-speed control systems used in industry.

15.1
MAGNETIC THEORY

DC motors operate on the principles of basic magnetism. You should remember that a coil of wire can be magnetized when current is passed through it. When this principle was applied in relay coils, the polarity of the current was not important. When the current is passed through a coil of wire to make a field coil for a motor, the polarity of the current will determine the direction of rotation for the motor. The polarity of the current flowing through the coil of wire will determine the location of the north and south magnetic poles in the coil of wire.

Another important principle involves the amount of current that is flowing through the coil. For a relay or solenoid, the amount of current was not important as long as enough current was present to move the armature of the relay or solenoid. In a DC motor, the amount of current in the windings will determine the speed (rpm) of the motor shaft and the amount of torque that it can produce.

You should remember from basic magnetic theory that the left-hand rule of current flow through a coil of wire helps you understand that the direction of current flow will determine the magnetic polarity of the coil. The left-hand rule is used to show you a principle from which several facts can be determined.

The first fact that you should understand is that the direction of current flow will determine which end of a coil of wire is negative or positive—in other words, which end of the coil will be the north pole of the magnet and which end will be the south pole. It is also easy to see that by changing the direction of the current flow in the coil of wire, the magnetic poles will be reversed in the coil. This is important to understand because the direction of the motor's rotation is determined by the changing magnetic field.

Another basic fact about magnets that you should remember is the relationship between two like poles and two unlike poles. When the north poles of two different magnets are placed close to each other, they will repel each other. When the north pole of one magnet is placed near the south pole of another magnet, the two poles will attract each other very strongly.

Still another fact that is important to understand with the coil of wire is that the strength of the magnetic field can be varied by changing the amount of current flowing through the wire in the coil. If a small amount of current is flowing, a small number of flux lines will be created, and the magnetic field will be relatively weak. If the amount of current is increased, the magnetic field will become stronger. The strength of the magnetic field can be increased to the point of saturation. A magnetic coil is said to be saturated when its magnetic strength cannot be increased by adding more current.

Saturation is similar to filling a drinking glass with water. You cannot get the level of the glass any higher than full. Any additional water that is put into the glass when it is full will not increase the amount of water in the glass. The additional water will run over the side of the glass and be wasted. The same principle can be applied to a magnetic coil. When the strength of the magnetic field is at its strongest point, additional electrical current will not cause the field to become any stronger.

15.2
DC MOTOR THEORY

The DC motor has two basic parts: the rotating part that is called the *armature* and the stationary part that includes coils of wire called the *field coils*. The stationary part is also called the *stator*. Figure 15–1 shows a typical DC motor, Fig. 15–2 shows a DC armature, and Fig. 15–3 shows a typical stator. From the picture in Fig. 15–2, you can see the armature is made of coils of wire wrapped around the core, and the core has an extended shaft that rotates on bearings. The ends of each coil of wire on the armature are terminated at one end of the armature. The termination points are called the *commutator*, and this is where the brushes make electrical contact to bring electrical current from the stationary part to the rotating part of the machine. The commutator segments are visible at the left end of the armature in the picture.

The picture in Fig. 15–3 shows the location of the coils that are mounted inside the stator. These coils will be referred to as field coils in future discussions, and they may be connected in series or parallel with each other to create changes of torque in the motor. You will find the size of wire in these coils and the number of turns of wire in the coil will depend on the effect that is trying to be achieved.

It will be easier to understand the operation of the DC motor from a basic diagram that shows the magnetic interaction between the rotating armature and the stationary field coils. Figure 15–4 shows three diagrams that explain the DC motor's operation in terms of the magnetic interaction. In Fig. 15–4a you can see that a bar magnet has been mounted on a shaft so that it can spin. The operation of this bar magnet will be similar to the armature. The field winding is one long coil of wire that has been separated into two sections. The top section is connected to the positive pole of the battery, and the bottom section is connected to the negative pole of the battery. It is important to understand that the battery represents a source of voltage for this winding. In the actual industrial-type motor this voltage will come from the DC

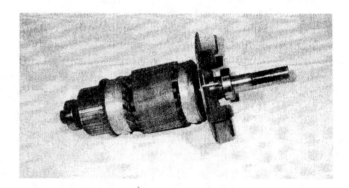

Figure 15–2 The armature (rotor) of a DC motor has coils of wire wrapped around its core. The ends of each coil are terminated at commutator segments located on the left end of the shaft. The brushes make contact on the commutator to provide current for the armature.

Figure 15–1 A typical DC motor.

Figure 15–3 The stationary part of a DC motor has the field coils mounted in it.

voltage source for the motor. The current flow in this direction makes the top coil the north pole of the magnet and the bottom coil the south pole of the magnet.

The bar magnet represents the *armature,* and the coil of wire represents the *field.* The arrow shows the direction of the armature's rotation. Notice that the arrow shows the armature starting to rotate in the clockwise direction. The north pole of the field coil is repelling the north pole of the armature, and the south pole of the field coil is repelling the south pole of the armature.

As the armature begins to move, the north pole of the armature comes closer to the south pole of the field, and the south pole of the armature comes closer to the north pole of the field. As the two unlike poles near each other, they begin to attract. This attraction becomes stronger until the armature's north pole moves directly in line with the field's south pole, and its south pole moves directly in line with the field's north pole (Fig. 15–4b).

When the opposite poles are at their strongest attraction, the armature will be "locked up" and will resist further attempts to continue spinning. For the armature to continue its rotation, the armature's polarity must be switched. Since the armature in this diagram is a permanent magnet, you can see that it would lock up during the first rotation and not work. If the armature is an electromagnet, its polarity can be changed by changing the direction of current flow through it. For this reason the armature must be changed to a coil (electromagnet), and a set of commutator segments must be added to provide a means of making contact between the rotating member (armature) and the stationary (stator) member. One commutator segment is provided for each terminal of the magnetic coil. Since this armature has only one coil, it will have only two terminals, so the commutator has two segments.

Since the armature is now a coil of wire, it will need DC current flowing through it to become magnetized. This presents another problem: Since the armature will be rotating, the DC voltage wires cannot be connected directly to the armature coil. A stationary set of carbon brushes is used to make contact with the rotating armature. The brushes ride on the commutator segments to make contact so that current will flow through the armature coil.

In Fig. 15–4c you can see that the DC voltage is applied to the field and to the brushes. Since negative DC voltage is connected to one of the brushes, the commutator segment that the negative brush rides on will also be negative. The armature's magnetic field causes the armature to begin to rotate. This time when the armature gets to the point where it becomes locked up with the magnetic field, the negative brush begins to touch the end of the armature coil that was previously positive, and the positive brush begins to touch the end of the armature coil that was negative. This action switches the direction of current flow through the armature, which also switches the polarity of the armature coil's magnetic field at just the right time so that the repelling and attracting continues. The armature continues to switch its magnetic polarity twice during each rotation, which causes it to continually be attracted to and repelled by the field poles.

This is a simple two-pole motor that is used primarily for instructional purposes. Since the motor has only two poles, the motor will operate rather roughly and not provide too much torque. Additional field poles and armature poles must be added to the motor for it to become useful for industry.

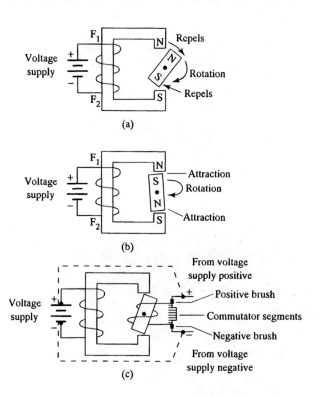

Figure 15–4 Magnetic diagram that explains the operation of a DC motor. (a) The rotating magnet moves clockwise because like poles repel. (b) The rotating magnet is being attracted because the poles are unlike. (c) The rotating magnet is now shown as the armature coil, and its polarity is determined by the brushes and commutator segments.

15.3
DC Motor Components

The armature and field in a DC motor can be wired three different ways to provide varying amounts of torque or different types of speed control. The armature and field windings are designed slightly differently for different types of DC motors. The three basic types of DC motors are the *series motor,* the *shunt motor,* and the *compound motor*. The series motor is designed to move large loads with high starting torque in applications such as a crane motor or lift hoist. The shunt motor is designed slightly differently since it is made for applications such as pumping fluids, where constant-speed characteristics are important. The compound motor is designed with some of the series motor's characteristics and some of the shunt motor's characteristics. This allows the compound motor to be used in applications where high starting torque and controlled operating speed are both required.

Figure 15–5   A cutaway picture of a DC motor.

It is important that you understand the function and operation of the basic components of the DC motor since motor controls will take advantage of these design characteristics to provide speed, torque, and direction of rotation control. Figure 15–5 shows a cutaway picture of a DC motor, and Fig. 15–6 shows an exploded view diagram of a DC motor. In these figures you can see that the basic components include the armature assembly, which includes all rotating parts; the frame assembly, which houses the stationary field coils, and the end plates, which provide bearings for the motor shaft and a mounting point for the brush rigging. Each of these assemblies is explained in depth so that you will understand the design concepts used for motor control.

15.3.1
Armature

The armature is the part of a DC motor that rotates and provides energy at the end of the shaft. It is basically an electromagnet since it is a coil of wire that has to be specially designed to fit around core material on the shaft. The core of the armature is made of laminated steel and provides slots for the coils of wire to be pressed onto. Figure 15–7a shows a sketch of a typical DC motor armature. Figure 15–7b shows the laminated-steel core of the armature without any coils of wire on it. This gives you a better look at the core.

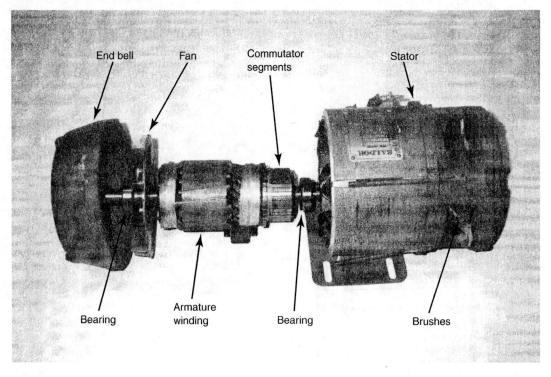

Figure 15–6   An exploded view of a DC motor. This diagram shows the relationship of all the components.

The armature core is made of laminated steel to prevent the circulation of eddy currents. If the core were solid, magnetic currents would be produced that would circulate in the core material near the surface and cause the core metal to heat up. These magnetic currents are called *eddy currents*. When laminated-steel sections are pressed together to make the core, the eddy currents cannot flow from one laminated segment to another, so they are effectively canceled out. The laminated core also prevents other magnetic losses called *flux losses*. These losses tend to make the magnetic field weaker so that more core material is required to obtain the same magnetic field strengths. The flux losses and eddy current losses are grouped together by designers and called *core losses*. The laminated core is designed to allow the armature's magnetic field to be as strong as possible since the laminations prevent core losses.

Notice that one end of the core has commutator segments. There is one commutator segment for each end of each coil. This means that an armature with four coils will have eight commutator segments. The commutator seg-

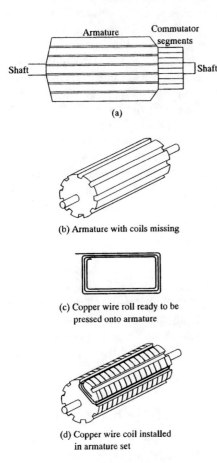

(a)

(b) Armature with coils missing

(c) Copper wire roll ready to be pressed onto armature

(d) Copper wire coil installed in armature set

Figure 15–7   (a) Armature and commutator segments; (b) armature prior to the coil wire being installed; (c) coil of wire prior to being pressed into the armature; (d) a coil pressed into the armature. The end of each coil is attached to a commutator segment.

ments are used as a contact point between the stationary brushes and the rotating armature. When each coil of wire is pressed onto the armature, the end of the coil is soldered to a specific commutator segment. This makes an electrical terminal point for the current that will flow from the brushes onto the commutator segment and finally through the coil of wire. Figure 15–7c shows the coil of wire before it is mounted in the armature slot, and Fig. 15–7d shows the coil mounted in the armature slot and soldered to the commutator segment.

The shaft is designed so that the laminated armature segments can be pressed onto it easily. It is also machined to provide a surface for a main bearing to be pressed on at each end. The bearing will ride in the end plates and support the armature when it begins to rotate. One end of the shaft is also longer than the other since it will provide the mounting shaft for the motor's load to be attached. Some shafts have a key way or flat spot machined into them so that the load that is mounted on it can be secured. You must be careful when handling a motor that you do not damage the shaft since it must be smooth to accept the coupling mechanism. It is also possible to bend the shaft or cause damage to the bearings so that the motor will vibrate when it is operating at high speed. The commutator is made of copper. A thin section of insulation is placed between each commutator segment. This effectively isolates each commutator segment from all others.

15.3.2
Motor Frame

The armature is placed inside the frame of the motor where the field coils are mounted. When the field coils and the armature coils become magnetized, the armature will begin to rotate. The field winding is made by coiling up a long piece of wire. The wire is mounted on laminated pole pieces called field poles. Similar to an armature, these poles are made of laminated steel or cast iron to prevent eddy current and other flux losses. Figure 15–8 shows the location of the pole pieces inside a DC motor frame.

The amount of wire that is used to make the field winding will depend on the type of motor being manufactured. A series motor uses heavy-gauge wire for its field winding so that it can handle very large field currents. Since the wire is a large gauge, the number of turns of wire in the coil will be limited. If the field winding is designed for a shunt motor, it will be made of very small-gauge wire, and many turns can be used. It is important to understand that the wire used in the coils looks like it is a bare wire without any insulation, in reality the wire is actually covered with a thin coating of clear insulation.

After the coils are wound, they are coated for protection against moisture and other environmental

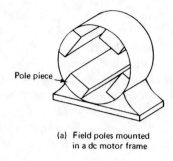

Pole piece

(a) Field poles mounted
in a dc motor frame

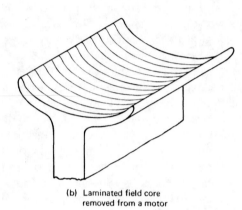

(b) Laminated field core
removed from a motor

Figure 15–8   (a) This diagram shows the location of the pole pieces in the frame of a DC motor. (b) This diagram shows an individual pole piece. You can see that it is made of laminated sections. The field coils are wound around the pole pieces.

elements. After they have been pressed onto the field poles, they must be secured with shims or bolts so that they are held rigidly in place. *Remember:* when current is passed through the coil, it will become strongly magnetized and attract or repel the armature magnetic poles. If the field poles are not rigidly secured, they will be pulled loose when they are attracted to the armature's magnetic field and then pressed back into place when they become repelled. This action will cause the field to vibrate and damage the outer protective insulation and cause a short circuit or a ground condition between the winding and the frame of the motor.

The ends of the frame are machined so that the end plates will mount firmly into place. An access hole is also provided in the side of the frame or in the end plates so that the field wires can be brought to the outside of the motor, where DC voltage can be connected.

The bottom of the frame has the mounting bracket attached. The bracket has a set of holes or slots provided so that the motor can be bolted down and securely mounted on the machine it is driving. The mounting holes will be designed to specifications by frame size. The dimensions for the frame sizes are provided in tables printed by motor manufacturers. Since these holes and slots are designed to a standard, you can predrill the mounting holes

in the machinery before the motor is put in place. The slots are used to provide minor adjustments to the mounting alignment when the motor is used in belt-driven or chain-driven applications. It is also important to have a small amount of mounting adjustment when the motor is used in direct-drive applications. It is very important that the motor be mounted so that the armature shaft can turn freely and not bind with the load.

15.3.3
End Plates

The end plates of the motor are mounted on the ends of the motor frame. Figures 15–5 and 15–6 show the location of the end plates in relation to the motor frame. The end plates are held in place by four bolts that pass through the motor frame. The bolts can be removed from the frame completely so that the end plates can be removed easily for maintenance. The end plates also house the bearings for the armature shaft. These bearings can be either sleeve or ball type. If the bearing is a ball-bearing type, it is normally permanently lubricated. If it is a sleeve type, it will require a light film of oil to operate properly. The end plates that house a sleeve-type bearing will have a lubrication tube and wicking material. Several drops of lubricating oil are poured down the lubrication tube, where they will saturate the wicking material. The wicking is located in the bearing sleeve so that it can make contact with the armature shaft and transfer a light film of oil to it. Other types of sleeve bearings are made of porous metal so that it can absorb oil to be used to create a film between the bearing and the shaft.

It is important that the end plate for a sleeve bearing be mounted on the motor frame so that the lubricating tube is pointing up. This position will ensure that gravity will pull the oil to the wicking material. If the end plates are mounted so that the lubricating tube is pointing down, the oil will flow away from the wicking, and it will become dry. When the wicking dries out, the armature shaft will rub directly on the metal in the sleeve bearing, which will cause it to quickly heat up, and the shaft will seize to the bearing. For this reason it is also important to follow lubrication instructions and oil the motor on a regular basis.

15.3.4
Brushes and Brush Rigging

The brush rigging is an assembly that securely holds the brushes in place so that they will be able to ride on the commutator. It is mounted on the rear end plate so that the brushes will be accessible by removing the end plate. An access hole is also provided in the motor frame so that the brushes can be adjusted slightly when the motor is initially set up. The brush rigging uses a spring to provide the proper amount of tension on the brushes so that they make proper contact with the commutator. If the tension is too light, the brushes will bounce and arc; and if the tension is too heavy, the brushes will wear down prematurely.

The brush rigging is shown in Figs. 15–5 and 15–6. Notice that it is mounted on the rear end plate. Since the rigging is made of metal, it must be insulated electrically when it is mounted on the end plate. The DC voltage that is used to energize the armature will pass through the brushes to the commutator segments and into the armature coils. Each brush has a wire connected to it. The wires will be connected to either the positive or negative terminal of the DC power supply. The motor will always have an even number of brushes. Half of the brushes will be connected to positive voltage, and half will be connected to negative voltage. In most motors the number of brush sets will be equal to the number of field poles. It is important to remember that the voltage polarity will remain constant on each brush. This means that for each pair, one of the brushes will be connected to the positive power terminal, and the other will be connected permanently to the negative terminal.

The brushes will cause the polarity of each armature segment to alternate from positive to negative. When the armature is spinning, each commutator segment will come in contact with a positive brush for an instant and will be positive during that time. As the armature rotates slightly, that commutator segment will come in contact with a brush that is connected to the negative voltage supply, and it will become negative during that time. As the armature continues to spin, each commutator segment will be alternately powered by positive and then negative voltage.

The brushes are made of carbon-composite material. Usually, the brushes have copper added to aid in conduction. Other material is also added to make them wear longer. The end of the brush that rides on the commutator is contoured to fit the commutator exactly so that current will transfer easily. The process of contouring the brush to the commutator is called *seating*. Whenever a set of new brushes is installed, the brushes should be seated to fit the commutator. The brushes are the main part of the DC motor that will wear out. It is important that their wear be monitored closely so that they do not damage the commutator segments when they begin to wear out. Most brushes have a small mark on them called a wear mark or wear bar. When a brush wears down to the mark, it should be replaced. If the brushes begin to wear excessively or do not fit properly on the commutator, they will heat up and damage the brush rigging and spring mechanism. If the brushes have been overheated, they can cause burn marks or pitting on the commutator segments and also warp the spring mechanism so that it will no longer hold the brushes with the proper amount of tension. Figures 15–5 and

15–6 show the location of the brushes riding on the commutator.

If the spring mechanism has been overheated, it should be replaced and the brushes should be checked for proper operation. If the commutator is pitted, it can be turned down on a lathe. After the commutator has been turned down, the brushes will need to be reseated so they make perfect contact with the commutator.

After you have an understanding of the function of each of the parts or assemblies of the motor, you will be able to understand better the operation of a basic DC motor. Operation of the motor involves the interaction of all the motor parts. Some of the parts will be altered slightly for specific motor applications. These changes will become evident when the motor's basic operation is explained.

15.4
DC Motor Operation

The DC motor you will find in modern industrial applications operates very similarly to the simple DC motor described earlier in this chapter. Figure 15–9 shows an electrical diagram of a simple DC motor. Notice that the DC voltage is applied directly to the field winding and the brushes. The armature and the field are both shown as a coil of wire. In later diagrams, a field resistor will be added in series with the field to control the motor speed.

When voltage is applied to the motor, current begins to flow through the field coil from the negative terminal to the positive terminal. This sets up a strong magnetic field in the field winding. Current also begins to flow through the brushes into a commutator segment and then through an armature coil. The current continues to flow through the coil back to the brush that is attached to other end of the coil and returns to the DC power source. The current flowing in the armature coil sets up a strong magnetic field in the armature.

The magnetic field in the armature and field coil causes the armature to begin to rotate. This occurs by the unlike magnetic poles attracting each other and the

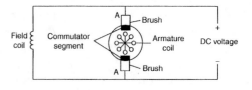

Figure 15–9 Simple electrical diagram of DC shunt motor. This diagram shows the field coil and armature connected in parallel.

like magnetic poles repelling each other. As the armature begins to rotate, the commutator segments will also begin to move under the brushes. As an individual commutator segment moves under the brush connected to positive voltage, it will become positive; and when it moves under a brush connected to negative voltage, it will become negative. In this way, the commutator segments continually change polarity from positive to negative. Since the commutator segments are connected to the ends of the wires that make up the field winding in the armature, it causes the magnetic field in the armature to change polarity continually from north pole to south pole. The commutator segments and brushes are aligned in such a way that the switch in polarity of the armature coincides with the location of the armature's magnetic field and the field winding's magnetic field. The switching action is timed so that the armature will not lock up magnetically with the field. Instead the magnetic fields tend to build on each other and provide additional torque to keep the motor shaft rotating.

When the voltage is de-energized to the motor, the magnetic fields in the armature and the field winding will quickly diminish, and the armature shaft's speed will begin to drop to zero. If voltage is applied to the motor again, the magnetic fields will strengthen, and the armature will begin to rotate again.

15.5
TYPES OF DC MOTORS

Three basic types of DC motors are used in industry today: the series motor, the shunt motor, and the compound motor. The series motor is capable of starting with a very large load attached, such as lifting applications. The shunt motor is able to operate with rpm control while it is at high speed. The compound motor, a combination of the series motor and the shunt motor, is able to start with fairly large loads and have some rpm control at higher speeds. In the remaining sections of this chapter we show a diagram for each of these motors and discuss their operational characteristics. As a technician you should understand methods of controlling their speed and ways to change the direction of rotation because these are the two parameters of a DC motor you will be asked to change as applications change on the factory floor. It is also important to understand the basic theory of operation of these motors because you will be controlling them with solid-state electronic circuits. You will need to know if problems that arise are the fault of the motor or the solid-state circuit.

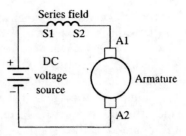

Figure 15–10   Electrical diagram of series motor. Notice that the series field is identified as S1 and S2.

15.5.1
DC Series Motors

The series motor provides high starting torque and is able to move very large shaft loads when it is first energized. Fig. 15–10 shows the wiring diagram of a series motor. From the diagram you can see that the field winding in this motor is wired in series with the armature winding. This is the attribute that gives the series motor its name.

Since the series field winding is connected in series with the armature, it will carry the same amount of current that passes through the armature. For this reason the field is made from heavy-gauge wire that is large enough to carry the load. Since the wire gauge is so large, the winding will have only a few turns of wire. In some larger DC motors, the field winding is made from copper bar stock rather than the conventional round wire used for power distribution. The square or rectangular shape of the copper bar stock makes it fit more easily around the field pole pieces. It can also radiate more easily the heat that has built up in the winding due to the large amount of current being carried.

The amount of current that passes through the winding determines the amount of torque the motor shaft can produce. Since the series field is made of large conductors, it can carry large amounts of current and produce large torques. For example, the starter motor that is used to start an automobile's engine is a series motor, and it may draw up to 500 A when it is turning the engine's crankshaft on a cold morning. Series motors used to power hoist or cranes may draw currents of thousands of amperes during operation.

The series motor can safely handle large currents since the motor does not operate for an extended period. In most applications the motor will operate for only a few seconds while this large current is present. Think about how long the starter motor on the automobile must operate to get the engine to start. This period is similar to that of industrial series motors.

15.5.2
Series Motor Operation

Operation of the series motor is easy to understand. In Fig. 15–10 you can see that the field winding is connected in series with the armature winding. This means that power will be applied to one end of the series field winding and to one end of the armature winding (connected at the brush).

When voltage is applied, current begins to flow from negative power supply terminals through the series winding and armature winding. The armature is not rotating when voltage is first applied, and the only resistance in this circuit will be provided by the large conductors used in the armature and field windings. Since these conductors are so large, they will have a small amount of resistance. This causes the motor to draw a large amount of current from the power supply. When the large current begins to flow through the field and armature windings, it causes a very strong magnetic field to be built. Since the current is so large, it will cause the coils to reach saturation, which will produce the strongest magnetic field possible.

15.5.3
Producing Back-EMF

The strength of these magnetic fields provides the armature shafts with the greatest amount of torque possible. The large torque causes the armature to begin to spin with the maximum amount of power. When the armature begins to rotate, it begins to produce voltage. This concept is difficult for some students to understand since the armature is part of the motor at this time.

You should remember from the basic theories of magnetism that any time a magnetic field passes a coil of wire, a current will be produced. The stronger the magnetic field is or the faster the coil passes the flux lines, the more current will be generated. When the armature begins to rotate, it will produce a voltage that is of opposite polarity to that of the power supply. This voltage is called *back voltage*, *back-EMF* (electromotive force), or *counter-EMF* (CEMF). The overall effect of this voltage is that it will be subtracted from the supply voltage so that the motor windings will see a smaller voltage potential.

When Ohm's law is applied to this circuit, you will see that when the voltage is slightly reduced, the current will also be reduced slightly. This means that the series motor will see less current as its speed is increased. The reduced current will mean that the motor will continue to lose torque as the motor speed increases. Since the load is moving when the armature begins to pick up

speed, the application will require less torque to keep the load moving. This works to the motor's advantage by automatically reducing the motor current as soon as the load begins to move. It also allows the motor to operate with less heat buildup.

This condition can cause problems if the series motor ever loses its load. The load could be lost when a shaft breaks or if a drive pin is sheared. When this occurs, the load current is allowed to fall to a minimum, which reduces the amount of back-EMF that the armature is producing. Since the armature is not producing a sufficient amount of back-EMF and the load is no longer causing a drag on the shaft, the armature will begin to rotate faster and faster. It will continue to increase rotational speed until it is operating at a very high speed. When the armature is operating at high speed, the heavy armature windings will be pulled out of their slots by centrifugal force. When the windings are pulled loose, they will catch on a field winding pole piece and the motor will be severely damaged. This condition is called *runaway*, and you can see why a DC series motor must have some type of runaway protection. A centrifugal switch can be connected to the motor to de-energize the motor-starter coil if the rpm exceeds the set amount. Other sensors can be used to de-energize the circuit if the motor's current drops while full voltage is applied to the motor. The most important part to remember about a series motor is that it is difficult to control its speed by external means because its rpm is determined by the size of its load. (In some smaller series motors, the speed can be controlled by placing a rheostat in series with the supply voltage to provide some amount of change in resistance to control the voltage to the motor.)

Figure 15–11 shows the relationship between series motor speed and armature current. From this curve you can see that when current is low (at the top left), the motor speed is maximum; and when current increases, the motor speed slows down (bottom right). You can also see from this curve that a DC motor will

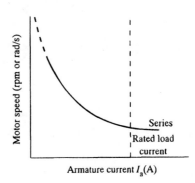

Figure 15–11   The relationship between series motor speed and the armature current.

run away if the load current is reduced to zero. (It should be noted that in larger series machines used in industry, the amount of friction losses will limit the highest speed somewhat.)

15.5.4
Reversing the Rotation of a Motor

The direction of rotation of a series motor can be changed by changing the polarity of either the armature or held winding. It is important to remember that if you simply changed the polarity of the applied voltage, you would be changing the polarity of both field and armature windings and the motor's rotation would remain the same.

Since only one of the windings needs to be reversed, the armature winding is typically used because its terminals are readily accessible at the brush rigging. Remember that the armature receives its current through the brushes, so that if their polarity is changed, the armature's polarity will also be changed. A reversing motor starter is used to change wiring to cause the direction of the motor's rotation to change by changing the polarity of the armature windings. Figure 15–12 shows a DC series motor that is connected to a reversing motor starter. In this diagram the armature's terminals are marked A1 and A2, and the field terminals are marked S1 and S2.

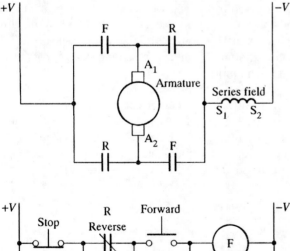

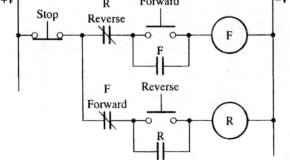

Figure 15–12 DC series motor connected to forward and reverse motor starter.

When the forward motor starter is energized, the top contact identified as F closes, so that the A1 terminal is connected to the positive terminal of the power supply, and the bottom F contact closes and connects terminal A2 and S1. Terminal S2 is connected to the negative terminal of the power supply. When the reverse motor starter is energized, terminals A1 and A2 are reversed. A2 is now connected to the positive terminal. Notice that S2 remains connected to the negative terminal of the power supply terminal. This ensures that only the armature's polarity has been changed, and the motor will begin to rotate in the opposite direction.

You will also notice that the normally closed (NC) set of R contacts is connected in series with the forward push button and that the NC set of F contacts is connected in series with the reverse push button. These contacts provide an *interlock* that prevents the motor from being changed from forward to reverse direction without stopping the motor. The circuit can be explained as follows: When the forward push button is depressed, current will flow from the stop push button through the NCR interlock contacts and through the forward push button to the forward motor-starter (F) coil. When the F coil is energized, it will open its NC contacts that are connected in series with the reverse push button. This means that if someone depresses the reverse push button, current could not flow to the reverse motor starter (R) coil. If the person depressing the push buttons wants to reverse the direction of the rotation of the motor, he or she will need to depress the stop push button first to de-energize the F coil, which will allow the normally closed F contacts to return to their normally closed position. You can see that when the R coil is energized, its normally closed R contacts that are connected in series with the forward push button will open and prevent the current flow to the F coil if the forward push button is depressed. You will see a number of other ways to control the F and R starter in later discussions and in Chapter 12 on motor controls.

15.5.5
Installing and Troubleshooting a Series Motor

Since a series motor has only two leads brought out of the motor for installation wiring this wiring can be accomplished rather easily. If the motor is wired to operate in only one direction, the motor terminals can be connected to a manual or magnetic starter. If the motor's rotation is required to be reversed periodically, it should be connected to a reversing starter.

Most DC series motors are used in direct-drive applications. This means that the load is connected directly to the armature's shaft. This type of load is generally used to get the most torque converted. Belt-drive applications are not recommended since a broken belt would allow the motor to run away. After the motor has been installed, a test run should be used to check it out. If any problems occur, the troubleshooting procedures should be used.

The most likely problem that will occur with the series motor is that it will develop an open in one of its windings or between the brushes and the commutator. Since the coils in a series motor are connected in series, each coil must be functioning properly or the motor will not draw any current. When this occurs, the motor cannot build a magnetic field, and the armature will not turn. Another problem that is likely to occur with the motor circuit is that circuit voltage will be lost due to a blown fuse or circuit breaker. The motor will respond similarly in both of these conditions.

The best way to test a series motor is with a voltmeter. The first test should be for applied voltage at the motor terminals. Since the motor terminals are usually connected to a motor starter, the test leads can be placed on these terminals. If the meter shows that full voltage is applied, the problem will be in the motor. If it shows that no voltage is present, you should test the supply voltage and the control circuit to ensure that the motor starter is closed. If the motor starter has a visual indicator, be sure to check to see that the starter's contacts are closed. If the overloads have tripped, you can assume that they have sensed a problem with the motor or its load. When you reset the overloads, the motor will probably start again, but remember to test the motor thoroughly for problems that would cause an overcurrent situation.

If the voltage test indicates that the motor has full applied voltage to its terminals but the motor is not operating, you can assume that you have an open in one of the windings or between the brushes and the armature and continue testing. Each of these sections should be disconnected from each other and voltage should be removed so that they can be tested with an ohmmeter for an open. The series field coils can be tested by putting the ohmmeter leads on terminals S1 and S2. If the meter indicates that an open exists the motor will need to be removed and sent to be rewound or replaced. If the meter indicates that the field coil has continuity, you should continue the procedure by testing the armature.

The armature can also be tested with an ohmmeter by placing the leads on the terminals marked A1 and A2. If the meter shows continuity, rotate the armature shaft slightly to look for bad spots where the commutator may have an open or the brushes may not be seated properly. If the armature test indicates that an open exists, you should continue the test by visually inspecting the brushes and commutator. You may also have an open in the armature coils. The armature must be removed from the motor frame to be tested further. When you have located the problem, you should remember that the commutator can be removed from the motor while the motor remains in place, and it can be turned down on a lathe. When the commutator is replaced in the motor, new brushes can be installed, and the motor will be ready for use.

It is possible that the motor will develop a problem but still run. This type of problem usually involves the motor overheating or not being able to pull its rated load. This type of problem is different from an open circuit because the motor is drawing current and trying to run. Since the motor is drawing current, you must assume that there is not an open circuit. It is still possible to have brush problems that would require the brushes to be reseated or replaced. Other conditions that will cause the motor to overheat include loose or damaged field and armature coils. The motor will also overheat if the armature shaft bearing is in need of lubrication or is damaged. The bearing will seize on the shaft and cause the motor to build up friction and overheat.

If either of these conditions occurs, the motor may be fixed on site or be removed for extensive repairs. When the motor is restarted after repairs have been made, it is important to monitor the current usage and heat buildup. Remember that the motor will draw DC current so that an AC clamp-on ammeter will not be useful for measuring the DC current. You will need to use an ammeter that is specially designed for very large DC currents. It is also important to remember that the motor can draw very high locked-rotor current when it is starting, so the ammeter should be capable of measuring currents up to 1,000 A. After the motor has completed its test run successfully, it can be put back into operation for normal duty. Any time the motor is suspected of faulty operation, the troubleshooting procedure should be rechecked.

15.5.6
DC Series Motor Used as a Universal Motor

The series motor is used in a wide variety of power tools such as electric hand drills, saws, and power screwdrivers. In most of these cases, the power source for the motor is AC voltage. The DC series motor will operate on AC voltage. If the motor is used in a hand drill that needs variable-speed control a field rheostat or other type of current control is used to control the speed of the motor. In some newer tools, the current control uses solid-state components to control the speed of the motor. You will notice that the motors used for these types of power tools have brushes and a commutator, and these are the main parts of the motor to wear out. You can use the same theory of operation provided for the DC motor to troubleshoot these types of motors.

15.6
DC SHUNT MOTORS

The shunt motor is different from the series motor in that the field winding is connected in parallel with the armature instead of in series. You should remember from basic electrical theory that a parallel circuit is often referred to as a shunt. Since the field winding is placed in parallel with the armature it is called a shunt winding and the motor is called a shunt motor. Figure 15–13 shows a diagram of a shunt motor. Notice that the field terminals are marked F1 and F2 and that the armature terminals are marked A1 and A2. You should notice in this diagram that the shunt field is represented with multiple turns using a thin line.

The shunt winding is made of small gauge wire with many turns on the coil. Since the wire is so small, the coil can have thousands of turns and still fit in the slots. The small-gauge wire cannot handle as much current as the heavy-gauge wire in the series field, but since this coil has many more turns of wire, it can still produce a very strong magnetic field. Figure 15–14 shows a picture of a DC shunt motor.

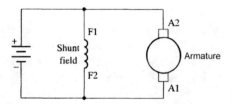

Figure 15–13 Diagram of DC shunt motor. Notice the shunt coil is identified as a coil of fine wire with many turns that is connected in parallel (shunt) with the armature.

Figure 15–14 Typical DC shunt motor. These motors are available in a variety of sizes. This motor is a 1 hp (approximately 8 in. tall).

15.6.1
Shunt Motor Operation

A shunt motor has slightly different operating characteristics than a series motor. Since the shunt field coil is made of fine wire, it cannot produce the large current for starting like the series field. This means that the shunt motor has very low starting torque, which requires that the shaft load be rather small.

When voltage is applied to the motor, the high resistance of the shunt coil keeps the overall current flow low. The armature for the shunt motor is similar to the series motor, and it will draw current to produce a magnetic field strong enough to cause the armature shaft and load to start turning. Like the series motor, when the armature begins to turn, it will produce back-EMF. The back-EMF will cause the current in the armature to begin to diminish to a very small level. The amount of current the armature will draw is directly related to the size of the load when the motor reaches full speed. Since the load is generally small, the armature current will be small. When the motor reaches full rpm, its speed will remain fairly constant.

15.6.2
Controlling the Speed of the Motor

When the shunt motor reaches full rpm, its speed will remain fairly constant. The reason the speed remains constant is due to the load characteristics of the armature and shunt coil. You should remember that the speed of a series motor could not be controlled since it was totally dependent on the size of the load in comparison with the size of the motor. If the load was very large for the motor size, the speed of the armature would be very slow. If the load was light compared with the size of the motor, the armature shaft speed would be much faster: and if no load was present on the shaft, the motor could run away.

The shunt motor's speed can be controlled. The ability of the motor to maintain a set rpm at high speed when the load changes is due to the characteristic of the shunt field and armature. Since the armature begins to produce back-EMF as soon as it starts to rotate, it will use the back-EMF to maintain its rpm at high speed. If the load increases slightly and causes the armature shaft to slow down, less back-EMF will be produced. This will allow the difference between the back-EMF and applied voltage to become larger, which will cause more current to flow. The extra current provides the motor with the extra torque required to regain its rpm when this load is increased slightly.

The shunt motor's speed can be varied in two different ways. These include varying the amount of current supplied to the shunt field and controlling the amount of current supplied to the armature. Controlling the current to the shunt field allows the rpm to be changed 10—20% when the motor is at full rpm.

This type of speed-control regulation is accomplished by slightly increasing or decreasing the voltage applied to the field. The armature continues to have full voltage applied to it while the current to the shunt field is regulated by a rheostat that is connected in series with the shunt field. When the shunt field's current is decreased, the motor's rpm will increase slightly. When the shunt field's current is reduced, the armature must rotate faster to produce the same amount of back-EMF to keep the load turning. If the shunt field current is increased slightly, the armature can rotate at a slower rpm and maintain the amount of back-EMF to produce the armature current to drive the load. The field current can be adjusted with a field rheostat or an SCR current control.

The shunt motor's rpm can also be controlled by regulating the voltage that is applied to the motor armature. This means that if the motor is operated on less voltage than is shown on its data plate rating, it will run at less than full rpm. You must remember that the shunt motor's efficiency will drop off drastically when it is operated below its rated voltage. The motor will tend to overheat when it is operated below full voltage, so motor ventilation must be provided. You should also be aware that the motor's torque is reduced when it is operated below the full voltage level.

Since the armature draws more current than the shunt field, the control resistors were much larger than those used for the field rheostat. During the 1950s and 1960s SCRs were used for this type of current control. The SCR was able to control the armature current since it was capable of controlling several hundred amperes. In Chapter 19 we provided an in-depth explanation of the DC motor drive.

15.6.3
Torque Characteristics

The armature's torque increases as the motor gains speed due to the fact that the shunt motor's torque is directly proportional to the armature current. When the motor is starting and speed is very low, the motor has very little torque. After the motor reaches full rpm, its torque is at its fullest potential. In fact, if the shunt field current is reduced slightly when the motor is at full rpm, the rpm will increase slightly, and the motor's torque will also increase slightly. This type of automatic control makes the shunt motor a good choice for applications where constant speed is required, even though the torque will vary slightly due to changes in the load.

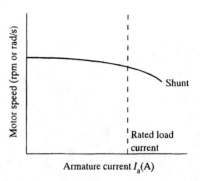

Figure 15–15   A curve that shows the armature current versus the armature speed for a shunt motor. Notice that the speed of a shunt motor is nearly constant.

Figure 15–15 shows the torque/speed curve for the shunt motor. From this diagram you can see that the speed of the shunt motor stays fairly constant throughout its load range and drops slightly when it is drawing the largest current.

15.6.4
Reversing the Rotation of the Motor

The direction of rotation of a DC shunt motor can be reversed by changing the polarity of either the armature coil or the field coil. In this application the armature coil is usually changed, as was the case with the series motor. Figure 15–16 shows the electrical diagram of a DC shunt motor connected to a forward and reversing motor starter. You should notice that the F1 and F2 terminals of the shunt field are connected directly to the power supply, and the A1 and A2 terminals of the armature winding are connected to the reversing starter.

When the F motor starter is energized, its contacts connect the A1 lead to the positive power supply terminal and the A2 lead to the negative power supply terminal. The F1 motor lead is connected directly to the positive terminal of the power supply, and the F2 lead is connected to the negative terminal. When the motor is wired in this configuration, it will begin to run in the forward direction.

When the R motor starter is energized, its contacts reverse the armature wires so that the A1 lead is connected to the negative power supply terminal and the A2 lead is connected to the positive power supply terminal. The field leads are connected directly to the power supply, so their polarity is not changed. Since the field's polarity has remained the same and the armature's polarity has reversed, the motor will begin to rotate in the reverse direction. The control part of the diagram shows that when the FMS coil is energized, the R motor starter coil is locked out.

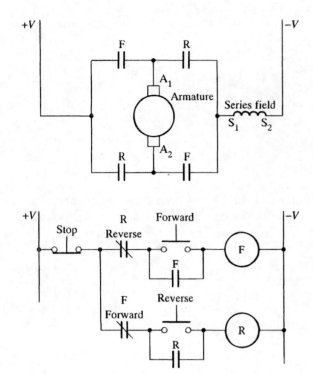

Figure 15–16 Diagram of a shunt motor connected to a reversing motor starter. Notice that the shunt field is connected across the armature and it is not reversed when the armature is reversed.

15.6.5
Installing a Shunt Motor

A shunt motor can be installed easily. The motor is generally used in belt-drive applications. This means that the installation procedure should be broken into two sections: (1) the mechanical installation of the motor and its load and (2) the installation of electrical wiring and controls.

When the mechanical part of the installation is completed, the alignment of the motor shaft and the load shaft should be checked. If the alignment is not true, the load will cause an undue stress on the armature bearing, and there is the possibility of the load vibrating and causing damage to it and the motor. After the alignment is checked, the tension on the belt should also be tested. As a rule of thumb, you should have about ½ to ¼ in. of play in the belt when it is properly tensioned.

Several tension measurement devices are available to determine when a belt is tensioned properly. The belt tension can also be compared with the amount of current the motor draws. The motor must have its electrical installation completed to use this method.

The motor should be started, and if it is drawing too much current, the belt should be loosened slightly but not enough to allow the load to slip. If the belt is slipping,

it can be tightened to the point where the motor is able to start successfully and not draw current over its rating.

The electrical installation can be completed before, after, or during the mechanical installation. The first step in this procedure is to locate the field and armature leads in the motor and prepare them for field connections. If the motor is connected to magnetic or manual across the line starter, the F1 field coil wire can be connected to the A1 armature lead and an interconnecting wire, which will be used to connect these leads to the T1 terminal on the motor starter. The F2 lead can be connected to the A2 lead and a second wire, which will connect these leads to the T2 motor-starter terminal.

When these connections are completed, field and armature leads should be put back into the motor and the field wiring cover or motor access plate should be replaced. Next the DC power supply's positive and negative leads should be connected to the motor starter's L1 and L2 terminals, respectively.

After all the load wires are connected, any pilot devices or control circuitry should be installed and connected. The control circuit should be tested with the load voltage disconnected from the motor. If the control circuit uses the same power source as the motor, the load circuit can be isolated so the motor will not try to start by disconnecting the wire at terminal L2 on the motor starter. Operate the control circuit several times to ensure that it is wired correctly and operating properly. After you have tested the control circuit, the lead can be replaced to the L2 terminal of the motor starter and the motor can be started and tested for proper operation. Be sure to check the motor's voltage and current while it is under load to ensure that it is operating correctly. It is also important to check the motor's temperature periodically until you are satisfied the motor is operating correctly.

If the motor is connected to a reversing starter or reduced-voltage starting circuit, the operation should also be tested. If the motor is not operating correctly or develops a fault, a troubleshooting procedure should be used to test the motor and locate the problem.

15.6.6
Troubleshooting the Shunt Motor

When a DC shunt motor develops a fault, you must be able to locate the problem quickly and return the motor to service or have it replaced. The most likely problems to occur with the shunt motor include loss of supply voltage or an open in either the shunt winding or the armature winding. Other problems may arise that cause the motor to run abnormally hot even though it continues to drive the load. The motor will show different symptoms for each of these problems, which will make the troubleshooting procedure easier.

When you are called to troubleshoot the shunt motor, it is important to determine if the problem occurs while the motor is running or when it is trying to start. If the motor will not start, you should listen to see if the motor is humming and trying to start. When the supply voltage has been interrupted due to a blown fuse or a de-energized control circuit, the motor will not be able to draw any current, and it will be silent when you try to start it. You can also determine that the supply voltage has been lost by measuring it with a voltmeter at the starter's L1 and L2 terminals. If no voltage is present at the load terminals, you should check for voltage at the starter's T1 and T2 terminals. If voltage is present here but not at the load terminals. It indicates that the motor starter is de-energized or defective. If no voltage is present at the T1 and T2 terminals, it indicates that supply voltage has been lost prior to the motor starter. You will need to check the supply fuses and the rest of the supply circuit to locate the fault.

If the motor tries to start and hums loudly, it indicates that the supply voltage is present. The problem in this case is probably due to an open field winding or armature winding. It could also be caused by the supply voltage being too low.

The most likely problem will be an open in the field winding since it is made from small-gauge wire. The open can occur if the field winding draws too much current or develops a short circuit between the insulation in the coils. The best way to test the field is to remove supply voltage to the motor by opening the disconnect or de-energizing the motor starter. Be sure to use a *lockout* when you are working on the motor after the disconnect has been opened. The lockout is a device that is placed on the handle of the disconnect after the handle is placed in the off position, and it allows it padlock to be placed around it so it cannot be removed until the technician has completed the work on the circuit. If the lockout has extra holes additional padlocks can be placed on it by other technicians who are also working on this system. This ensures that the power cannot be returned to the system until all technicians have removed their padlocks. The lockout will be explained in detail in Chapter 20 later in this text.

After power has been removed, the field terminals should be isolated from the armature coil. This can be accomplished by disconnecting one set of leads where the field and armature are connected together. Remember that the field and armature are connected in parallel, and if they are not isolated, your continuity test will show a completed circuit even if one of the two windings has an open.

When you have the field coil isolated from the armature coil, you can proceed with the continuity test. Be sure to use the R × 1-k or R × 10-k setting on the ohmmeter because the resistance in the field coil will be very high since the field coil may be wound from sev-

eral thousand feet of wire. If the field winding test indicates the field winding is good you should continue the procedure and test the armature winding for continuity.

The armature winding test may show that an open has developed from the wire in the coil burning open or from a problem with the brushes. Since the brushes may be part of the fault, they should be visually inspected and replaced if they are worn or not seating properly. If the commutator is also damaged. The armature should be removed so the commutator can be turned down on a lathe.

If either the field winding or the armature winding has developed an open circuit, the motor will have to be removed and replaced. In some larger motors it will be possible to change the armature by itself rather than remove and replace the entire motor. If the motor operates but draws excessive current or heats up, the motor should be tested for loose or shorting coils. Field coils may tend to come loose and cause the motor to vibrate and overheat, or the armature coils may come loose from their slots and cause problems. If the motor continues to overheat or operate roughly, the motor should be removed and sent to a motor rebuilding shop so that a more in-depth test may be performed to find the problem before the motor is permanently damaged by the heat

15.7
DC COMPOUND MOTORS

The DC compound motor is a combination of the series motor and the shunt motor. It has a series field winding that is connected in series with the armature and a shunt field that is in parallel with the armature. The combination of series and shunt winding allows the motor to have the torque characteristics of the series motor and the regulated speed characteristics of the shunt motor Figure 15–17 shows a diagram of the compound motor. Several versions of the compound motor are also shown in this diagram.

15.7.1
Cumulative Compound Motors

Figure 15–17a shows a diagram of the cumulative compound motor. It is so called because the shunt field is connected so that its coils are aiding the magnetic fields of the series field and armature. The shunt winding can be wired as a *long shunt* or as a *short shunt* parts a and b of Fig. 15–7 show the motor connected as a short shunt where the shunt field is connected in parallel with only the armature. Figure 15–17c shows the motor connected as a long shunt where the shunt field is connected in parallel with the series field, the interpoles, and the armature.

Figure 15–17 (a) Diagram of a cumulative compound motor; (b) diagram of a differential compound motor; (c) diagram of an interpole compound motor.

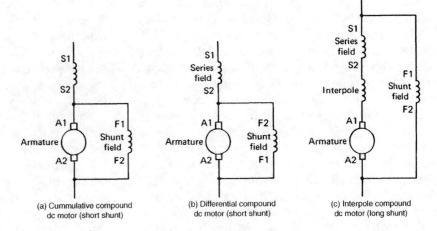

(a) Cummulative compound dc motor (short shunt)

(b) Differential compound dc motor (short shunt)

(c) Interpole compound dc motor (long shunt)

Figure 15–17a also shows the short shunt motor as a cumulative compound motor, which means the polarity of the shunt field matches the polarity of the armature. You can see in this figure that the top of the shunt field is positive polarity and that it is connected to the positive terminal of the armature. In Fig. 15–17b you can see that the shunt field has been reversed so that the negative terminal of the shunt field is now connected to the positive terminal of the armature. This type of motor is called a differential compound because the polarities of the shunt field and the armature are opposite.

The cumulative compound motor is one of the most common DC motors because it provides high starting torque and good speed regulation at high speeds. Since the shunt field is wired with similar polarity in parallel with the magnetic field aiding the series field and armature field, it is called cumulative. When the motor is connected this way, it can start even with a large load and then operate smoothly when the load varies slightly.

You should recall that the shunt motor can provide smooth operation at full speed, but it cannot start with a large load attached, and the series motor can start with a heavy load, but its speed cannot be controlled. The cumulative compound motor takes the best characteristics of both the series motor and shunt motor which makes it acceptable for most applications.

15.7.2
Differential Compound Motors

Differential compound motors use the same motor and windings as the cumulative compound motor, but they are connected in a slightly different manner to provide slightly different operating speed and torque characteristics Figure 15–17b shows the diagram for a differential compound motor with the shunt field connected so its polarity is reversed to the polarity of the armature. Since the shunt field is still connected in parallel with only the armature, it is considered a short shunt.

In this diagram you should notice that F1 and F2 are connected in reverse polarity to the armature. In the differential compound motor the shunt field is connected so that its magnetic field opposes the magnetic fields in the armature and series field. When the shunt field's polarity is reversed like this, its field will oppose the other fields, and the characteristics of the shunt motor are not as pronounced in this motor. This means that the motor will tend to overspeed when the load is reduced just like a series motor. Its speed will also drop more than the cumulative compound motor when the load increases at full rpm. These two characteristics make the differential motor less desirable than the cumulative motor for most applications.

15.7.3
Compound Interpole Motors

The compound interpole motor is built slightly differently from the cumulative and differential compound motors. This motor has interpoles connected in series with the armature (Fig. 15–17c). The interpoles are connected in series between the armature and series winding. The interpole is physically located beside the series coil in the stator. It is made of wire that is the same gauge as the series winding, and it is connected so that its polarity is the same as the series winding pole it is mounted behind. Remember that these motors may have any number of poles to make the field stronger.

The interpole prevents the armature and brushes from arcing due to the buildup of magnetic forces. These forces are created from counter-EMF called *armature reaction.* They are so effective that normally all DC compound motors that are larger than ½ hp will utilize them. Since the brushes do not arc, they will last longer, and the armature will not need to be cut down as often. The interpoles also allow the armature to draw heavier currents and carry larger shaft loads.

When the interpoles are connected, they must be tested carefully to determine their polarity so that it can be matched with that of the main pole. If the polarity of the interpoles does not match that of the main pole it is mounted behind, it will cause the motor to overheat and may damage the series winding.

15.7.4
Reversing the Rotation of the DC Compound Motor

Each of the compound motors shown in Fig. 15–17 can be reversed by changing the polarity of the armature winding. If the motor has interpoles, the polarity of the interpole must be changed when the armature's polarity is changed. Since the interpole is connected in series with the armature, it should be reversed when the armature is reversed. The armature winding is always marked as A1 and A2, and these terminals should be connected to the contacts of the reversing motor starter.

15.7.5
Controlling the Speed of the Motor

The speed of a compound motor can be changed very easily by adjusting the amount of voltage applied to it. In fact, it can be generalized that prior to the late 1970s any industrial application that required a motor to have a constant speed would be handled by an AC motor and any application that required the load to be driven at variable speeds would automatically be handled by a DC motor. This statement was true because it was easier to change the speed of a DC motor than an AC motor. Since the advent of solid-state components and microprocessor controls this circumstance is no longer true. In fact today a solid-state AC variable-frequency motor drive can vary the speed of an AC motor as easily as that of DC motors. This means you now must understand methods of controlling the speed of both AC and DC motors. Information about AC motor speed control is provided in Chapter 19.

Figure 15–18 shows the characteristic curves of speed versus armature current for the compound motors. From this diagram you can see that the speed of a differential compound motor increases slightly when the motor is drawing the armature's highest cur-

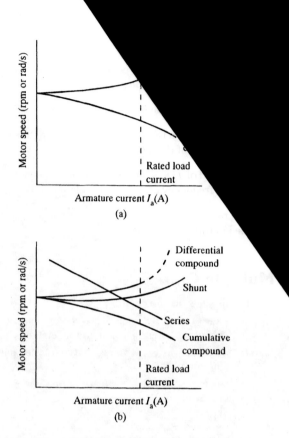

Figure 15–18 (a) Characteristic curve of armature current versus speed for the differential compound motor and cumulative compound motor; (b) composite of the characteristic curves for all the DC motors.

rent. The increase in speed occurs because the extra current in the differential winding causes the magnetic field in the motor to weaken slightly because the magnetic field in the differential winding opposes the magnetic in series field. As you learned earlier in regard to the speed control of shunt motors, the speed of the motor will increase if the magnetic field is weakened.

Figure 15–18 also shows the characteristic curve for the cumulative compound motor. This curve shows that the speed of the cumulative compound motor decreases slightly because the field is increased. This slows the motor because the magnetic field in the shunt winding aids the magnetic field of the series field.

QUESTIONS

Short Answer

1. What is the function of the armature in a DC motor?

2. What is the function of the field in a shunt-type DC motor?

3. Explain how the forward and reverse motor-starter circuits shown in Fig. 15–12 provide lockout protection so the motor cannot be switched directly from the forward direction to the reverse direction.

4. What is the function of the brushes and commutator segments in the DC motor?

5. Explain how you can change the speed of a DC shunt motor.

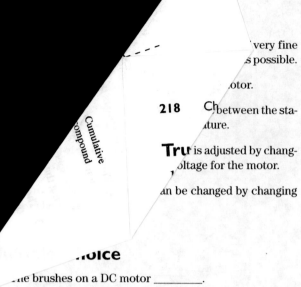

very fine
s possible.

.otor.

between the sta-
ature.

Tru is adjusted by chang-
oltage for the motor.

an be changed by changing

.noice

...he brushes on a DC motor _____.

a. make contact between the stationary part of the motor and the rotating, commutator segments

b. provide a way to get voltage to the rotating part of the generator

c. cause the polarity of each armature segment to alternate from positive to negative

d. all of the above

2. The armature _____.

a. is the part of the DC motor that rotates

b. has a core that is made of laminated steel and has windings pressed into it

c. has commutator segments on the end of its shaft

d. all of the above

3. The motor end plates _____.

a. are mounted on the end of the motor frame and have bearings to support the shaft

b. have commutator segments connected to them

c. produce flux lines for the motor

d. all of the above

4. The DC shunt motor has _____.

a. an armature, a series field, and a shunt field

b. an armature and only a series field

c. an armature and only a shunt field

d. all of the above

5. Torque is _____.

a. rotational force

b. flux lines that are created by the field

c. flux lines that are created by the armature

d. all of the above

CHAPTER 16

DC Generators

OBJECTIVES

After reading this chapter you will be able to:

1. Identify the basic parts of a generator

2. Explain the operation of each part of a DC generator.

3. Explain how a DC generator produces voltage.

4. Explain what happens when the direction of rotation or the polarity of the field coil is changed on a DC generator.

16.0

INTRODUCTION

By definition a generator produces voltage when its shaft (armature) is rotated. The DC generator has the same parts as a DC motor. The major difference is that a motor takes electrical energy in and produces rotating energy (torque) and the electric generator creates electrical current when its shaft receives rotating energy. The DC motors explained in Chapter 15 could be used as DC generators if an energy source were applied to turn their shaft rather than voltage being applied to their terminals. When the DC machine is used as a generator, electrical current is available at the armature terminals when the armature is rotated rapidly.

The voltage that is produced by a DC generator is used to power DC loads such as welding equipment, emergency lighting, or emergency pumps. At one time before the advent of AC variable-frequency drives, DC generators were used to produce variable voltage that was supplied to large DC motors, which enabled the load connected to the motor shaft to run at variable speeds. Today these are not as common, because variable-frequency drives for AC motors provide a

means to accurately adjust AC motor speeds on a large variety of loads. Even though you may not work on many DC generators, it is important to understand their theory of operation because it applies to a large variety of other electrical devices such as AC generators (alternators) and tach generators.

16.1

BASIC PARTS AND BASIC OPERATION OF A DC SHUNT GENERATOR

The basic parts of a DC shunt generator include the field winding, which is mounted in the stationary part of the generator called the stator; the armature, which is the rotating part with its commutator segments; and the shaft. The brushes make contact between the stationary part of the generator and the commutator segments, which are electrically connected to the armature winding. The end plate also has the bearings that allow the armature shaft to rotate more easily. The armature shaft is extended out one end of the generator to provide a place to mount the pulley sheave or coupling that is used to transfer mechanical rotating energy to the armature shaft. Figure 16–1 shows a diagram of the two stator segments for the DC generator. The dashed lines between two stator segments represent the flux lines of the magnetic field. In this figure the small magnetic field is produced by residual magnetism from the iron in the stator. The magnetic field in the stator can be increased by the current flowing through the field winding. At this point we need to clarify some simple electrical theory. A generator will produce voltage that is a potential energy. Current flow will depend on the amount of resistance (load) that is connected to the potential voltage source. In this section you will see that at times the generator will-produce a voltage; and when a load is connected to the generator, it will cause current to flow. When the generator is not loaded, it will not provide a current flow.

When a wire passes through a magnetic field, current is caused to flow in the wire. In the case of the generator, the wire is part of the armature, which is the rotating part of the generator. When the armature begins to rotate, the wire in its windings will pass through the magnetic field, which will create current flow in the winding. The stronger the magnetic field or the faster the armature turns, the more current is produced in the armature.

Figure 16–2 shows a diagram of wire moving through a magnetic field. Since the armature rotates in a circular motion, you can assign values 0-360 to represent the degrees of a circle. You can then plot the output waveform from the generator's armature as the armature turns through one rotation. As the wire in the armature moves from $0°$ through $90°$, the sine wave displays 0 V to peak positive. As the wire in the armature moves from 90° to 180°, the sine wave displays peak voltage back to zero. Each armature winding has one commutator segment connected to each end of the winding. The generator in Fig. 16–2 shows only one armature winding, and two commutator segments are shown connected to the ends of the armature winding, and two brushes are shown making contact with the commutator segments. This means that the commutator segments rotate in the same 360° pattern as the armature and the brushes, which are permanently mounted so that they make contact with the commutator segments for $180°$, or one-half

a revolution. The important point to remember about the relationship between the brushes and commutator segments is that the brush identified with a positive sign (+) makes contact with the end of the armature coil and commutator segment that is positive at any instant of time and the brush that is identified with a negative sign (−) always touches the commutator segment that is connected to the end of the armature wire that is negative at that instant in time. At first this is difficult to understand, but on closer inspection, you find that as the armature rotates and the wire in its coil moves down through the magnetic field, a positive half of a sine wave is produced. When the armature rotates through the remaining 180°, the commutator segments have also moved 180°, which results in another positive half of a sine wave being produced and sent through the positive brush. This causes the generator to produce a series of positive half-waves with respect to the positive brush. Since all the half sine waves are the same polarity, the output voltage of the generator is identified as DC voltage.

The brushes and commutator segments provide two important functions. First, the brushes and commutator segments provide an electrical contact between a stationary part (the brushes) and a rotating part (the commutator segments). The second function the brushes and commutator provide is timing function so that the voltage produced by the armature is always pulsing DC.

16.2
SEPARATELY EXCITED SHUNT GENERATOR

Figure 16–3 shows the diagram of a separately excited shunt generator. In this diagram you can see that the armature is now represented as a circle with the word *Arm* inside it, and a brush is shown on each side of the

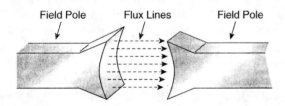

Figure 16–1   Field poles and flux lines of a simple DC generator.

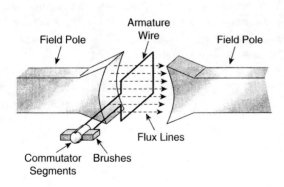

Figure 16–2   Armature wire cutting through the flux lines of a DC generator.

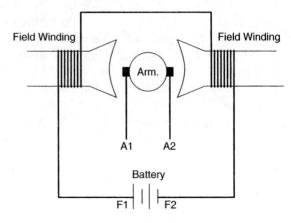

Figure 16–3   Separately excited generator.

armature. From this point on, the armature for each type of generator will be shown with this symbol. The two wires connected to the brushes will be identified as Al and A2 to indicate they are armature wires.

In the separately excited generator, an outside voltage (called a battery voltage) is used to increase the current flow in the field, which in turn creates a stronger magnetic field. The separately excited generator exists only where a second power source is available. The point to remember about the separately excited generator is that the amount of voltage that is used to control the magnetic field is relatively small when compared with the amount of voltage and current produced at the armature of the generator.

16.3

SELF-EXCITED SHUNT GENERATOR

Figure 16–4 shows a diagram of a self-excited shunt generator. In this diagram you can see that the Fl terminal for the field winding is connected to the Al terminal of the armature and the F2 terminal for the field winding is connected to the A2 terminal of the armature. This connection allows a small amount of voltage from the armature to be routed back through the field winding to create the magnetic field. Again as in the separately excited generator, the actual amount of voltage that the field needs is usually less than 2% of the output of the generator at the armature terminals.

Since the self-excited generator requires field current to create armature voltage, a small amount of residual magnetism is required to get the generator to begin producing voltage when the armature first begins to turn. The small amount of residual magnetism will provide sufficient magnetic flux lines to produce a small amount of armature voltage. As the armature voltage increases, it produces more field current, which in turn allows the generator to produce more armature voltage. You should

remember that the field winding is essentially a fixed resistance, so that any increase in voltage applied to the field winding will result in an increase in field current. The increase in field current will cause the magnetic field to become stronger to the point where the magnetic field becomes saturated. Saturation is defined as the amount of field current that causes the strongest magnetic field, such that any further increase in field current will not cause an increase in magnetic strength.

If a self-excited generator has gone for a long period of time without generating a voltage, it may lose its residual magnetism. If this occurs, an outside source of voltage (such as a battery) must be used to provide the voltage for the field to begin to create the magnetic field again. When an outside voltage source is used to establish the field current, it is called flashing the field. It is important to observe the polarity of the external voltage and the field polarity so that you do not reverse the polarity of the generator field, which will cause the armature voltage to be reversed with respect to the markings on the generator terminals. If this occurs, you can reestablish the proper polarity by flashing the field again with a voltage source that has the reverse polarity of the previous voltage. It is important to control the polarity in the field winding with respect to the armature windings by ensuring that the positive field terminal is connected to the positive armature terminal. This means that Fl should be connected to Al and F2 should be connected to A2.

In Fig. 16–4 you should notice that the field winding and the armature winding are connected in parallel. This type of configuration is called a shunt generator. The term *shunt* means that the windings are connected in parallel with each other. It also means that if this type of generator is turned at a constant speed, the armature voltage will remain rather constant. Since the field is connected in parallel with the armature, the voltage applied to the field will remain rather constant too. This equilibrium will continue as long as the load is constrained. If the load resistance goes down, it will cause the armature amperage to increase and cause the generator to produce more horsepower, so the electrical load is said to go up. When this occurs in a shunt generator, the output voltage will decrease slightly. This means the shunt generator produces the most voltage when its electrical load is low (electrical resistance is high at this time), and it produces the least voltage when its load is high.

16.4

COMPOUND GENERATOR

A diagram of a compound generator is shown in Fig. 16–5. In this figure you can see that the generator has both a series field and a shunt field connected.

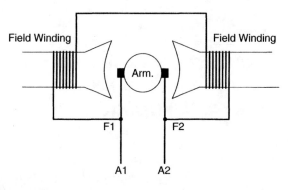

Figure 16–4   Self-excited generator.

The series field consists of a few turns of very large wire, and it is connected in series with the armature so that all the current the armature produces when a load is connected to the generator will also flow through the series field winding, which creates a stronger magnetic field. The main feature of this type

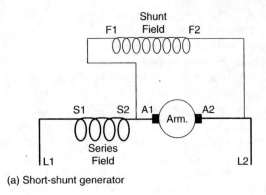

(a) Short-shunt generator

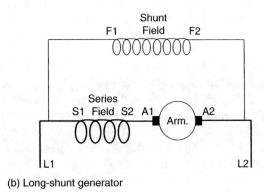

(b) Long-shunt generator

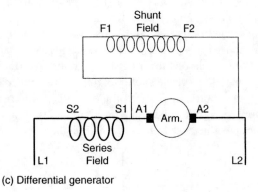

(c) Differential generator

Figure 16–5   (a) An example of a compound generator connected as a short-shunt generator; (b) an example of a compound generator connected as a long-shunt generator; (c) an example of a compound generator connected as a differential generator.

of generator is that it will produce a rather constant voltage regardless of its load.

Figure 16–5 shows several examples of how a compound generator can be connected. Figure 16–5a shows an example of a short-shunt generator, Fig. 16–5b shows a long shunt generator, and Fig. 16–5c shows a short-shunt differential generator. When the shunt field winding (Fl-F2) is connected in parallel with only the armature in the compound generator, it is called a *short shunt*. If the field winding is connected in parallel with both the armature and the series field (S1-S2), it is called a *long shunt*. If the series field is connected so that its polarity is reversed in reference to the armature, the generator is called a differential generator. The short shunt is the most popular way to connect these windings because it provides the most constant voltage under varying loads. The differential generator will provide fairly constant voltage when its load is low (electrical resistance high), but its voltage will drop significantly when a load is applied.

16.5
SERIES GENERATOR

If the DC machine is connected so that only the series field winding is connected in series with the armature and the shunt field is not used, it is called a series generator. In this generator the series field, armature, and electrical load connected to the generator are all connected in series. This means that as the electrical resistance in the load decreases, electrical current will increase, so the generator will need to produce more electrical current. When the electrical resistance in the load is high, the amount of electrical current is low, so the load on the generator is relatively low. This also means that when the current flow from the generator is low, its voltage is also low; and as its current increases, its voltage will increase. Since this produces a variable voltage that can range from near 0 V to the maximum of the generator, the series generator is not very useful in most applications.

16.6
CONTROLLING THE AMOUNT OF VOLTAGE AND ITS POLARITY IN A DC GENERATOR

The output voltage of a DC generator can be controlled by either of two means. First the speed of rotation can be controlled at a constant rate. This is usually accomplished by controlling the speed of the diesel or gasoline motor that is turning the generator. If the DC generator is turned by an AC motor, its speed will also tend to be constant. The second way to control the voltage of the DC generator is to control the field current. This can be automatically controlled by an electrical circuit called a *voltage regulator.* When the speed is kept constant and a voltage regulator is used, the output voltage of the DC generator can be accurately controlled. If the voltage regulator uses solid-state devices such as transistors or op amps, the voltage can be controlled automatically, even as the load changes. If the voltage regulator is an older style, you may need to manually adjust the field current to set the voltage to the level that the application requires.

The polarity of a DC generator can be controlled in either of two ways. First, the direction of rotation can be changed, and the polarity of the armature voltage will be reversed. If the generator is driven by a diesel or gasoline engine, this may not be an option. If the generator is driven by a three-phase AC motor, the motor can be reversed fairly easily by swapping any two of the three incoming voltage lines to the motor.

If the direction of rotation cannot be changed, you can change the polarity of the field winding with respect to the armature winding. In some cases you will have trouble getting the polarity the way you want it, and you may need to flash the field with an external voltage to cause the polarity to change.

16.7

TROUBLESHOOTING A DC GENERATOR

The main problem that will occur with a DC generator is that it will produce no voltage or low voltage when it is turning. Since the generator consists of relatively few and simple parts—such as the field winding, armature winding, with its commutator segments, and brushes—these components can all be easily tested. You can test the field winding with an ohmmeter for continuity by placing the ohmmeter leads across the FI and F2 terminals. You do not need to be concerned with the exact amount of resistance in the winding, just that it has continuity (some resistance). You can also test the armature for resistance in the same way by connecting an ohmmeter across the AI and A2 terminals. You must remember that you are reading the resistance of the brushes as they touch the commutator segments and the total resistance of each individual part of the armature winding. Since the armature is made of many windings, each having its own set of commutator seg-

ments, you must turn the armature shaft very slowly to allow the brushes to come into contract with each set of commutator segments. When you do this test, you will notice the amount of resistance will change slightly as the brushes move across the different commutator segments. Do not spin the generator quickly or try this test while the generator is turning, since a voltage will be produced when the generator turns and the voltage becomes larger as the generator turns faster. Remember, too much voltage at the terminals of the ohmmeter will damage it, since it has its own internal battery.

If you have tested the field windings and the armature and find that they are good, you can try to apply external field voltage to the generator to force it to produce an armature voltage. In some generators, the residual magnetism is too weak to get the generator to build up to a voltage. In these cases the external voltage will easily cause the generator to produce a voltage. Once the generator is producing a voltage, you can remove the external voltage, and it should continue to produce a voltage from its armature voltage. If the generator does not continue to produce a voltage at this point, you should suspect the voltage regulator on the generator.

Another common problem with DC generators is that their brushes become worn and the commutator segments become excessively dirty. In fact, this problem is so common that DC generators require considerable preventative maintenance, so they are not the generator of choice. If the commutator segments are moderately dirty, you can clean them with a commutator stone; sometimes a large eraser used to remove pencil marks will also work. You need to be aware that if you use sandpaper or emery paper to clean the commutator, small particles of copper from the commutator will become wedged between the segments and cause the segments to short out, which will cause the generator to produce less and less voltage and current. If the commutator segments are too dirty, you may need to pull the armature shaft and have it professionally turned down on a lathe. The brushes will continue to wear down until they are so small that the spring tension that holds them tight to the commutator segments is not sufficient to do the job; then the brushes will begin to bounce and arc as they move over the commutator segments. The brushes have a mark on them called a *wear mark,* which indicates when the brush is worn down too far and must be changed out. When you replace a set of brushes, always change all the brushes. Pay close attention to the angle of the face of each brush and place it in the brush holder so the face angle matches the angle of the commutator segments. Also be sure that the springs that hold the brush in place allow the brush to move up and down slightly.

QUESTIONS

Short Answer

1. What is the function of the armature in a DC generator?

2. What is the function of the field in a shunt-type DC generator?

3. Identity three things that need to occur for a DC generator to produce voltage.

4. What is the function of the brushes and commutator segments in the DC generator?

5. Explain how you can change the voltage output of a DC shunt generator.

True or False

1. The series field of a DC series generator is made of very fine wire so that its magnetic field will be as strong as possible.

2. The armature is the rotating part of a DC generator.

3. The brushes in a DC generator make contact between the stationary part of the generator and the armature.

4. The voltage output of a DC shunt-type generator is adjusted by changing the speed of the rotation of the generator armature.

5. The polarity of the voltage of a DC generator can be changed by changing the direction of rotation of the generator.

Multiple Choice

1. The brushes on a DC generator _____.
   a. make contact between the stationary part of the generator and the rotating commutator segments
   b. provide a way to get voltage to the rotating part of the generator
   c. cause the polarity of each armature segments to alternate from positive to negative
   d. all of the above

2. The armature of a generator _____.
   a. is the part of the DC generator that rotates
   b. has a core that is made of laminated steel and has windings pressed into it
   c. has commutator segments on the end of its shaft
   d. all of the above

3. The generator end plates _____.
   a. are mounted on the end of the generator frame and have bearings to support the shaft
   b. have commutator segments connected to it
   c. produce flux lines for the generator
   d. all of the above

4. The DC shunt generator has _____.
   a. an armature, a series field, and a shunt field
   b. an armature and only a series field
   c. an armature and only a shunt field
   d. all of the above

5. The amount of voltage a DC generator can produce can be increased by _____.
   a. turning the armature faster
   b. supplying more field current
   c. increasing the size of the field or armature windings
   d. all of the above

CHAPTER 17

AC Alternators

OBJECTIVES

After reading this chapter you will be able to:

1. Identify the basic parts of an alternator

2. Explain the operation of each part of an alternator.

3. Explain how the alternator produces voltage.

4. Explain what happens when the direction of rotation or the polarity of a field coil is changed on an alternator.

17.0

INTRODUCTION

An AC alternator is similar to a DC generator in that it produces a voltage when a coil of wire is passed through a magnetic field. The major difference is that the alternator has several design features that compensate for the problems that are common in the DC generator. One design difference is that the part of the alternator that produces the large current is located in the stationary part of the machine called the *stator*, so that no brushes are needed to transfer the large amount of electrical energy that is produced. This means that the part of the alternator that uses a small amount of current to produce the magnetic field must be located in the rotating part of the machine, called the *rotor*. Since the rotor needs a small amount of constant DC voltage, this voltage can be supplied through two slip rings and two brushes rather than commutator segments. Figure 17–1 shows the basic parts of the AC alternator. You should notice that slip rings are mounted on the rotor, and they run completely around the armature shaft like a collar, instead of the commu-

tator segments in the DC generator. Because the DC voltage supplied to the rotor by the slip rings is constant, the slip rings do not need to be segmented to provide a timing function like the commutator. Since the slip rings are continuous, they have a very high polished finish, which makes the brushes last longer. The brushes also tend to last much longer in the alternator because only the small amount of current needed to produce the magnetic field must pass through the brushes and slip rings. This current is usually less than 5 A, and it is generally called the *exciter voltage*. Another point to remember is that the source of DC voltage for the field can be an outside DC source on larger alternators, or it can easily be obtained from the AC output voltage at the stator windings by placing a diode in the circuit for rectification. In large alternators that produce the voltage that you use in factories or in your home, a separate DC generator is mounted on the rotor shaft in addition to the field winding, and this separate DC generator produces the voltage for the field.

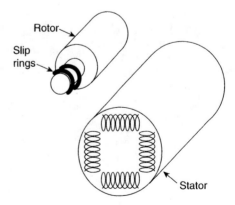

Figure 17–1 Basic parts of an alternator.

17.1

THEORY OF OPERATION
OF THE ALTERNATOR

An alternator produces voltage when a coil of wire passes through a magnetic field. In an alternator, a magnetic field is produced in the rotor when DC current is supplied through the brushes to the slip rings. The alternator needs only two slip rings because the rotor has only one coil of wire with two ends. One slip ring is connected to each end of the coil. The magnetic field in the rotor increases its magnetic field as the small amount of current is supplied to it until the field reaches saturation. Less than 5 A is usually needed. A field regulator is used to control the field current and keep the output voltage of the alternator constant.

When DC current flows through the rotor winding, it produces the magnetic field. Since the rotor is turned by an outside energy source, it produces a rotating magnetic field. When the rotating field passes the stationary coils of the stator, a large voltage is produced in the stator. When a load is applied to the stator terminals, the alternator begins to produce current. The amount of voltage is controlled by varying the amount of current in the rotating field and by the speed of the rotor. Since the voltage produced by the alternator is alternating current (AC), it will have a frequency. The frequency is a result of the speed of the rotor's rotation. This means that the rotor speed must be kept constant to keep the frequency constant, so the voltage regulation is controlled mainly by adjusting field current.

If the alternator relies on AC voltage from its stator to provide the rectified DC field voltage for the rotor, the rotor must first provide some residual magnetism to create the initial flux lines in the rotor. This residual magnetism occurs naturally because the rotor is made of ferrous material that readily maintains a small amount of permanent magnetism.

17.2

AC VOLTAGE IN THE STATOR

When the rotating magnetic field passes the stator winding, the polarity of the magnetic field will determine the polarity of the voltage being produced in the field. Figure 17–2 shows the rotating field as it passes the stator winding and the resulting AC voltage that is produced. You can see in Fig. 17–2a that when the flux lines of the positive end of the magnetic field pass

Figure 17–2 (a) The rotor's magnetic field moving passed the stationary field in an alternator to create the positive half of the sine wave; (b) the rotor's magnetic field moving passed the stationary field in an alternator to create the negative half of the sine wave.

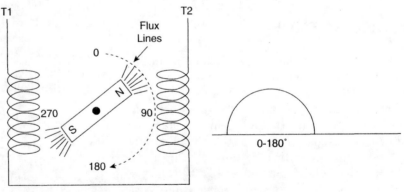

(a) Creating the positive half of an AC sine wave 0-180

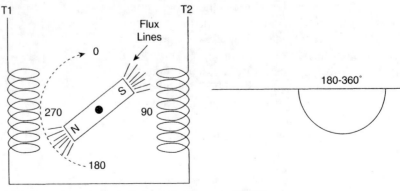

(b) Creating the negative half of an AC sine wave 180-360

across the wires in the stator winding, a positive voltage will begin to be produced. Since only a few flux lines are passing through the winding at first, the amount of voltage will be small. As the rotor continues to move the magnetic field more and more, flux lines will continue to cut across the stator coil until the maximum number of flux lines are moving across the coil and the amount of voltage produced is at its maximum. When the rotor completes its movement from 90° to 180°, fewer and fewer positive flux lines cut across the stator winding. This causes the amount of voltage being produced to decrease until the amount of voltage reaches zero again. You should recognize the waveform that is produced as the positive part of the AC sine wave.

When the rotor moves from 180° to 270°, the flux lines from the opposite end of the rotating magnetic field will begin to cut across the stator winding. Since these flux lines are created from the south magnetic pole, the resulting voltage that is produced in the stator will be negative. This voltage will start at zero and begin to build continually until the maximum number of flux lines are cutting across the stator winding when the rotor reaches 270°. At this point the voltage will be at its negative peak. As the rotor continues to rotate from 270° to 360°, the voltage will continually be reduced until it reaches zero again. In Fig. 17–2b you can see that this voltage waveform is the negative half of the sine wave. This means that when the rotor moves in one complete rotation (360°), the alternator will produce a waveform that is a sine wave with one positive and one negative peak. The time it takes the rotor to make one revolution: which produces one sine wave, is called the *period*. If the alternator is to produce voltage at 60 cycles, it must make one rotation in 0.016 s. This can also be described in terms of the alternator rpm. If the alternator produces 60 sine waves in 1 s, it must be turned at 2,600 rpm (60 Hz/s times 60 s/min). Since the frequency for an AC alternator is important, the speed of the alternator must be closely controlled.

17.3
PRODUCING THREE-PHASE VOLTAGE

The alternator produces some amount of AC voltage, which will cause current to flow in a load. The amount of AC power that is created by the voltage and current can be calculated. It is important to understand that the amount of energy needed to turn the alternator shaft must be larger than the amount of electrical power that is produced by the alternator. This energy can be produced by a gasoline engine, a diesel engine, a steam turbine, or other means. One way to design an alternator so that it produces the most energy for the smallest

physical size is to create three individual stator windings so that each produces its own set of sine waves. The stator windings are placed in the alternator so that the three waveforms are produced 120° apart. This type of voltage is called three-phase voltage. Figure 17–3 shows an example of the three-phase voltage waveform. Each winding may be made of multiple coils of wire to provide more power.

17.4
USING AN ALTERNATOR TO PRODUCE DC VOLTAGE

Since a DC generator has many design problems that require a large amount of preventative maintenance, the AC alternator is now commonly used to provide DC voltage and current. This is easily accomplished by placing two diodes in each winding of the alternator. If the alternator is a three-phase alternator, it will have six diodes (three sets) to rectify the AC voltage to DC. The diodes must be sized to carry the maximum working voltage and current of the alternator. If one diode in a set becomes faulty, the amount of voltage produced by that winding will be approximately half of the rating of that coil. If two diodes in any set burn out, all the voltage produced by that coil will be lost. Bad diodes are the primary cause of these types of alternator failures.

17.5
TROUBLESHOOTING AN ALTERNATOR

The main problem you will encounter with an alternator is the total loss of voltage. If an alternator is used to produce AC voltage, you should first suspect the loss of the field voltage that is supplied to the motor. You can test for this by placing a DC ammeter in the field terminals. If the current is zero, it means that you have an open somewhere in that circuit or that you have lost voltage to the field. The simplest test is for voltage at the terminals. Remember, this voltage is DC, and it comes from a separate DC source. Hence, it is easy to check the amount of DC voltage that is present and determine if there is a voltage loss.

This problem becomes more complex when the alternator uses AC voltage from its stator and rectifies it to produce the DC voltage for the field. In this type of alternator, the rotor must have some amount of residual magnetism to provide the initial flux lines. If you suspect the alternator has been sitting for some time, you

Figure 17–3 Example of three-phase AC voltage.

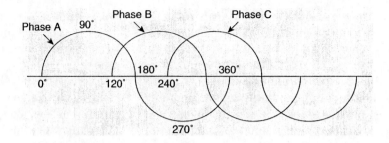

can apply a small amount of DC voltage to the rotor from an external DC source such as a battery.

If the field does not show any current flow when the DC voltage is supplied from an external power source, you must begin to suspect that the brushes or slip rings are bad or that you have an open in the rotor winding. You can test this circuit for continuity by removing all wires to the rotor terminals and checking for continuity. Be sure to slowly rotate the alternator shaft so that the brushes come into contact with all 360° of the slip ring's surface during this test. In some cases, the slip rings will pick up a spot or begin to pit in one area, and this will not show up if you do not rotate the shaft. A better test is to remove the inspection cover and view the slip rings and brushes with a flashlight to ensure they are operating correctly and are not worn.

If you find a problem with brushes or slip rings, you can remove and replace the brushes rather easily. Be careful, however, when you put the brushes back into the brush holder that you have the bevel on the face of the brush pointing in the correct direction so that it makes perfect contact with the slip rings. Also be sure to check the brush after you have returned the spring to its proper place to ensure there is proper tension on the brush and that it can freely move as it rides on the slip rings. It is important that as much of the brush face makes contact with as much of the slip ring as possible. Be sure to check for misalignment. After you have checked the brushes and slip rings and are sure they are operating correctly, you should test the field circuit again for continuity. If you still do not have continuity in this circuit, you should suspect that the rotor has an open in its winding. You can test the winding for conti-

nuity by putting the ohmmeter on the rotor conductors at a point that is beyond the slip rings and brushes.

If your test indicates that you have field current, the next test is to be sure to test each stator terminal for voltage. Depending how the alternator windings are connected, you can do this with a voltage test from terminal to terminal or from terminal to ground. If you do not have voltage at the stator terminal and you do have field current, you should suspect that you have one or more open windings in the stator. You can test the stator windings for continuity by disconnecting any load wires from the stator terminals and testing between the terminals for resistance. The amount of resistance is not important at this test. You are just trying to determine that the windings do not have an open. Be sure that the rotor does not turn during this test so that it does not produce voltage that would make the ohmmeter readings incorrect. If you determine any of the windings have an open, you will need to send the alternator out to a rewind shop or remove the alternator and replace it with a new one.

Another problem you may encounter with an alternator is that it will produce some amount of voltage but not the proper amount. When this occurs, you can start by trying to increase the field current to the amount listed on the data plate as the maximum field current. You can use an ammeter to verify the amount of current flowing through the field. If the amount of field current will not reach the maximum, you should suspect that the slip rings are dirty or that the brushes are damaged. If the field current is correct, you can test the amount of voltage at each terminal on the stator next. Since the stator is made of multiple windings, you may find that you are getting the correct amount of voltage at one or two of the windings and have low voltage or no voltage at the other winding.

QUESTIONS

Short Answer

1. Identify the basic parts of an AC alternator.

2. Explain the operation of the AC alternator.

3. Explain how voltage is directed into the rotating part of the machine.

4. Explain how you would get more voltage from the alternator.

5. Explain what happens when the speed of the rotor in an alternator is increased.

True or False

1. The alternator uses slip rings and brushes to get voltage from the rotor.

2. The alternator will produce more voltage when the field current is decreased.

3. The speed of the alternator will affect the frequency of the AC voltage it produces.

4. If you change the direction of rotation of the alternator shaft, you will change the polarity of the AC voltage the alternator produces.

5. If the field of the alternator has an open, the alternator will not produce any voltage.

Multiple Choice

1. The moving part of the alternator is called the _____.
   a. stator
   b. rotor
   c. field
   d. slip rings

2. The alternator uses _____ to get voltage from its rotating part.
   a. brushes and commutator segments
   b. brushes and stator
   c. brushes and slip rings
   d. all of the above

3. A three phase alternator _____.
   a. has three individual stator windings that each produces a sine wave that is 120° out of phase with each other
   b. has three separate rotors that spin simultaneously and produce sine waves that are 120° out of phase with each other
   c. has three separate shafts for the pulley turn
   d. all of the above

4. A voltage is produced in the alternator when _____.
   a. the windings in the rotor make contact with the slip rings
   b. the windings in the rotor make contact with the brushes
   c. the windings in the rotor pass through the magnetic flux lines
   d. all of the above

5. If you encounter an alternator that does not work, the first thing you should do is _____.
   a. test the field for voltage
   b. check for continuity in the field winding
   c. check for continuity in the rotor
   d. all of the above

CHAPTER 18

Programmable Controllers

OBJECTIVES

After reading this chapter you will be able to:

1. Describe the four basic parts of any programmable controller

2. Explain the four things that occur when the PLC processor scans its program.

3. Explain the function of an input module and describe the circuitry used to complete this function.

4. Explain the function of an output module and describe the circuitry used to complete this function.

5. Explain the function of an internal control relay.

6. Explain the terms *power flow* and *transition* in regard to contacts in a PLC program.

7. Describe what happens when a PLC is in run mode.

8. Identify the input instruction and output instruction for a PLC.

9. Explain the operation of a PLC on-delay timer.

10. Explain the operation of a PLC up-counter and down-counter.

18.0

INTRODUCTION

The programmable controller is the most powerful change to occur in factory automation in the past twenty years. The programmable controller (P/C) is also called the programmable logic controller (PLC). Since the personal computer is also called a PC, the programmable controller is now usually referred to as a PLC to prevent confusion. As a maintenance technician, you will run into the PLC in a number of places, such as on the factory floor. In this chapter you are going to study PLCs from two different perspectives. First, you will see how the PLC is programmed to perform logic functions that control machines much the same way that motor controls do. Second, you will see how simple it is to troubleshoot large machine controls and automation with a programmable controller. You will see how to use the status indicators on PLC modules to help you troubleshoot systems. Finally, you will learn to wire switches to PLC inputs and motor-starter coils to PLC outputs.

Today in industry you may find PLCs as stand-alone controls or as part of a complex computer-integrated manufacturing (CIM) system. In these large integrated manufacturing systems the PLC will control individual machines or groups of machines. They may also provide the interface between machines and robots or between machines and color graphics systems, called man-machine interfaces (MMI) or human machine interfaces (HMI).

In previous years companies were able to hire both an electrician and an electronics technician to install, interface, program, and repair PLCs. Now the number of PLCs has grown so large that many companies have found that it is better to hire one individual as a maintenance technician to make minor programming changes and troubleshoot. If you have never heard of PLCs or have had only a minor introduction to them, this chapter will provide you with all the information necessary to work successfully with them. Several typical types of PLCs are shown in Fig. 18–1. This chapter will use generic PLC addresses wherever possible. When specific applications are provided, the Allen-Bradley MicroLogix, PLC5, and SLC500 will be used as example systems. The early examples in this chapter will not be specific to any brand of PLC so that you may understand the functions that are generic to all controllers.

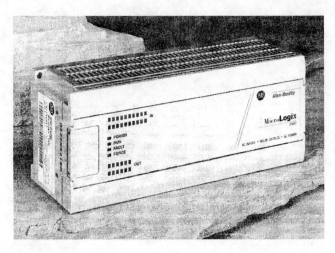

Figure 18–1 Typical programmable controller. (Courtesy of Rockwell Automation's Allen-Bradley Business)

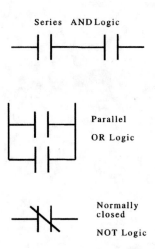

Figure 18–2 Logic for AND, OR, and NOT functions.

18.1

THE GENERIC PROGRAMMABLE LOGIC CONTROLLER

The programmable logic controller is a computer that is designed to connect input switches such as start and stop push buttons to outputs such as solenoids and motor starters. When the switches open and close, they control the on-off action of the output just as in a hardwired circuit where the switches and motor starter are connected directly to each other. The PLC allows the user to determine how the switches should be connected to each other such as in series or in parallel. This is called *logic*, and when this circuit is put into the PLC, it is called a program. Figure 18–2 shows examples of AND, OR, and NOT logic. Switches that are connected in series are called AND logic in a PLC, and switches that are connected in parallel are called OR logic. If closed contacts are used, they are called NOT logic. At this point do not let the term *logic* confuse you. *Logic* is the word that describes the action that switches and contacts will perform when they are wired in series or parallel, and *program* is the name we call the circuit when it is entered into the PLC to be controlled.

The PLC is designed to have industrial-type switches, such as 110-V push-button and 110-V limit switches, connected to it through an optically isolated interface called an input module. The PLC can also control 480-V three-phase motors through 120-V motor-starter coils with a similar interface called an output module. The computer part of the PLC, called a central processing unit (CPU), allows a program to be entered into its memory that will represent the logic functions. The program in the PLC is not a normal computer programming language like BASIC. Instead, the PLC program uses contact and coil symbols to indicate which switches should control which outputs. These symbols look very similar to the typical relay ladder diagram that was discussed earlier in this text. You can view the program through a hand-held terminal, or the program can be displayed on a computer screen where the actions of switches and outputs are animated as they turn motor-starter coils and other outputs on or off. The animation includes highlighting the input and output symbols in the program when they are energized.

In short, the PLC allows the logic program that is stored inside it to connect switches and output such as push buttons and motor starters just as you would if you were connecting a copper wire between the components. This allows the start and stop push buttons in the circuit to execute machine control functions such as starting and stopping the motor just like the hardwired circuits you used in earlier chapters.

An additional function the PLC provides is simpler troubleshooting of the switches and outputs that are connected to it. Each switch is connected to its own input circuit at the input module, and the input module provides a status light for each circuit. If power is passing through the switch to its individual input circuit, the status light for that circuit will be lighted. This allows you to determine if the switch contacts are open or closed and if the switch is operating correctly without using a voltmeter. The output module also provides status indicators for each output circuit. When the PLC energizes an output such as the motor starter, you can check the output status indicator for that circuit to see that the PLC is sending power to that output device. This feature allows you, as a maintenance technician, to quickly determine the condition of each switch and output device on the PLC by looking at each circuit's status indicator.

The PLC was designed to provide the technology of programmable circuit control (logic) with the simplicity of an electrical diagram so that its inputs and outputs

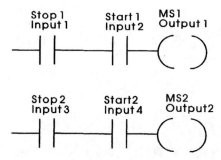

Figure 18–3 Example of a typical programmable controller program that is called ladder logic. This program uses a generic address for input 1, input 2, and output 1.

are easy to troubleshoot. Because the circuit logic is programmable, it is easily changed. Since each line of the program looks exactly like an electrical diagram, you can quickly understand the control function it is providing. As you know, it may take several hours to make a change in a hardwired circuit, such as adding a limit switch to a circuit. In a PLC, you will be able to make changes to the operation of a machine in a matter of minutes. The logic circuits can also be easily stored on a disk. This way they can be loaded into a PLC on the factory floor when the program logic for a controller needs to be changed as different parts are made or when the machine function needs to be changed.

Today, in most factories, it is important for an expensive machine to be able to make more than one part or different sizes or styles of the same part. This requires the machine to have a different set of switches and different control circuits. The PLC is designed to make these changes as simple as possible.

When PLCs were first introduced, they allowed the factory electrician to use them to help troubleshoot large automated systems. PLCs have evolved and now provide many of the more complex circuit control (logic) functions, such as timing, up/down counting, shift registers, and first-in, first-out (FIFO) functions.

Figure 18–3 shows an example of a PLC program. The program looks exactly like an electrical relay ladder diagram that you use when you connect components directly to each other. As the program (diagram) gets larger, additional lines will be added. This gives the program the appearance of a wooden ladder that has multiple rungs. For this reason the program is called a *ladder diagram*, or it may be referred to as *ladder logic*.

The original function of the PLC was to provide a substitute for the large number of electromechanical relays that were used in industrial control circuits. Early automation used large numbers of relays, which were difficult to troubleshoot. The operation of the relays was slow, and they were expensive to rewire when changes were needed in the control circuit. In 1969 the

automotive industry designed a specification for a reprogrammable electronic controller that could replace the relays while providing the same control function. This original PLC was actually a sequencer device that executed each line of the ladder diagram in a precise sequence. Later in the 1970s the PLC evolved into the microprocessor controller used today. Several major companies, such as Allen-Bradley, Texas Instruments, and Modicon, produced the earliest versions. These early PLCs allowed a program to be written and stored in memory; and when changes were required to the control circuit, changes were made to the program rather than changing the electrical wiring. It also became feasible for the first time to design one machine to do multiple tasks by simply changing the program in its controller. Today, the major controller companies are Allen-Bradley, Modicon, GE, Siemens (formerly Texas Instruments), Omron, and Square D.

18.1.1
Basic Parts of a Simple Programmable Controller

All PLCs, regardless of brand name, have four basic major parts: the *power supply*, the *processor*, *input modules*, and *output modules*. A fifth part, a programming device, is not considered a basic part of the PLC, since the program can be written on a laptop computer and downloaded to the PLC or the program can be written to an EPROM (erasable programmable read-only memory) chip. Figure 18–4 shows a block diagram of the typical PLC. You can see that input devices such as push-button switches are wired directly to the input module, and the coil of the motor starter or solenoids are connected directly to the output module. Before the PLC was invented, each switch would be directly connected (hard-wired) to the coil of the motor starter or solenoid it was controlling. In the PLC, the switches and output devices are connected to the modules, and the program in the PLC will determine which switch will control which output. In this way, the physical electrical wiring needs to be connected only one time during the installation process, and the control circuitry can be changed unlimited times through simple changes to the ladder logic program.

18.1.2
The Programming Device

At the bottom of the diagram in Fig. 18–4, you can also see a programming device. A programming device is necessary to program the PLC, but it is not considered one of the parts of a PLC because it can be disconnected

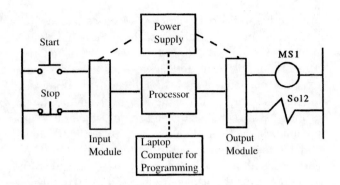

Figure 18-4 Block diagram showing the four major parts of a programmable controller. The programming panel is the fifth part of the system, but it is not considered a basic part of the PLC because it can be disconnected when it is not needed.

after the program is loaded and the PLC will run by itself. The programming device is used by technicians to make changes to the program, to troubleshoot the inputs and outputs by viewing the status indicators for the inputs and outputs to see if they are energized or deenergized, and to save programs to disk or load programs from disk.

The programming device in smaller PLCs is usually a hand-held programmer, and in larger systems it is a laptop computer with PLC programming software loaded on it. The ladder logic program is able to be displayed on the programming device just like a wiring diagram, and it can become animated. Animation shows whether an input switch is on or off and indicates if the switch is passing power to the switch located next to it. This feature is unique to the PLC, and it helps the technician troubleshoot very large logic circuits that control complex equipment by watching each circuit line by line. When a switch that is connected to the PLC input module is turned on, the program display can show this on the screen by highlighting the switch symbol everywhere it shows up in the program. The display can also highlight each output when it becomes energized by the ladder logic program. When the on-off state of a switch on the display is compared with the electrical state of the switch in the real world, you can quickly decide where the problem exists.

Another feature that makes the PLC so desirable is that it has a wide variety of logic functions. This means that functions such as timing and counting can be executed by the PLC rather than using electromechanical or electronic timers and counters. The advantage to using timer and counter functions in the PLC is that they do not involve any electronic parts that could fail, and their preset times and counts are easily altered through changes in the program. Additional PLC instructions provide a complete set of mathematical functions as well as manipulation of variables in memory files. Mod-

ern PLCs such as the Allen-Bradley PLC-5 have the computing power of many small mainframe computers as well as the ability to control several hundred inputs and outputs.

18.2

UNDERSTANDING HOW THE PLC CONTROLS A SIMPLE CIRCUIT WITH A SELECTOR SWITCH AND MOTOR-STARTER COIL

The simplest way to understand how the PLC operates is to see how it controls a simple circuit with a selector switch that energizes the motor-starter coil. We will compare the PLC-controlled circuit with the hardwired circuit with a selector switch that energizes the motor-starter coil. Figure 18-5a shows an example of a traditional circuit where the selector switch controls a motor-starter coil. Figure 18-5b shows how this same circuit is controlled by a PLC. Instead of the selector switch and the motor-starter coil being connected to each other, the selector switch is connected to the input module of the PLC, and the motor-starter coil is connected to the output module of the PLC. You should notice that each input circuit on the PLC input and output module has an indicator lamp called the status indicator. The selector switch is connected to 110-VAC voltage; and when the switch is closed and passing 110 V to its input circuit on the PLC module, its status light will illuminate. This feature allows you to troubleshoot each switch without using a voltmeter. You can simply look at the status indicator and open and close each switch and determine if they are working correctly. If the status indicator illuminates for a switch, you know that it is working correctly. Each input circuit also has an address number that the PLC recognizes and uses in its program. The selector switch is connected to address I1. In this diagram you can also see that the motor-starter coil is connected to the output module at address O1. The output module also has a status indicator for each circuit to indicate when the circuit is energized.

18.2.1
The Scan Cycle

The third part of the PLC to understand at this time is the scan cycle. A diagram of the PLC scan cycle is shown in Figure 18-5c. The scan cycle consists of three distinct functions: the *read cycle*, which allows the PLC to check the status of every input; the *solve logic cycle*, where the PLC determines if each line of logic in its

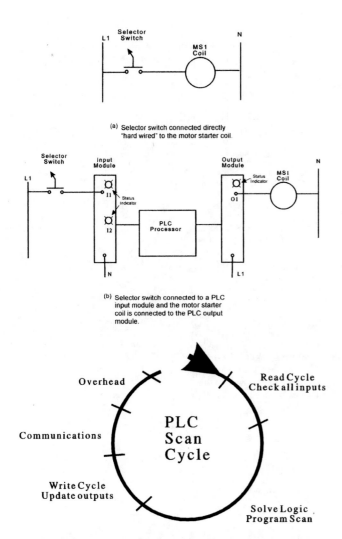

(a) Selector switch connected directly "hard wired" to the motor starter coil.

(b) Selector switch connected to a PLC input module and the motor starter coil is connected to the PLC output module.

Figure 18–5   (a) Ladder diagram of a hardwired selector switch connected to a motor-starter coil; (b) hardware diagram of a selector switch wired to a PLC input module and a motor-starter coil connected to PLC output module; (c) scan cycle for a PLC.

cycle and the *overhead cycle.* During the communications part of the scan cycle, the PLC sends information to and receives information from its programming device such as the hand-held programmer or laptop computer. A copy of the image register is sent to the hand-held programmer or laptop during the communications part of the scan so that you can use the hand-held to view the status of each line of the program and the condition of each input or output for troubleshooting purposes. The final part of the scan cycle is the overhead part of the cycle, which the PLC uses for internal purposes.

The PLC executes a scan cycle about every 10–20 ms. This means that every 10/1,000 of a second the PLC checks the status of every switch to see if it is on or off and then writes the condition of each line of logic (each circuit in the program) to the outputs in the image register to tell each circuit in the output module whether it should turn on or off. The speed of the scan cycle slows down as more lines of logic are added to the program.

If an input is in the "on state" and is providing voltage to the input circuit that it is connected to, two things occur. First the status indicator for that input is illuminated, and second the PLC places a 1 in the image register in the address for that input. When the switch changes state to its off state, the status indicator turns off and the PLC places a 0 in the address for the input to tell the PLC that input is now turned off. Since the PLC scan cycle is so quick, it will pick up any change in the condition of any input switch instantaneously. It will also send a signal to the output connected in the program to the switch within the time of the next scan cycle so it looks like the switch is controlling the output instantaneously.

18.2.2
The Operation of the Program in the PLC

When the PLC is set to the "run mode," it begins to execute the scan cycle, and it checks all of its inputs, solves logic, and writes a 1 or 0 to all its outputs. Figure 18–6a shows a diagram of the selector switch connected to the input and the motor-starter coil connected to the output module. Figure 18–6b shows the program for the circuit with a selector switch that energizes the motor-starter coil that is controlling a motor starter. From this figure you can see the program. It has a set of normally open contacts that have the PLC address I1 and the name "selector switch." The output is identified as O1, and it has the name "MS1 Coil."

The PLC program cannot use standard electrical symbols for different types of switches such as limit switches, push buttons, or selector switches. Instead the PLC program uses only the symbols for normally open or normally closed contacts to represent all the switches in the

program should be energized or de-energized; and the *write cycle,* where the PLC writes the condition, on or off, to every output in its image table. If the condition of the switches in a line of logic in the PLC program indicates the output should be energized, the PLC writes a "1" in the image register for that output address. If the PLC determines the output should be de-energized the PLC writes a "0" in the image register for that output address. This means that the PLC writes a 1 or a 0 in every output address in the output image register during the write-cycle part of the PLC scan cycle. When a 1 is placed in the image register address for an output such as O1 where the motor-starter coil is connected, the electronic circuit in the output module for that address is energized and 110-VAC is sent to the motor-starter coil. The diagram of the scan cycle shows two additional segments of the scan cycle: the *communications*

Figure 18–6 (a) The selector switch in its normally open position connected to the PLC input module and the motor-starter coil connected to the PLC output module; (b) PLC diagram of a selector switch controlling motor-starter coil; (c) image table in the PLC for the selector switch and the motor-starter coil.

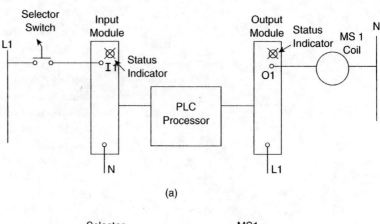

(a)

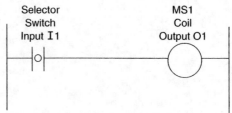

O Signifies that contacts have not (changed state) "transitioned" and are still open in the program and they are not passing power to the next part of the circuit so the motor starter coil remains in the off state.

(b)

Image Register

Select SW	I1	0
MS1	O1	0

Selector switch OFF
and MS1 coil is OFF.

(c)

program. Even though we know the selector switch has a standard electrical symbol that is shown in the hardware diagram in Fig. 18–6a, when it is used in the PLC program as in figure 18–6b it is always shown as a set of contacts –][— The PLC only uses the symbol –][— for normally open switches and the symbol ––]\[— for normally closed switches. The coil will use the the symbol ( ) for all of its outputs in its program.

18.2.3
The Image Register

Another diagram that is needed to understand how the PLC works is the diagram of the *image register*, which is also called the *image table* in some PLCs. Figure 18–6c shows an example of an image register with just

the address for each input and output shown. This example is only showing one input and one output. If the PLC has twenty inputs and ten outputs, the image table will show all the addresses. The column on the right side of the image table provides a location to place a 1 or 0 to indicate the condition or status of the input or output. The PLC uses the image register to store or keep track of the status of each of the inputs and outputs in regard to whether they are on or off. The PLC assigns an "address" that contains a letter of the alphabet and a number. For example, all inputs have the letter I, and all outputs have the letter O so the PLC can tell the input addresses from the output addresses. Some PLCs start the first number for inputs and outputs with zero (0) and continue through the maximum number of inputs and outputs provided. In our examples we will use inputs I1 and I2 and output O1. The stop push-button is connected to the input address I1, and the start push button is connected to input I2. The motor-starter coil is connected to output O1. You can see in Fig. 18–6c that these names have been added to the image table. The PLC will place a 1 or 0 in the address for these inputs at all times to indicate the current state of each component. The PLC does not know the names of the switches and outputs that are connected to the addresses; it only knows the addresses.

Figure 18–6c shows a copy of the image register for the selector switch and motor-starter coil in the PLC. You can see the selector switch is assigned address I1, and the output has address O1 for the motor-starter coil. You should remember that the selector switch is wired normally open in the hardware section of the PLC where it is connected to the input module as input I1. The image register for this point in time is shown in Fig. 18–6c where input I1 (the selector switch) is open and the status light on the input module for this input is not illuminated. The PLC has placed a value of 0 in the image register address for address I1. When the PLC solves the logic for this line of the program, it shows the selector switch is off, so the output is not energized and the PLC will place a 0 in the O1 address of the image register.

There are two additional concepts you will need to review in order to fully understand the operation of the PLC. One is transitioning contacts, and the other, is power flow.

18.2.4
Transitioning Contacts

The first concept, transitioning contacts, deals with identifying when contacts *change state* or *transition*. The contacts in the program of a PLC will change state any time the hardware connected to that input address changes state. For example in Fig. 18–7a you can see that the selector switch connected to the input module address I1 is in the closed position. When this occurs, 110 VAC is sent to the PLC input module address for input I1. The PLC recognizes this during its read cycle and places a 1 in the image register for address I1, Figure 18–7b shows the program for the selector switch controlling the motor-starter coil, and Fig. 18–7c shows the image table for this program. When the 1 is placed in the image register, every set of contacts with the I1 address on them will "change state," or "transition." This means that contacts that are programmed as normally open will change state and act in the program as closed contacts. Any contacts that are programmed normally closed will change state and act in the program as open contacts. You may hear this referred to as *transitioning the contacts*. One troubleshooting

method includes drawing, on a separate sheet of paper, the contacts and outputs for the line of logic you are working on just as they are programmed into the PLC. You could also use a printout for the line of logic that is affected. This paper copy of the program line of logic is called the *working copy* of the program. You can identify that a set of contacts in a program has *changed state* or *transitioned* from open to closed by placing a black dot between the contacts in the working copy of the program. If the contacts are shown in a program as normally closed and they transition, we place an open circle to indicate the contacts have changed state and are now in the open state.

18.2.5
Power Flow

The second concept is *power flow*. We will again refer to Fig. 18–7a, in which the selector switch is in the closed position and the status indicator for input 1 is illuminated. We will also refer to Fig. 18–7b again, where the diagram shows a line of a PLC program where the contacts with address I1 (selector switch) are shown located against the power rail on the left side of the program. When this first set of contacts in the program is transitioned to the closed position, the contacts will *pass power* to output O1, which is the next element in the program that is located to its right. In the previous discussion we placed a black dot in the I1 contacts on the working copy of the program to indicate the contacts have changed state.

An arrow pointing to the right is then added under the contacts to indicate that power is flowing through the contacts to the next component that is located to the right of the contacts. In this line of the program, the next component located to the right is output O1, which is the motor-starter coil. Since this component is receiving power, you would put a black dot inside the output instruction to indicate it is receiving power and it is transitioned (changed state) to be energized in the on state. Remember when you are troubleshooting a PLC system, you will need to develop a working diagram and add *black dots* to indicate any place where contacts have transitioned, and an arrow should be added below the contacts to indicate power flow to the next component. When power flow reaches the output as in the diagram with the motor-starter coil, you can switch your troubleshooting to the output module and the component that is connected to it, because all the switches in the circuit have done their job in passing power to the output.

You should understand that when you are trying to troubleshoot an electrical system that is controlled by a PLC, you will have to check each line in the program that is affected by the problem and determine which

Figure 18–7   (a) Status lights shown illuminated on the input module when the selector switch is in the closed position. You can also see that the status light for the output module address for the motor-starter coil is also illuminated; (b) PLC program with black dots showing contacts transitioning and an arrow showing power flow through the contacts; (c) image table for the PLC program when the selector switch is closed. Input I1 is now a "1" and out O1 is also a "1."

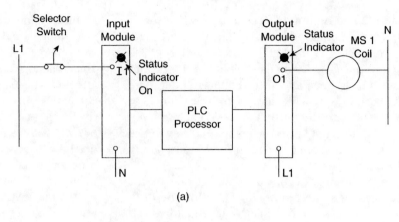

(a)

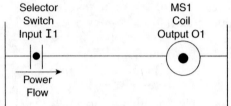

Power
Flow

● Signifies that contacts have "transitioned" and are now closed in the program and they are passing power to the next part of the circuit. The dot in the output signifies it is receiving power and the output is energized.

──→ The arrow signifies that the contact is now passing power to the next component in the circuit, so the coil is receiving power and it is energized.

(b)

Image
Register

Select SW	I1	1
MS1	O1	1

Selector switch ON
and MS1 coil is ON.

(c)

inputs and which outputs are energized. If an output is not energized, you would look to the contacts in the program that are located to the left of the output in that line and determine which ones are in the on state and passing power to the components to the left. Some lines of the program will be more complex and have several sets of normally open or normally closed contacts that you need to verify to see where the power flow stops when an output is not energized. It is important to learn how to make the marks on the working copy of the program so you can determine how the machine logic for that line of the program is supposed to work.

18.3

USING CHANGE OF STATE AND POWER FLOW TO UNDERSTAND THE STOP-START CIRCUIT

Now that you have seen contacts transition and power flow demonstrated in a simple circuit with a selector switch and motor-starter coil, you are ready to see the same concepts in a more complex stop-start circuit that controls a motor-starter coil. Figure 18–8a shows the hardware diagram of a stop-start circuit that is control-

Figure 18–8 (a) A stop-and-start switch is connected in a hardwired circuit where each component is connected directly to each other; (b) a stop-and-start switch is connected to a PLC input module, and a motor-starter coil is connected to an output module; (c) the image table for the PLC-controlled stop-start circuit. The stop push button (I1) is on so a "1" is written in its address in the image register. The start push button (I2) and the motor starter coil (O1) are both off so an "O" is written in their addresses in the image register.

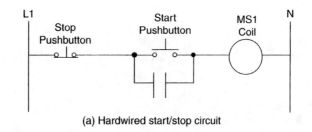

(a) Hardwired start/stop circuit

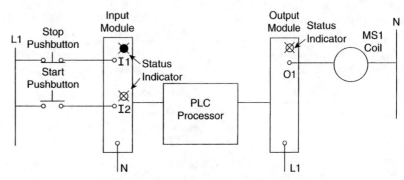

(b) Start and stop push buttons connected to PLC input module and motor starter coil connected to PLC output module.

Image
Register

Stop PB	I1	1
Start PB	I2	0
MS1	01	0

(c) Image register for stop, start push buttons and motor starter coil with the stop push button normally closed (on) and start push button and motor start coil are off.

ling a motor-starter coil. This is a motor control circuit that you have seen previously in earlier chapters. The stop push button is wired as a normally closed switch, and the start push button is wired as a normally open push button. A set of normally open auxiliary contacts from the motor starter is wired in parallel around the start push button. When the start push button is depressed, the voltage flows from the power source at line 1, through the normally closed stop push button, through the closed contacts of the normally open push button, and on to the coil. When the coil becomes energized, closes its normally open auxiliary contacts that seal in the start push button to provide a path for current even after the start push button is released and returns to its normally open state. The coil remains energized until the stop push button is depressed.

Figure 18–8b shows the stop-start circuit controlled by a programmable controller (PLC). In this diagram

you can see that the normally closed stop push button is connected to the PLC input module at address I1. Since the stop push button is wired normally closed, the status light for the input at address I1 is illuminated. The normally open start push button is wired to the input module at address I2. Since this switch is normally open, the status light for address I2 is not on.

Figure 18–8c shows the image register for this PLC, and you can see that the stop push button address I1 has a 1 in its image register location because the normally closed stop push button is passing power to the input address at I1. The image register address for I2 has a 0 in its location because the normally open push button is not depressed and its input module address is not receiving power. During the "solve logic" part of the PLC scan cycle, the logic indicates that the output should not be turned on. During the "write" part of the PLC scan cycle, the output image register receives a 0

Figure 18–9 (a) Example of a working diagram of the PLC program of the stop-start circuit. The black dot indicates contacts that have transitioned, and the arrow shows where contacts have passed power; (b) a working diagram of the PLC stop-start circuit with the start push button depressed. The black dot shows where contacts have transitioned, and the arrows indicate power flow to the next component; (c) the image table for the PLC-controlled stop-start circuit.

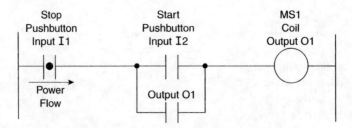

● Signifies that contacts have "transitioned" and are now closed in the program and they are passing power to the next part of the circuit.

⟶ The arrow signifies that the contact is now passing power to the next component in the circuit.

(a)

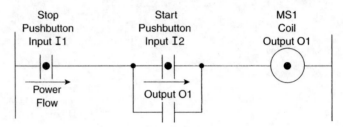

● Signifies that contacts have "transitioned" and are now closed in the program and they are passing power to the next part of the circuit.

● Signifies that the coil is receiving power and is energized.

⟶ The arrow signifies that the contact is now passing power to the next component in the circuit.

(b)

Image Register

Stop PB	I1	1
Start PB	I2	1
MS1	01	1

(c)

for the output address O1 which controls the motor-starter coil, so the motor-starter coil is not energized.

18.4
THE PLC PROGRAM FOR THE STOP-START CIRCUIT

Figure 18–9a shows the program for the stop-start circuit in the PLC. In this diagram you can see that the I1 contacts for the stop push button are programmed normally open. This is not a mistake; we will explain the reason why at the end of this section. For now just review the status of the circuit and it will make sense. You can see that a black dot has been added between the normally open contacts for the stop push button, which indicates the push button is normally closed in the hardwired portion of the system and it is passing power to the input module address at I1. An arrow has been added to indicate that power flow is occurring through the I1 contacts and passing power to the normally open I2 contacts. Since the start push button has not been depressed yet, this is as far as the power flow can pass power.

Figure 18–9b shows the program at the point where the start push button has been depressed. In this diagram of the program you can see that a black dot has been added to the I2 contacts because the start push button has been depressed and is passing power to the I2 input. A second arrow has been added to the program now under the I2 contacts, indicating that power is now flowing through the I2 contacts to the motor-starter coil, which is output address O1. You should notice that a black dot has also been added to the output coil O1 to indicate it is receiving power and has transitioned and is now energizing the output circuit in the output module so that 110 VAC is now being passed to the motor-starter solenoid. If you were to view the status indicator for output O1 on the output module, you would notice that it is illuminated because the output at O1 is now energized by the PLC program.

As soon as the person operating this circuit removes his or her finger from depressing the start push button, the start push button would return to its normally open state, and its I1 contacts in the program would return to their open state. The output would remain on because power in the PLC program would continue to flow through the motor-starter auxiliary contacts (O1) that are programmed in parallel around the start push button's I1 contacts. This would function much like the hardwired circuit where the auxiliary contacts from the motor starter are connected around the start push button. In the PLC circuit, the auxiliary contacts are programmed in parallel around the start contacts rather than use hardwired contacts.

18.4.1
The Image Register for the PLC-Controlled Stop-Start Circuit

Figure 18–9c shows the image register for this circuit when the stop button I1 is closed, the start button is being depressed, and contacts I2 are closed and the output O1 is energized. You can see that the image register address 11 for the stop push button is showing a 1 because the stop push button is normally closed. The I2 address is also showing a 1 because the start push button is being depressed, and you should also notice that the output O1 is showing a 1 because the output is energized.

Any time the stop push button is depressed, the normally open stop push-button contacts will open, and a 0 would be written into the image register for address I1. This would cause the I1 contacts in the program to transition back to the open condition, it would interrupt power flow to the circuit, the output O1 would become de-energized, the auxiliary O1 contacts that are in parallel with the start push button I1 contacts would also open, and the circuit would return to its off state. At this point, the O1 address in the image register would return to a 0 to indicate it is off.

18.4.2
Why the Stop Push Button Is Wired Normally Closed and Programmed Normally Open

In the previous example, you can see that the stop push button is wired normally closed when it is connected to address I1 on the input module in the hardwired portion of the circuit and it is programmed normally open in the PLC program. You know that the black dots were added to show when the program contacts transition and the arrows were added to show that power flow through the circuit works correctly with the stop push button hardwired normally closed and programmed normally open.

The other way this circuit could operate is by wiring the stop push button normally open in the hardwired section where it is connected to the input module and then programming it as a set of normally closed contacts when it is in the program. If you analyze the circuit with the stop push button wired normally open and programmed normally closed, you would see it would work the same as in the previous example, where it is wired normally closed and programmed normally open. The problem is created when a wire comes loose on the normally open stop push button. If this occurred, the operator would depress the stop push button to stop the machine, but no power would be sent to address I1 on the input module, and the machine could not be stopped. This would create an unsafe condition. You should always examine the hardware and program conditions to be sure that the machine will always operate in the "fail-safe" condition. In the case of the stop-start circuit, the fail-safe condition is to have the machine stop if anything goes wrong in the circuit. You can see that when the stop push button is hardwired to the input module as a normally closed switch, and it is programmed as a normally open set of contacts, the output would become de-energized if a wire came loose on the stop push-button switch or if the switch malfunctioned.

SAFETY NOTICE!
You will encounter many occasions when you are working with PLCs that you have several different ways to connect hardwired switches as normally open or normally closed to input modules, and you will have the option of programming the contacts as either normally open or normally closed. You will find that several combinations will actually perform the same logic function for the circuit. You must always review the circuit to ensure it acts the way you expect it to when it has a fault and goes to a *fail-safe* condition. Always be sure you can determine the machine action when the *fail-safe* condition occurs so that the machine will not cause damage to persons or the equipment.

18.5

AN EXAMPLE PROGRAMMABLE CONTROLLER APPLICATION

It may be easier to understand the operation of a PLC when you see it used in a simple application of sorting boxes according to height as they move along a conveyor. The address numbering schemes for inputs and outputs of some PLCs may be difficult to understand when you are first exposed to them. For this reason, the first examples of the PLC presented in this section will use generic numbering that does not represent any particular brand name. This will allow the basic functions of the PLC that are used by all brand names of PLCs to be examined in a simplified manner. Later sections of this chapter go into specific numbering systems for several major brands of PLCs.

Figure 18–10 shows a top view of the conveyor sorting system. The location of the photoelectric switches that detect the boxes and the pneumatic cylinders used to push the boxes off the conveyor are also shown.

Each air cylinder is controlled by a separate solenoid. The conveyor is powered by a three-phase electrical motor that is connected to a motor starter. Start and stop push buttons are used with the photoelectric switches to energize the coil of the motor starter, which turns the conveyor motor on and off. Figure 18–11 shows all the switches connected to an input module and solenoids and the motor starter connected to an output module. The inputs are generically numbered input 1, input 2, and so on. The outputs are numbered output 10, output 11, and so on. The ladder logic program that determines the operation of the system is shown in Fig. 18–12. In the diagram, the standard electrical symbol is used for each switch and output device.

In the ladder diagram you can see that each switch is represented as a set of open contacts or closed contacts (-] [-or-]\[-). Each motor starter or solenoid in the hardware diagram is represented by an output symbol -( )-. The same numbers that are assigned to the switches, motor starters, and solenoids in the hardware diagram are used in the logic diagram so that you can determine which switch is being used.

Figure 18–10 Diagram of the overhead view of a conveyor sorting system that is controlled by a generic programmable controller.

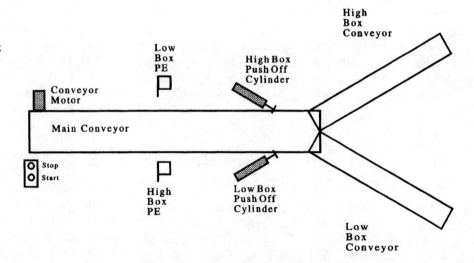

Figure 18–11 Input and output diagram that shows (left) switches that are connected to the PLC input module and (right) the coil of a motor starter connected to the PLC output module.

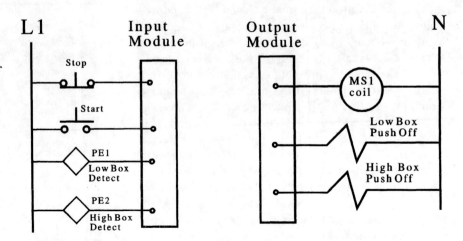

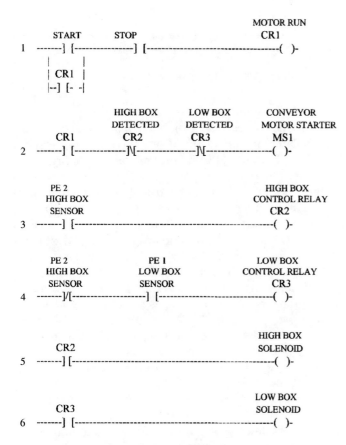

```
                                                  MOTOR RUN
        START          STOP                          CR1
1  -------] [---------------] [-------------------------------(  )-
        |            |
        | CR1 |
        |--] [- -|


                  HIGH BOX        LOW BOX         CONVEYOR
                  DETECTED        DETECTED      MOTOR STARTER
        CR1         CR2             CR3             MS1
2  -------] [--------------]\[---------------]\[----------------(  )-


        PE 2                                    HIGH BOX
      HIGH BOX                               CONTROL RELAY
      SENSOR                                     CR2
3  -------] [--------------------------------------------(  )-


        PE 2                PE 1               LOW BOX
      HIGH BOX            LOW BOX           CONTROL RELAY
      SENSOR             SENSOR                 CR3
4  -------]/[-------------------] [-----------------------(  )-


                                             HIGH BOX
         CR2                                  SOLENOID
5  -------] [--------------------------------------------(  )-


                                             LOW BOX
         CR3                                  SOLENOID
6  -------] [--------------------------------------------(  )-
```

Figure 18–12 Ladder logic diagram showing the logic control for the conveyor sorting system. The diagram is generic and does not represent any particular PLC brand. Notice that all switches are represented by the contact symbol and all outputs are represented by the coil symbol.

18.5.1
Operation of the Conveyor Sorting System

When the conveyor system is ready for operation, the start push button must be depressed. Since the stop push button is wired as a normally closed switch, power will pass through its contacts in the program to the start push-button contacts. When the start push button is depressed, the "motor-run" condition will be satisfied in the first rung of the program and the "motor-run" control relay in the program will be energized. (The PLC can have hundreds of control relays in its program like the motor-run relay. The advantage of having control relays in a PLC program is that it provides the logic functions of a control relay, which can have multiple sets of normally open and normally closed contacts; and yet they exist only inside the PLC, so they can never fail or break.) A set of normally open contacts from CR1 is connected in parallel with the start push button to seal it in because the start push button is only momentarily closed. The photoelectric switches are used to energize

and de-energize the conveyor motor starter in the second line of the logic. Line 2 of the diagram is designed so the start-stop circuit can turn the motor starter on or off without affecting the stop-start circuit. The photoelectric switches send out a beam of light that is focused on a reflector that is mounted on the opposite side of the conveyor. As long as a box does not interrupt the beam of light, the photoelectric switches will be energized.

When the motor-run control relay coil in the first rung of the program is energized, its normally open contacts in the second rung of the program will become "logically" closed. Since no boxes are being sensed by the photoelectric switches, they will be energized, and their contacts in the program—low box PE and high box PE—will pass power to the output in the second rung, called the motor starter, which directly controls the coil of motor starter 1. When motor starter 1 is energized, the conveyor motor is energized and the conveyor belt begins to move.

As boxes are placed on the conveyor, they will travel past the photoelectric switches. If the box is a low box, photoelectric switch PE1, which is mounted to detect the low box, will become energized. This will cause three things to happen. First, the normally open contacts of the PE1 low box sensor in rung 4 will close. Since the box is not a high box, the normally closed contacts of the PE2 high box sensor in rung 4 will remain closed to pass power to the PE1 low box sensor contacts (that are now closed) and on to the coil of CR3 (low box control relay). Second, the coil of CR3 is energized, and all the contacts in the program that are identified as CR3 will change state. This means the normally closed CR3 contacts in rung 2 will open and de-energize MS1, the conveyor motor starter. When MS1 is de-energized, the motor starter will become de-energized and the conveyor motor will stop. The third event to happen in this program when the coil of CR3 is energized by the PE1 low box sensor is that the output for the low box solenoid is energized, the rod of the low box push-off cylinder is extended, and the low box is pushed off the main conveyor onto the low box conveyor.

After the low box is pushed off the main conveyor, it will no longer activate PE1, and the coil of CR3 will become de-energized in rung 3. When the coil of CR3 is de-energized, the normally closed CR3 contacts in rung 2 will return to their closed state, which will energize the output for MS1 and start the conveyor motor again.

When a high box moves into position on the conveyor, it will block both the high box photoelectric switch PE2 and the low box photoelectric switch PE1. This could create a problem of having both the low box control relay CR3 and the high box control relay CR2 energized at the same time. The logic in rung 4 is specifically designed to prevent this. Since the contacts of the high box sensor PE2 are programmed normally closed in rung 4, they will open when the high box is detected and not allow power to pass to the coil of CR3. This type of logic is called an exclusion since it prevents both control relay coils from energizing at the same time even though both photoelectric switches are activated by the high box.

When the high box is detected by PE2, the normally closed PE2 contacts in rung 3 will close and energize the coil of the high box control relay CR2. When the coil of CR2 is energized, all the contacts in the program that are identified as CR2 will change state. The normally closed CR2 contacts in rung 2 will open and de-energize the coil of MS1 in the conveyor motor starter. The normally open CR2 contacts in rung 5 will close at this time and energize the high box solenoid, air will be directed to the high box pneumatic cylinder, and the high box will be pushed off the conveyor.

After the box has been pushed off the conveyor, the photoelectric switch that detected the high box will return to its normal state, high box contacts in rung 2 will return to their normally closed logic state, and the motor starter will become energized again and start the conveyor motor. The high box contacts in rung 3 will return to their normally open logic state, and the push-off cylinder will be retracted. This allows the conveyor to return to its normal operating condition.

18.6
THE RUN MODE AND THE PROGRAM MODE

When the PLC is in the *run mode*, it is continually executing its scan cycle. This means that it monitors its inputs, solves its logic, and updates its outputs. When the PLC is in the *program mode*, it does not execute its scan cycle. Perhaps a better name for the program mode is the "not-run mode" or the "no-scan mode." The term *program mode* was originally given to this mode because older PLCs could not have their programming changed while they were executing their scan cycle. Modern PLCs have the ability to have their program edited and changed while the processor is in the run mode. At first glance, changing the PLC program while it is in the run mode looks dangerous, especially when the machine the PLC is controlling is in automatic operation. But you will find in larger control applications, such as in factories where glass is being poured continually or in continuous steel rolling mills, it is not practical to stop the machine process to make minor program changes such as the changes to a timer. In these types of applications the changes are made while the PLC is in the run mode. As a rule, you should switch the PLC to the program mode if possible when any program changes are being made.

18.7
ON-LINE AND OFF-LINE PROGRAMMING

The programming software for a PLC may allow you to write the ladder logic program in a personal computer and later download the program from the personal computer to the PLC. If you are connected directly to a PLC and you are writing the ladder logic program in the PLC memory, it is called on-line programming. If you are writing the ladder logic program in a personal computer or programming panel that is not connected to a PLC, it is called off-line programming. Most modern PLC programming software allows the program to be written on a personal computer or laptop computer without being connected to the PLC. This allows new programs or program changes to be written at a location away from the PLC. For example, you may write a program in Detroit at an office, save the changes on a floppy disk, and mail the disk to a company in Chicago where the PLC is connected to automated machinery.

If you are using a hand-held programmer, it must be connected directly to a PLC so that you are writing the program in the PLC memory. Some small PLCs like the Allen-Bradley SLC100 do not have a microprocessor chip, so you cannot write the PLC program on-line to the processor with a laptop computer. Since the SLC100 does not use a microprocessor chip, all programming must be completed off-line in the laptop, and then you must download the program changes to the PLC memory. An alternative method would be to use a hand-held programmer to program the PLC on-line.

18.8
FEATURES OF THE PROGRAMMABLE CONTROLLER

The examples in Fig. 18–10 through 18–12 show several unique features of the PLC. First, the ladder diagram allows you to analyze the program in the sequence that it will occur. Second, the PLC allows you to program more than one set of contacts in the program for each switch. For example, the photoelectric switch may be in the program both normally open and normally closed and placed in the program as many times as the logic requires. The PLC gives you the capability of solving AND, OR, and NOT logic by connecting contacts in series or parallel and by using normally open and normally closed contacts as needed. Figure 18–13 shows examples of the five basic logic symbols: AND, OR, NOT, NAND, and NOR. This figure also shows the truth tables

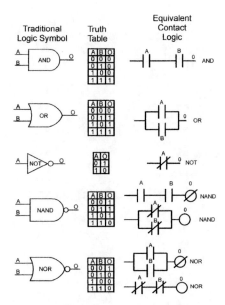

Figure 18–13 Traditional logic symbols for AND, OR, NOT, NAND, and NOR functions.

for these logic functions and the equivalent contact arrangements that provide the same function as you would see them in ladder logic. Notice that the NAND and NOR functions can be provided with either series contacts of parallel contacts. As complex as some ladder logic programs look, the logic program can consist only of combinations of these five basic functions.

Another feature of the PLC is the control relay or internal coils or memory relays. As stated in the discussion of image registers, the control relay acts like the control relays that were used in relay panels years ago. The relay has one coil, and it can have any number of normally open or normally closed contacts. When the coil is activated, the contacts in the program that have the same address number as the coil will change state. The control relay exists only inside the PLC, which means it can never break or malfunction physically. Also, since control relays exist only in the PLC memory, you can have as many of them as the memory size of the PLC can accommodate. This means you can add control relays to a PLC program to accommodate changes in the machine control logic as long as the PLC has space remaining in its memory.

It may help to see a case where a control relay is used in an actual circuit to better understand its function. For example, in Fig. 18–12 you can see that the low box detected control relay (CR3) is used to indicate that the logic has determined that a low box has been detected. The control relay in this rung of the program does nothing more than "represent" the condition that indicates that a low box has been detected. This condition exists when low box PE1 is on and the high box PE2 is off. Any time CR3 is energized, a low box has been detected, and it will change the state of all the contacts

that have its number throughout the remainder of the program, to show that the low box has been detected.

In many cases, each control relay in a program represents a condition for the machine operation, such as "ready to go to auto," "safety doors closed," or "ready to jog." If you think of the condition the control relay represents each time you see the control relay contacts, you will find it easier to read and interpret the program.

Another feature of the PLC that makes complex factory automation easier to troubleshoot is the way that each input switch or each output load is electrically connected to its interface module. In Fig. 18–11 you can see that when a switch is connected to an input module, only two wires are involved. The outputs also use only two wires to connect the load device like the coil of a motor starter to the module. This means that when there is a malfunction in the electrical part of the circuit, the troubleshooter has only the switch and two wires to check for any individual input and the coil and two wires for any output. When you compare this type of wiring with a complex series parallel circuit in an older-style relay panel, you can see it is much easier to troubleshoot two wires and a switch, or two wires and a solenoid. In the older-style relay panels, a typical signal may pass through 10 to 15 relay contacts and switches and all the wiring that connects them together before it gets to a solenoid. This means that an open could occur in this complex circuit when any set of contacts opens, when any switch opens, or if a wire has a fault where it becomes open.

To make troubleshooting the switch and two wires that connect it to an input module even easier, a small indicator light called a status indicator is connected to each input circuit, and it is mounted on the face of the module so it is easy to see. The troubleshooter can look at the status indicator and quickly determine if power is passing through the contacts or not. If the switch contacts are closed, but the status indicator is not illuminated, the troubleshooter needs to check only the two wires and the switch for that circuit to see where the problem is. This feature is particularly valuable when the automated system the PLC is controlling has several hundred switches that could cause a problem. A status indicator is also connected to each output circuit and mounted on the face of the output module. Any time the PLC energizes an output, the status indicator for that circuit will be illuminated.

18.9

CLASSIFICATIONS OF PROGRAMMABLE CONTROLLERS

PLCs are generally classified by size. A small system costs around $500 to $1,000 and will have a limited number of inputs and outputs. As a general rule, the small

PLCs have less than 100 inputs and outputs with approximately 12 inputs and 8 outputs mounted locally with the processor. In some systems additional inputs and outputs can be added through remote I/O racks to accommodate the remaining inputs and outputs. In other systems, the number of inputs and outputs is limited, but several small PLCs can be networked together to provide a larger number of I/Os that are controlled together. The smaller PLCs generally have 2 to 10 K of memory that can be used to store the user's logic program.

Medium-size PLCs cost between $1,000 and $3,000 and have extended instruction sets that include math functions, file functions, and Pill process control. These PLCs are capable of having up to 4,000 to 8,000 inputs and outputs. They also support a wide variety of specialty modules, such as ASCII communication modules, BASIC programming modules, 16-bit multiplexing modules, analog input and output modules that allow interfaces to both analog voltages and currents, and communication modules or ports that allow the PLC to be connected to a local area network (LAN).

Large PLCs were very popular before networks were perfected. The concept for the large PLC was to provide enough user memory and input and output modules to control a complete factory. Problems occurred when minor failures in the system brought the complete factory to a halt.

The advent of local area networks brought about the concept known as distributive control, where small and medium-size PLCs are connected together through the network. In this way, the entire factory is brought under the control of a number of PLCs, but a failure in one system will not disturb any other system. The majority of PLCs that you will encounter today will be small and medium-size systems that are used as stand-alone controls or as part of a network.

18.10

REVIEW THE OPERATION OF A PROGRAMMABLE CONTROLLERS

Now that you have a better understanding of how the basic PLC operates, you are ready to understand some of the more complex operations. You may need to review some of the previous diagrams to help you get a better understanding of these functions. In Fig. 18–14 you can see that the PLC has a processor, input module, output module, and a power supply. The processor has several sections, including resident memory that contains the instruction set, user memory reserved for programs, and addressable data memory that has room to store a timer, counter, and other variables. The variables may be stored as binary data, as an integer (whole

Inputs	Outputs
I0	O0
I1	O1
I2	O2
I3	O3
I4	O4
I5	O5
I6	O6
I7	O7
I8	O8
I9	O9
I10	O10
I11	O11
I12	
I13	
I14	
I15	
I16	
I17	
I18	
I19	

Figure 18–14   Input and output addresses for the Allen-Bradley MicroLogix PLC.

numbers), or as a floating point (numbers that have decimal points and can be expressed with exponents).

When the processor is in the run mode, it will continually scan the inputs to see if they are on or off and then update the input image register. Next it will sequentially solve the ladder logic that is stored in its user program area. It completes this part of its cycle by reading the ladder diagram. The processor goes through a fetch cycle to get each address found in the ladder diagram and puts it into the logic solving area. Each address is compared with the value found in its image register address, and this value is used to solve the logic. During the next part of the cycle, the processor writes the results of the logic to the output part of the image register. If the output in the ladder logic has contacts assigned to it, like the run CR1 in the second rung of the diagram in Fig. 18–12, the contacts will reflect the status of the output image register address when the processor solves that line of logic. If the output address is connected to the output module, the signal is sent to the module to ensure that the device connected to it is energized.

The final section of each processor scan cycle is called overhead. During the overhead part of the scan, the processor does all its work outside of solving logic, such as updating the computer screen (monitor) if one is connected, sending messages to its printer, and interacting with the network. Since the main purpose of the PLC is to read inputs, solve logic, and write to its outputs, less than 10% of the scan cycle is devoted to the overhead. The amount of time for the typical scan cycle for a generic PLC is approximately 20–60 ms. Newer enhancements to the PLCs try to lower the scan cycle to less than 10 ms.

18.11

ADDRESSES FOR INPUTS AND OUTPUTS FOR THE MICROLOGIX PLC

Up to this point in our discussion we have identified inputs and outputs in generic terms by name only. In reality the processor must keep track of each input and output by an address or number. Each brand name of PLC has devised its own numbering and addressing scheme. These addressing schemes generally fall into one of two categories: (1) sequentially numbering each input and output or (2) numbering according to location of input and output modules in the rack. The sequential numbering system is used in most small PLCs and about half of the medium-size systems. Allen-Bradley, which is one of the most widely used brands, uses two different addressing systems. In its medium-size PLCs (PLC2, PLC5, and SLC500) the address for an input or output uses numbers that indicate the location of the input or output module in the rack. In the smaller PLCs like the MicroLogix a sequential numbering system for addresses is used.

From this point on, the applications will use the address and numbering scheme of the Allen-Bradley MicroLogix. If you are using a different brand name of PLC, you will need to convert the addresses to your system if you wish to try the program examples. Figure 18–14 shows an example of the numbering system used in the Allen-Bradley MicroLogix small PLC. You can see that the input addresses start with address I0 (I stands for input and zero is the first address) and continue through address I11. Output addresses start with address O0 (O stands for output and zero is the first address) and continue through address O7. The first group of I/Os are mounted locally on the processor. Each I/O address refers to a specific circuit in the hardware module and to a specific address in the processor's memory. The addresses in memory are assigned to memory blocks. The diagram that displays all the addresses in a memory block is called a memory map.

The MicroLogix PLC is available in three sizes: a unit that has 10 inputs and 6 outputs, one that has 12 inputs and 8 outputs, and one that has 20 inputs and 12 outputs. At this time the MicroLogix does not support expansion modules; instead, it provides a means to network multiple PLCs so that they can provide common control to a larger group of inputs and outputs.

The MicroLogix also provides up to 512 internal relays. As you know, the internal relays act like a control relay in that they perform logic, but they do not directly energize an output.

You can view the entire User Manual for the Allen-Bradley MicroLogix PLC by visiting the Allen-Bradley site on the Internet at http://www.ab.com/ and looking at Publications On Line and Manuals On Line.

18.12

EXAMPLE INPUT AND OUTPUT INSTRUCTIONS FOR THE MICROLOGIX

Figure 18–15 shows an example of the input and output instructions the MicroLogix uses. From this figure you can see that normally open and normally closed contacts are provided for input instructions under a number of different names to indicate if they are connected in series or parallel in the circuit. The first column in this figure shows a symbol for each instruction. The symbol shows the way the instruction looks when it is displayed in a program as you would see it with a hand-held terminal display or when the program is printed out. The second column shows the mnemonic (pronounced ni-mon-ic) for each instruction. A mnemonic is the code that you will enter in the hand-held programmer, and it will be similar to an abbreviation for each instruction. The function code (column 3) is the code you can use with the mnemonic. The fourth column is the name of the instruction. Notice that the name of the instruction is very similar to the mnemonic. The last column in this table is the purpose of the instruction, which explains the function of each instruction.

From the table you can see that there are several different instructions for normally open and normally closed contacts. If the normally open contact is the first contact in the line of logic, it is identified by the term *LOAD* and has the mnemonic LD. The LD mnemonic is used to indicate that the contact is the start of a new rung. The AND instruction is used for all other contacts in the line of logic that are connected in series. If a contact is connected in parallel in the line of logic, it will be identified by the OR instruction and mnemonic.

If the contacts are normally closed, the term *inverted* will be added to the AND instruction, so that a normally closed set of contacts that is connected in series will have

Basic Instructions

HHP Display	Mnemonic	Code	Name	Purpose
⊢⊢	LD	20	Load	Examines a bit for an ON condition. It is the first normally open instruction in a rung or block
⊢/⊢	LDI	21	Load Inverted	Examines a bit for an Off condition. It is the first normally closed instruction in a rung or block.
⊣⊢	AND	22	And	Examines a bit for an On condition. It is a normally open instruction placed in series with any previous input instruction in the current rung or block..
⊣/⊢	ANI	23	And Inverted	Examines a bit for an Off condition. It is a normally closed instruction places in series with any previous input instruction in the current rung or block.
⊔⊢	OR	24	Or	Examines a bit for an On condition. It is normally open instruction placed in parallel with any previous input instruction in the current rung or block.
⊔/⊔	ORI	25	Or Inverted	Examines a bit for an Off condition. It is a normally closed instruction placed in parallel with any previous input instruction in the current rung or block.
⊢LDT⊢	LDT	26	Load True	Represents a short as the first instruction in a rung or block.
⊢ORT⊢	ORT	27	Or True	Represents a short in parallel with the previous instruction in the current rung or block.
⊢OSR⊢	LD OSR	28	One-Shot Rising	Triggers a one time event
⊣OSR⊢	AND OSR	29	One-Shot Rising	Triggers a one time event
─( )─	OUT	40	Output (Output Energize)	Represents an output driven by some combination of input logic. Energized (1) when conditions preceding it pemit power continuity in the rung, and de-energized after the rung is false.
─(L)─	SET	41	Set (Output Latch)	Turns a bit on when the rung is executed, and this bit retains its state when the rung is not executed or a power cycle occurs.
─(U)─	RST	42	Reset (Output Unlatch)	Turns a bit off when the rung is executed, and this bit retains its state when the rung is not executed or when power cycle occurs.
	TON	0	Timer-On Delay	Counts timebase intervals when the instruction is true.
	TOF	1	Timer-Off Delay	Counts timebase intervals when the instruction is false.
	RTO	2	Retentive Timer	Counts timebase intervals when the instruction is true and retains the accumulated value when the instruction goes false or when power cycles occur.
	CTU	5	Count Up	Increments the accumulated value at each false-to-true transition and retains the accumulated value when the instruction goes false or when power cycle occurs.
	CTD	6	Count Down	Decrements the accumulated value at each false-to-true transition and retains the accumulated value when the instruction goes false or when power cycle occurs.
	RES	7	Reset	Resets the accumulated value and status bits of a timer or counter. Do not use with the TOF timers.

Figure 18–15   Example input, output, timer, and counter instructions available in the Allen-Bradley MicroLogix PLC.

the mnemonic ANI and a normally closed contact that is connected in parallel will have the mnemonic ORI.

Sometimes Allen-Bradley uses the terms *examine on* for the normally open contacts -] [- and *examine off* for the normally closed contacts -]\[-. The term *examine* was chosen as part of this name to remind you that the processor looks at (examines) the electronic circuit in the input module to see if it is energized (on) or de-energized (off). The processor will keep track of the status of the circuit by placing a 1 in the image register address if it is energized and a 0 if it is de-energized. The word *examine* is also used to remind you to look at the status light on the front of the module so that you can also determine if the input circuit is energized or de-energized. The branch instructions are used to allow contacts to be connected in parallel to each other in the ladder diagram.

The output instruction in the table in Fig 18–15 includes the traditional coil, which uses the symbol -( )-. This instruction is named "Output" or "Output Energize." One specialized output instruction is called the "Output Latch" -(L)-, and it has a mnemonic called SET. Its companion is called the Output Unlatch -(U)-, which has the mnemonic RST. The latch output will maintain its energized state after it is energized by an input even if the input condition becomes de-energized. The unlatch output must be energized to toggle the output back to its reset (off) state. The latch coil and unlatch coil must have the same address to operate as a pair.

Latch coils are used in industrial applications where a condition should be maintained even after the input that energized the coil returns to its de-energized state. The latch coil bit will remain HI even if power to the system goes off and comes back on. This feature is important in several applications such as fault detection. Many times a system has a fault such as an over-temperature or over-current condition. After the fault is detected, the machine is automatically shut down by the logic. By the time the technician or operator gets to the machine, the condition may return to near normal, and it would be difficult to determine why the system shut down. If a latch coil is used in the detection logic, the fault bit will remain HI until the fault reset switch energizes the unlatch instruction. Another application for the latch coil is for feed water pumps or fans that are located in remote locations in the building, such as in the ceiling. A latch coil is used in these circuits because the motors must return to their energized condition automatically after a power loss. In some parts of the country, it is common for power to be lost for several seconds during severe lightning storms. If a regular type of coil is used in the program, a technician would have to go to each fan or pump and depress the start push button to start the motor again. If a latch coil is used, the coil will remain in the state it was in when the power went off. This means that if the latch coil was energized and the motor was running when power went off, the latch coil will remain energized when power is returned, and the motor will begin to run again automatically. The latch coil is very useful in these circumstances, especially when a factory has dozens of fans and pumps and it would take several hours to reset them all.

Other output instructions have a rather specific function that is beyond traditional relay logic. These instructions include the timer, counter, sequencer, and zone control. Each of these instructions will be covered in detail.

18.13
MORE ABOUT MNEMONICS: ABBREVIATIONS FOR INSTRUCTIONS

In the previous figure you can see that each instruction has a mnemonic, which is an abbreviation for the function of the instruction. The mnemonic for each instruction is selected because it sounds like the name of the function that the instruction provides. It also is limited to two or three letters so that it is easy to program. The mnemonic is the letters the PLC's microprocessor recognizes as instructions. You will need to learn the mnemonic for each instruction so that you can read and write PLC programs. The mnemonics for different PLC brand names may be slightly different but they will always describe the instruction they represent.

18.14
EXAMPLE OF A PROGRAM WITH INPUTS CONNECTED IN SERIES

Figure 18–16 shows a typical PLC program with the inputs and output identified. In this program the first input is number 1, and it is identified as I:1/1. The second input is I3, and it is identified as I:1/3. The I indicates the contact is controlled by an input. The :1 indicates the processor keeps the status of this contact (on or off) in

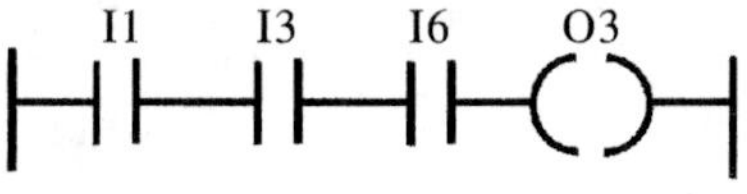

1. LD I1
2. AND I3
3. AND I6
4. OUT O3

Figure 18–16 Example of a PLC program with contacts in series using Allen-Bradley MicroLogix addressing.

File Type	Identifier	Number
Output	O	0
Input	I	1
Status	S	2
Bit	B	3
Timer	T	4
Counter	C	5
Control	R	6
Integer	N	7

Figure 18–17   File-type identifiers and numbers used in the Allen-Bradley MicroLogix PLC.

file number 1, and the /3 indicates this is input number 3. The output is identified in the program as O:0/3. The O:0 indicates this is an output whose status is stored in file 0, and the /3 indicates this is output number 3. The format for numbering files is used because it is the format that Allen-Bradley uses in its other PLCs like the PLC5.

When you enter inputs with the hand-held programmer, you would simply enter I1 for the first input, I3 for the second input, and I6 for the third input. You would use O3 for the output. The PLC automatically does the remainder of the formatting, such as adding the file number.

Figure 18–17 shows the format Allen-Bradley uses for file types, identifiers, and numbers to store information about inputs, outputs, timers, and counters in the MicroLogix PLC.

18.15
USING TIMERS IN A PLC

Many industrial applications require time delay. The MicroLogix uses the mnemonic TON to represent the timer on-delay instruction. The timer will operate like the traditional hardware-type motor-driven timer. Figure 18–18 shows an example of a timer instruction used in a line of logic. The timers can use addresses T0–T39. Each timer has a preset value (PRE value) and an accumulative value (ACC value). The preset value is the amount of time delay the timer will provide, and the accumulative value is the actual time delay that has accumulated since the timer has been energized. The timer accumulative value increments (counts up) at a rate determined by the time base. The maximum preset value for timers in the MicroLogix is 32767.

The time base for the MicroLogix timers can be 1.0 or 0.01 second. When the timer accumulative value equals the preset value, the timer times out. When the timer times out, it can cause a set of contacts to change state. These contacts need to have the same address as the timer, and they should be identified with the

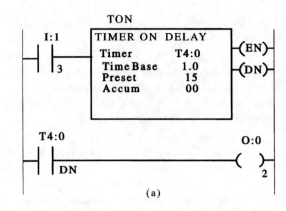

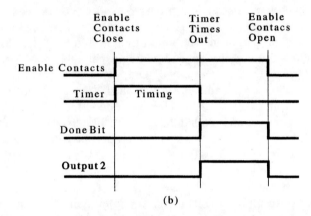

Figure 18–18   Example of a **TON** (delay-on) timer.

mnemonic DN, indicating they are controlled by the done bit for the timer.

The timer can control normally open or normally closed contacts with the done bit or two other status bits. Each of these bits is identified with a number as well as a mnemonic. The done bit is bit 13, and it is assigned the mnemonic DN. It will change state when the timer times out. The timer timing bit is bit 14, and it is assigned mnemonic TT. It will be set (ON) any time the timer is timing. The enable bit is bit 15, and it is assigned the mnemonic EN. It will be set (ON) any time the timer is enabled. You can use either the bit number or its mnemonic when you are entering a program. In Fig. 18–18 the timer is timer T0, so it has the number T4:0 in the program. In the second line of the program you can see a set of contacts that are controlled by T4:0 DN, which is the done bit for this timer. When the timer times out, these contacts will close and energize output O:0/2.

Another important part of each timer is the reset function. When a timer is reset, its accumulative value is reset to 0. Each timer must be reset at some point to make it ready to be used for the next timing cycle. The TON timer is reset any time power provided to it is interrupted. In the example in Fig. 18–18, contacts I:1/3 control power to the timer; and any time these contacts are open, the timer will reset its accumulative value. Any time these contacts are closed, the timer is enabled; and its accumulative value will increment (count up) at the rate determined by the time base, which in this example is a 1-s interval. This may also be described by saying that the timer's accumulative value will increase 1 s with every tick of the timer time base. The contacts I:1/3 in this example will be called enable/reset contacts since they control the enable and reset functions of the timer.

18.15.1
Retentive Timer

Another type of timer provided in the PLC is called the retentive timer, and it has the mnemonic RTO. A retentive timer is different from the TON timer in that its accumulative value will not reset when the enable contacts are opened. In the retentive timer, the accumulative value is retained; and when the enable contacts are closed again, the accumulative value in the timer will begin accumulating values again. This type of timer operates similarly to an hour meter on a piece of equipment. For example, this type of timer can be enabled by a set of contacts that control an air compressor motor. If the contacts are closed and the motor runs for 20 min, the timer will keep track of this time. When the contacts open and the motor stops, the timer also stops, but the accumulative value (20 min) does not reset. When the contacts close again and the motor runs an additional 30 min, the timer will show the total value of 50 min.

18.15.2
Resetting the RTO Timer

Since the enable contacts will not cause the RTO timer to reset, it will need a special reset instruction whose mnemonic is RES. Figure 18–19 shows an example of the RTO timer and the RES instruction that will cause its accumulative value to reset. In this example, the timer will add time to its accumulative value any time the enable contacts I:1/4 are closed. The timer will reset its accumulative value any time contacts I:1/5 are closed and pass power to the RES instruction. Notice the RES instruction must have the same address as the timer it is resetting.

18.15.3
The TON as an Automatic
Resetting Timer

A TON timer can be made to automatically reset itself. The automatic reset part of this circuit is accomplished by using a TON timer instruction, placing a set of the normally closed timer contacts in series with it, or by

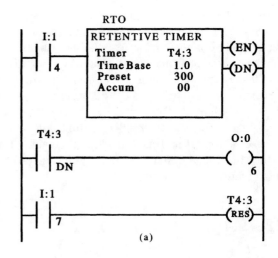

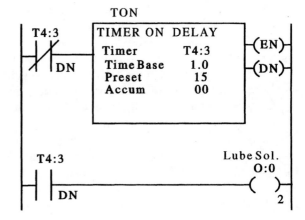

Figure 18–19 Example of an RTO (retentive) timer with a reset instruction.

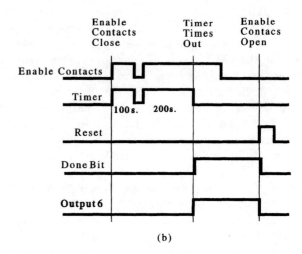

Figure 18–20 Example of a TON timer programmed as an automatic resetting timer that controls a lube solenoid.

placing a set of normally open timer contacts in series with the timer reset instruction. Figure 18–20 shows an example of a timer programmed as an auto resetting timer with its contacts controlling a lubrication solenoid. In this example, the TON timer T4:3 instruction has a set of normally closed contacts controlled by the DN bit connected in series with it so that the timer begins to time any time the PLC is in the run mode. After 60 s it times out, and it will close the normally open T4:3/DN contacts in line 2 of the program to energize the lube solenoid. The normally closed T4:3/DN contacts that enable the timer will open and cause the timer to reset and start its timing cycle over again automatically. The lube system only needs to be solenoid pulsed to activate the lubrication system.

18.15.4
Off-Delay Timers

The off-delay timer is used in applications where the amount of time delay is timed from the time a condition turns off. For example, in a heating system, at the end of a cycle, the fuel is turned off, and the fan remains running for several minutes to ensure that all the residual heat in the heat exchanger is removed. The time-delay-off function of the heating system fan can be controlled by an off-delay timer. For the off-delay timer to operate, its enable contacts must be energized and then turned off. At the point where the enable contacts are turned off, the timer will begin its delay cycle, which will run until the timer's accumulative value becomes equal to the preset value.

The off-delay timer in the PLC is called timer-off-delay instruction, and it has the mnemonic TOF. The TOF timer and the TON timer are similar in that they both can use any of the 40 timers (T0–T39), and they both have preset values, accumulative values, time base values, and status bits. An example of a TOF instruction is shown in part a of Fig. 18–21. The important difference between the TOF and TON timers is their timing cycle of when the done bit is on or off. Part b of Fig. 18–21 shows an example of the timing cycle of the done bits (DN) for the on-delay timer and the off-delay timer. At the start of a cycle when the TON instruction is first energized by its enable contacts, the timer's done bit is off and remains off until the accumulative value becomes equal to the preset value and the timer times out. At that point, the done bit becomes energized and remains energized until the enable contacts to the timer are opened and the timer is reset. The sequence for the TOF timer shows that its done bit (DN) contacts are off until the enable contacts for the timer are closed. At that point, the done bit is energized, and it remains energized until the timer cycle is complete. The TOF timer does not start its timing cycle until the enable contacts are opened again. After the enable contacts are opened, the timer runs its cycle and times out, and the done bit will be turned off.

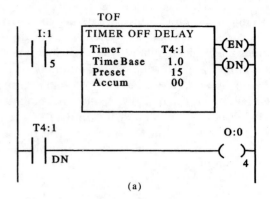

(a)

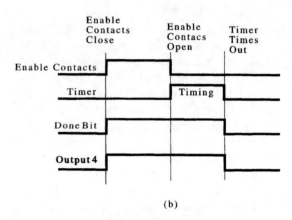

(b)

Figure 18–21 (a) A TOF (timer-off delay) instruction with its done bit controlling output O4; (b) the timing diagram for an off-delay timer.

18.16
COUNTER APPLICATIONS FOR INDUSTRY

Many industrial applications use counters. These applications can be separated into two basic operations: The first is called totalizing and is used when a machine is making a large number of parts and the counter must keep track of the total. The second type of application is called sequencing. In this type of application the counter is used to keep track of a number of steps and then it resets itself to start the sequence over again. An example of the sequence-type application is where a counter is used to keep track of the number of boxes being placed on each layer of a pallet and the number of layers the pallet has. For example, the pallet may require 4 boxes for each layer, which are stacked 3 layers high. One counter would count 1, 2, 3, 4 to keep track of the 4 boxes that are placed on each layer, and then it would reset to start the next layer. A second counter would count 1, 2, 3 and then reset to keep track of the layers on the pallet.

18.16.1
Counter Operation in a PLC

The PLC has two types of counter instructions available for use. They include the up-counter, whose mnemonic CTU stands for "count up," and the down-counter, whose mnemonic CTD stands for "count down." The counters are similar to the timer instruction in that they have a preset and an accumulative value and they each can control a number of status bits. The up-counter instruction increments (adds 1 to) *its* accumulative value when the enable contacts that allow power to flow to it transition from open to closed. This means that the enable contacts must return to their open state before they can transition for a second time to add another count to the counter accumulative value. Unlike the timer instruction, the counter does not care how long the contact stays closed; rather the counter looks only for the transition of the enable contacts. The down-counter instruction decrements (subtracts 1) from its accumulative value each time power is transitioned to it.

Both types of counters require a reset instruction (RES) to set the accumulative value back to zero. The reset instruction must have the same address as the instruction it is resetting. Figure 18–22 shows the instruction for an up-counter that includes the reset operation. The enable contacts (I4) are not part of the counter, but they must be present for the counter to operate.

The preset value is the desired value for the counter and may represent how large a value the counter is expected to count. The maximum preset value for the up-counter is +32767 and for the down-counter is −32768. Each counter also has several status bits under its control. Both types of counters use bit I3 as a done bit. The done bit is energized. (set) any time the accumulative value is equal to the preset value. The up-counter uses bit I2 as an overflow bit. The overflow bit is set any time the accumulative value is larger than +32767. The down-

counter uses bit I1 as an underflow bit. The underflow bit is set any time the accumulative value goes below −32768. The up-counter uses bit I5 as an enable bit, and the down-counter uses bit I4 as an enable bit. The bit number is used with the counter address (number) to create control logic. For example, if you want counter number 6 (C5:6) to control a solenoid that places a piece of cardboard between layers on a pallet, you would use contacts that have the number C5:6/13 on them. The /13 indicates that the done bit for the timer controls the contacts, and they will become energized each time the counter accumulative value reaches the preset value.

18.16.2
Up-Down Counters

The PLC can use an up-counter or a down-counter. Some applications are easier to understand when the up- and down-counters are used together in applications such as a good parts/bad parts counter. In this type of application the up-counter (CTU) and the down-counter (CTD) instructions are assigned to the same counter address. Each time the CTU instruction is transitioned, the counter will count up 1 value; and each time the CTD instruction is transitioned, the counter will count down 1 value. In this application both counters have the same accumulative value since they use the same counter. At the end of the day, the number displayed as the accumulative value will be the total difference between the good parts and the bad parts.

18.17
ENTERING PROGRAMS WITH THE HAND-HELD PROGRAMMER

After you understand how inputs, outputs, timers, and counters work, you are ready to enter a program into the PLC using the hand-held programmer. Figure 18–23 shows the hand-held programmer for the Allen-Bradley MicroLogix PLC. You can see that it has an LCD display, a set of number keys (0–9), and a set of instruction keys, such as contacts and coils. The programmer also has a set of function keys that are used for diagnostics and troubleshooting. These keys are named Menu, Mode, Force On, Force Off, Trace, Search, and Fault. Other keys are named Enter, Escape, Monitor, Overwrite, and Delete, and they are used for general editing of the PLC program.

The Menu function has eight choices, which are listed in the table in Fig. 18–24. Figure 18–25 shows the six choices listed under the MODE function.

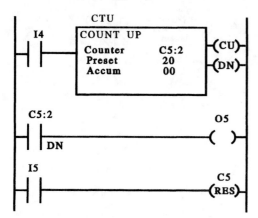

Figure 18–22   An example of a CTU (up-counter) with a reset instruction.

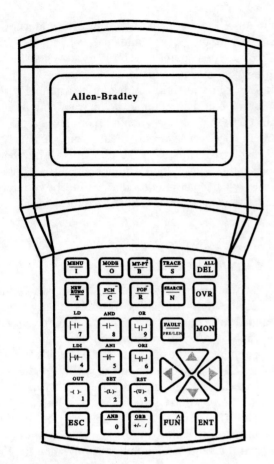

Figure 18–23   Hand-held programmer for the Allen-Bradley MicroLogix PLC.

1. LANGUAGE
2. ACCEPT EDITS
3. PROG. CONFIG
4. MEM MODULE
5. CLEAR FORCES
6. CLEAR PROG.
7. COMMS.
8. CONTRAST

Figure 18–24   Choices listed under the MENU function.

1. RPRG   Remote Program
2. RRUN   Remote Run
3. RCSN   Remote Test-Continuous Scan
4. RSSN   Remote Test-Single Scan
5. RSUS   Remote Suspend
6. FLT   Fault

Figure 18–25   Functions provided under the MODE function.

18.18
PROGRAMMING BASIC CIRCUITS INTO THE PLC

At times you will be requested to program simple programs into the PLC through the hand-held programmer. These simple programs include contacts that are connected in series or parallel, timers, and counters. Figures 18–26 through 18–32 show six simple programs and all the keystrokes required to enter these programs. If you have a MicroLogix PLC available, you can use these examples to try these programs.

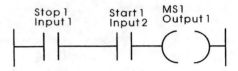

1. MON
2. ENT
3. NEW RUNG
4. ⊣⊢ I1 ENTER
5. ⊣⊢ I2 ENTER
6. ⊣(⟩ O1 ENTER
7. Change MODE to RRUN and try switches and view input and output status lights.

Figure 18–26   A simple circuit with two normally open contacts.

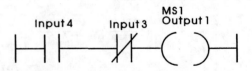

1. MON
2. ENT
3. NEW RUNG
4. ⊣⊢ I4 ENTER
5. ⊣⊬ I3 ENTER
6. ⊣(⟩ O1 ENTER
7. Change MODE to RRUN and try switches and view input and output status lights.

Figure 18–27   A simple circuit with one normally open and one normally closed contact.

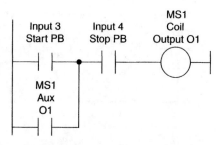

Note: The start push button and auxiliary contacts are shown before the stop push button.

1. MON
2. ENT
3. NEW RUNG
4. ⊢⊢ I3 ENTER
5. ⊔⊔ O1 ENTER
6. ⊣⊢ I4 ENTER
7. ⊸◯⊸ O1 ENTER
8. Change MODE to RUN and try switches and view input and output status lights.

Figure 18–28 A basic start-stop circuit for a PLC.

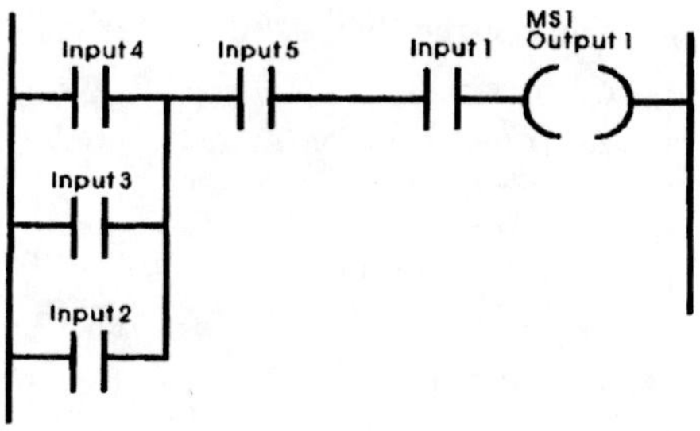

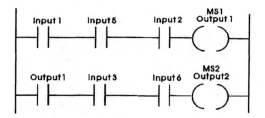

1. MON
2. ENT
3. NEW RUNG
4. ⊢⊢ I1 ENTER
5. ⊣⊢ I5 ENTER
6. ⊣⊢ I2 ENTER
7. ⊸◯⊸ O1 ENTER
8. NEW RUNG
9. ⊢⊢ O1 ENTER
10. ⊣⊢ I3 ENTER
11. ⊣⊢ I6 ENTER
12. ⊸◯⊸ O2 ENTER
13. Change MODE to RRUN and try switches and view input and output status lights.

Figure 18–30 Two sets of series circuit creating a starting sequence where motor 1 must start before motor 2 can be started.

1. MON
2. ENT
3. NEW RUNG
4. ⊢⊢I4 ENTER
5. ⊔⊔ I3 ENTER
6. ⊔⊔ I2 ENTER
7. ⊣⊢ I5 ENTER
8. ⊣⊢ I1 ENTER
9. ⊸◯⊸ O1 ENTER
10. Change MODE to RRUN and try switches and view input and output status lights.

Figure 18–29 A simple parallel circuit for a PLC.

○ ## 18.19

WIRING A STOP-START CIRCUIT TO INPUT AND OUTPUT TERMINALS ON THE PLC

All the inputs and outputs used in the PLC system must be connected to the input or output module hardware. The simplest circuit to understand is a start-stop circuit that controls a motor-starter coil. Figure 18–33 shows an example of this circuit, and you can see where the start push-button switch and the stop push-button switch are wired to two input addresses and to a motor-starter coil connected to an output address. You can also see the location of the screws on the MicroLogix PLC where the wires from the switch and coil for the motor starter are connected.

The start push button is wired as a normally open switch, and the stop push button is wired as a normally closed switch, just as in a hardwired control system. The major difference in the start-stop circuit in the PLC from a hardwired system is in the way the start and stop push buttons are programmed into the PLC. In the PLC program the stop push button is programmed normally open. At first this seems to be incorrect, but if you look at the line of logic in the PLC, you will notice that since the stop push button is wired normally closed, it will pass power to the input module, which in turn sends a 1 to the image table. Since the stop contacts in the program are programmed normally open, they will change state and pass power to the start contacts whenever the input module sends a 1 to the image table. If the stop contacts were programmed normally closed, they would open when the input module received power from the hardwired stop switch and the circuit would not operate.

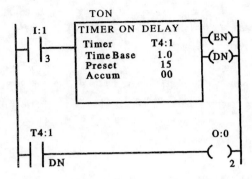

1. MON

2. ENT

3. NEW RUNG

4. ⊣⊢ I3 ENTER

5. FUN

6. ANB 0 ENTER   (0=TON 1=TOF 2=RTO)

7. T1 ENTER

8. 15 (FOR PRESET VALUE)
   0 (FOR ACC VALUE)

9. △ ENTER TO GET 1.0 TIME BASE

10. NEW RUNG

11. ⊣⊢ T1 ORB ENTER   THIS GIVES DN BIT

12. ⊣( )⊢ O2 ENTER

13. Change MODE to RRUN and
    try switches and view input
    and output status lights.

Figure 18–31 A simple timer program that delays the time the motor starts after the input switch is closed.

IMPORTANT NOTICE:
The PLC program would work if you wired the stop push button normally open and programmed its contacts normally closed. This creates a severe problem in that the stop push button becomes a safety hazard if it is wired normally open and a wire comes loose from its terminal or if the stop button fails, because it cannot send a signal to the PLC input module and stop the circuit. It is important that you always wire the stop push button as a normally closed switch so that if it fails, it will interrupt power to the input module and stop the circuit.

The hardwired circuit of a PLC must also be protected by a real relay called a master control relay. This relay will be identified as CRM (control relay master), and all the voltage to the input and output module is controlled by the contacts of the CRM. The coil of the CRM is connected to a hardwired start-stop push-button station that is not part of the PLC program. This allows the power to be shut off to the inputs and output by depressing the stop button. Power is connected directly to the processor so that it stays energized regardless of the state of the CRM.

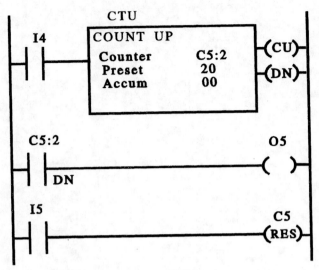

1. MON

2. ENT

3. NEW RUNG

4. ⊣⊢ I4 ENTER

5. FUN

6. ANB 5 ENTER   (5=CTU 6=CTD 7=RES)

7. C2 ENTER

8. 20 (FOR PRESET VALUE)
   0 (FOR ACC VALUE)

9. NEW RUNG

10. ⊣⊢ C2 ORB ENTER   THIS GIVES DN BIT

11. ⊣( )⊢ O5 ENTER

12. NEW RUNG

13. ⊣⊢ I5 ENTER

14. FUN

15. ANB 7 ENTER (5=CTU 6=CTD 7=RES)

16. C2 ENTER

17. Change MODE to RRUN and
    try switches and view input
    and output status lights.

Figure 18–32 A simple counter program.

18.20
TYPICAL PROBLEMS A TECHNICIAN WILL ENCOUNTER WITH PLCs

As a technician you may encounter problems with PLCs at either the software or hardware level. At the software level, you may be required to lay out and enter the original ladder logic program. You may also be involved at the start-up of a machine when a program has been written for the PLC and a technician is required to test the program against the machine operation. In other cases, you may be involved in using a program to trou-

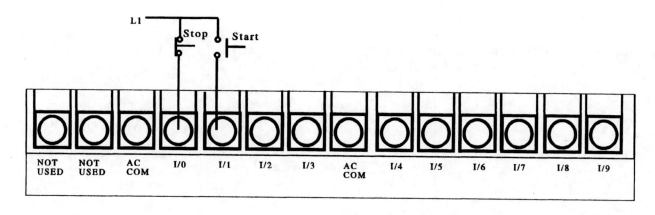

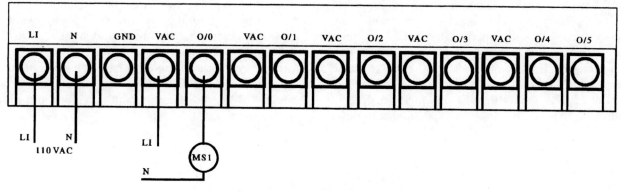

(a) Start/stop hardwired to PLC

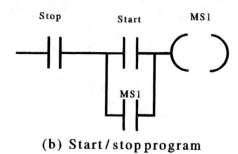

(b) Start/stop program

Figure 18–33   (a) Start and stop push buttons hardwired to a PLC; (b) program for start and stop circuit.

bleshoot the system when a fault occurs. Sometimes you will be requested to make changes in the documentation that describes the operation of the system and identifies all the inputs and outputs. In each of these cases, you may need additional information to satisfactorily accomplish these jobs. This additional information may be provided in manufacturers' manuals or in short four- to five-day training courses that are provided on specific PLCs.

You may also be involved with the hardware modules and I/O devices that are connected to PLCs. In some systems you will be requested to read the specification sheets for the signals that each module sends or receives so that you can select the proper types of electronic devices to interface with the PLC. Sometimes you will be requested to analyze a signal to see

that it meets the proper criteria for data transmission. If system component failure occurs, you will be required to use all the knowledge you have about electronics to test the hardware and make decisions. You now have a better idea of how electronics has been blended with the programmable controller to provide industrial automation.

18.21
TROUBLESHOOTING THE PLC

When you are called to troubleshoot the PLC, the first thing you need to understand is that very seldom is the

Figure 18–34 Example of the status indicators on input and output modules for the MicroLogix PLC.

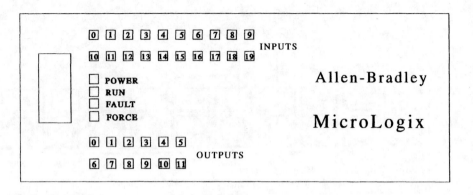

PLC the problem; rather it is indicating a problem with hardware, such as switches and coils. Each input module has a status indicator (light) for each input switch, and there is an output status indicator for each output circuit. If the input switch is in the "on state" and passing 110 VAC to the module, the status light will be on. The status light is provided on the PLC module so that you do not need a voltmeter to deter-mine if power is at the terminals. You can tell if the input switch is passing power to the PLC by observing the status light. On the output module, the status light will be on whenever the output circuit is receiving power. Again, the status light can be used instead of a voltmeter to quickly tell if the circuit has power. Figure 18–34 shows an example of input and output status lights.

QUESTIONS

Short Answer

1. Describe the four basic parts of any programmable controller.

2. Explain the operation of a PLC up-counter and down-counter.

3. Explain the four things that occur when the PLC processor scans its program.

4. Explain why you should use a black dot to show transi-tioned contacts and an arrow to indicate power flow when you are trying to determine if contacts are provid-ing power to a specific output in a program.

5. The input module has a status indicator for each input circuit. Explain how the status indicator is used in trou-bleshooting.

True or False

1. One advantage of all PLCs is that you can use contacts with the same address more than once in the program.

2. When the PLC processor is in the run mode, it does not execute its scan cycle.

3. A mnemonic is a timer whose time delay is controlled by air pressure.

4. When a timer in a PLC program will not time, you should suspect that the timer is broken and that you will need to program a replacement timer.

5. A control relay (memory coil) in a PLC program is differ-ent from an output instruction in that its signal does not control a circuit in an output module and it is used mainly to determine logic conditions.

Multiple Choice

1. When the input contact that enables a TON-type timer is opened, the accumulated value in the timer will _____.
   a. reset (go to zero)
   b. freeze (remain at its present value)
   c. go to an undetermined value, and the timer must be reset manually

2. An RTO-type timer has its accumulative value reset _____.
   a. any time the enable contacts are opened
   b. only when the reset for the timer is "HI"
   c. any time the accumulative value in the timer exceeds 99

3. When the PLC processor is in the PROGRAM mode, _____.
   a. it executes its scan cycle
   b. it does not execute its scan cycle
   c. it is impossible to tell whether the processor is exe-cuting its scan cycle because you are not on-line

4. When it is used as part of a start-stop circuit for a PLC system, the stop push button should be wired _____ and programmed _____.

   a. normally closed, normally open

   b. normally open, normally closed

   c. It doesn't matter.

5. You can use the _____ on the face of a PLC to determine whether an input switch is on or off.

   a. power light

   b. input status indicators

   c. output status indicators

PROBLEMS

1. Enter the conveyor box-sorting program in Figure 18–12 in a programmable controller and execute the inputs to watch the program operate. Describe the operation of each output.

2. Enter an on-delay timer and an off-delay timer in your PLC and describe the following: preset time, accumulative time, what must occur for the timer to run, what must occur for the timer to time out, and what contacts change when the timer times out.

3. Enter an up-counter and a down-counter in your PLC and describe the following: preset value, accumulative value, what must occur for each counter to count, what occurs when the counter's preset and accumulative values are equal, what happens when the counter's accumulative value exceeds the overflow value, what contacts change when the counter reaches its preset value, and what contacts change when the counter exceeds the overflow.

4. Program an auto resetting timer and describe its operation.

5. Program an auto resetting counter into your PLC and describe its operation.

CHAPTER 19

Electronics for Maintenance Personnel

OBJECTIVES

After reading this chapter you will be able to:

1. Explain P-type and N-type material.

2. Identify the terminals of a diode and explain its operation.

3. Explain the operation of PNP and NPN transistors.

4. Identify the terminals of a silicon-controlled rectifier (SCR) and explain its operation.

5. Explain the operation of a triac.

19.0

OVERVIEW OF ELECTRONICS USED IN INDUSTRIAL CIRCUITS

Electronic devices such as diodes, transistors, and SCRs have become commonplace in industrial circuits because they provide better control than electromechanical devices and are less expensive to manufacture. When you are troubleshooting, you will find solid-state components in control boards and other controls. In this chapter you will gain an understanding of P-type material and N-type material, which are the building blocks of all electronic components. After you have a basic understanding of P-type and N-type material, you will be introduced to diodes, transistors, SCRs, and triacs, and you will see application circuits of each of these types of components. You will also learn the theory of operation and troubleshooting techniques for each type of device.

19.1

CONDUCTORS, INSULATORS, AND SEMICONDUCTORS

In Chapter 3 you learned that atoms have protons and neutrons in their nuclei and electrons that move around the nucleus in orbits, which are also called shells. The number of electrons in the atom is different for each element. For example, you learned that copper has 29 electrons, 3 of which are located in the outermost shell. The outermost shell is called the valence shell, and the electrons in that shell are called valence electrons. The atoms of the most stable material have eight valence electrons, and these eight valence electrons are found as four pairs. As an atom for materials that are used as conductors and insulators fills its valence shell with eight electrons, it will open a new orbit with the additional electrons. This means that an atom may have five, six, or seven atoms, and it will take less energy to add electrons to get a full shell (eight); or it may have one, two, or three electrons, and it will take less energy to give up these electrons to get down to the previous full shell that has eight electrons.

A *conductor* is a material that allows electrons (electrical current) to flow easily through it, and an *insulator* does not allow current to flow through it. An example of a conductor is copper, which is used for electrical wiring. An example of an insulator is rubber or plastic. The atomic structure of a conductor makes it easier for electrons to flow through it, and the atomic structure of an insulator makes it nearly impossible for any electrons to flow through it.

Figure 19–1 shows the atomic structure of a conductor. Atoms of conductors can have one, two, or three valence electrons. The atom in this example has

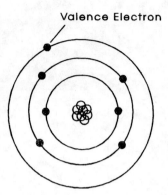

Figure 19–1   Atomic structure of a conductor.

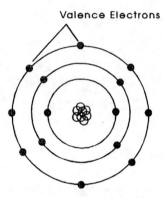

Figure 19–2   Atomic structure of an insulator.

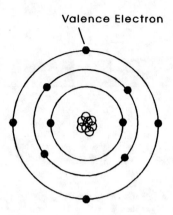

Figure 19–3   Atomic structure for semiconductor material.

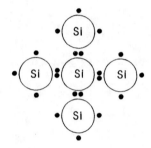

Figure 19–4   Atoms of silicon semiconductor material are combined to create a lattice structure.

one valence electron. Because all atoms try to get eight electrons (four pairs) in their valence shell, it takes less energy for conductors to give up these electrons (one, two, or three), so the valence shell will become empty. At this point the atom becomes stable because the previous shell becomes the new valence shell, and it has eight electrons. The electrons that are given up are free to move as current flow.

Figure 19–2 shows the atomic structure of an insulator. Insulators have five, six, or seven valence electrons. In this example you can see that the atom has seven valence atoms. This structure makes it easy for insulators to take on extra electrons to get eight valence electrons. The electrons that are captured to fill the valence shell are electrons that would normally be free to flow as current.

Semiconductors are materials whose atoms have exactly four valence electrons. Because these atoms have exactly four valence electrons, they can take on four new valence electrons, like an insulator, to get a full valence shell; or they can give up four valence electrons, like a conductor, to get an empty outer shell; then the previous shell, becomes the valence shell. Figure 19–3 shows an example of the atomic structure for semiconductor material.

19.2
COMBINING ATOMS

When solid-state materials or other materials are manufactured, large numbers of atoms are placed together. The structure that becomes most stable at this point is called a *lattice structure.* Figure 19–4 shows an example of the lattice structure that occurs when atoms are combined. In this diagram you can see that atoms of silicon (Si), which is a semiconductor material with four valence electrons, are combined so that one valence electron from each of the neighbor atoms is shared so that all atoms look and act as if they each have eight valence electrons.

19.3
COMBINING ARSENIC AND SILICON TO MAKE N-TYPE MATERIAL

Other types of atoms can be combined with semiconductor atoms to create the special material that is used in solid-state transistors and diodes. Figure 19–5 shows

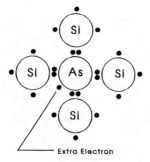

Figure 19–5 N-type material formed by combining four silicon atoms with a single arsenic atom.

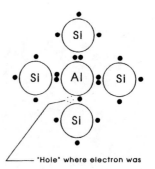

Figure 19–6 P-type material formed by combining four silicon atoms with a single aluminum atom.

a diagram of four silicon atoms that are combined with one atom of arsenic. From this figure you can see that arsenic has five valence electrons, and when the silicon atoms are combined with it, they create a very strong lattice structure. You can see that each silicon atom donates one of its valence electrons to pair up with each of the valence electrons of arsenic atom. Because the arsenic atom has five valence electrons, one of the electrons is not paired up and becomes displaced from the atom. This electron is called a *free electron,* and it can go into conduction with very little energy. Because this new material has a free electron and electrons have a negative charge, it is called *N-type material.*

19.4

COMBINING ALUMINUM AND SILICON TO MAKE P-TYPE MATERIAL

An atom of aluminum can also be combined with semiconductor atoms to create a special material that is called *P-type material.* Figure 19–6 shows a diagram of four silicon atoms that are combined with one atom of aluminum. From this figure you can see that aluminum has three valence electrons, and when the silicon atoms are combined with it, they create a very strong lattice structure. You can see that each silicon atom donates one of its valence electrons to pair up with each of the valence electrons of the aluminum atom. Because the aluminum atom has three valence electrons, one of the four aluminum electrons will not be paired up, and it will have a space that any free electron can move into to combine with the single electron. This free space is called a *hole,* and it is considered to have positive charge, because it is not occupied by a negatively charged electron. Because this new material has an excess hole that has a positive charge, it is called *P-type material.*

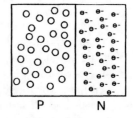

Figure 19–7 An example of a piece of P-type material connected to a piece of N-type material.

19.5

THE PN JUNCTION

One piece of P-type material can be combined with one piece of N-type material to make a *PN junction.* Figure 19–7 shows a typical PN junction. The PN junction forms to make an electronic component called a *diode.* The diode is the simplest electronic device. When DC voltage is applied to a PN junction with the proper polarity, it causes the junction to become a very good conductor. If the polarity of the voltage is reversed, the PN junction becomes a good insulator.

19.6

FORWARD-BIASING THE PN JUNCTION

When DC battery voltage is applied to the PN junction so that positive voltage is connected to the P-type material and negative voltage is connected to the N-type material, the junction is forward biased. Figure 19–8a shows a battery connected to the PN junction so that it is forward biased. In this diagram you can see that the positive battery voltage causes the majority of holes in the P-type

Figure 19–8 (a) A forward-biased PN junction; (b) a reversed-biased PN junction.

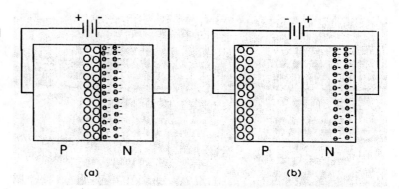

material to be repelled, so the free holes move toward the junction, where they come into contact with the N-type material. At the same time the negative battery voltage also repels the free electrons in the N-type material toward the junction. Because the holes and free electrons come into contact at the junction, the electrons recombine with the holes and cause a very low resistance junction, which allows current to flow freely through it. When a PN junction has low resistance, it allows current to pass just as if the junction were a closed switch.

It is important to understand that up to this point in this text, all current flow has been described in terms of *conventional current flow*, which is based on a theory that electrical current flows from a positive source to a negative return terminal. At this point you can see that this theory will not support current flow through electronic devices. For this reason *electron current flow theory* must be used when discussing electronic devices. In electron current flow theory, current is the flow of electrons, and it flows from the negative terminal to the positive terminal in any electronic circuit.

19.7
REVERSE-BIASING THE PN JUNCTION

Figure 19–8b shows a battery connected to the PN junction so that it is reversed biased. In this diagram you can see that the positive battery voltage is connected to the N-type material and the negative battery voltage is connected to the P-type material. The negative voltage on the P-type material attracts the majority of holes in the P-type material so that they move away from the junction and cannot come into contact with the N-type material. At the same time the positive battery voltage that is connected to the N-type material attracts the free electrons away from the junction. Because the holes and free electrons are both attracted away from the

junction, no electrons can recombine with any holes, so a high-resistance junction is formed that will not allow any current flow. When the PN junction has high resistance, it will not allow any current to pass, just as if the junction were an open switch.

19.8
USING A DIODE FOR RECTIFICATION

The electronic diode is a simple PN junction. Figure 19–9 shows the symbol for a diode. You can see that the symbol for the diode looks like an arrowhead that is up against a line. The part of the symbol that is the arrowhead is called the *anode;* it is also the P-type material of the PN junction. The other terminal of the diode is called the *cathode*, and it is the N-type material. Because the anode is made of positive P-type material, it is identified with a + sign. The cathode is identified with a − sign because it is made of N-type material.

Figure 19–10 shows a diode connected in a circuit that has an AC power source. The AC power source produces a sine wave that has a positive half-cycle and then a negative half-cycle. The diode converts the AC sine wave to half-wave DC voltage by allowing current to pass when the AC voltage provides a forward bias to the PN junction, and it blocks current when the AC voltage provides a reverse bias to the PN junction. The

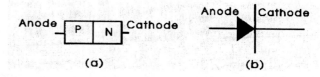

Figure 19–9 (a) PN junction for a diode; (b) electronic symbol of a diode. The anode is the arrowhead part of the symbol, and the cathode is the other terminal.

Figure 19–10 A single diode half-wave rectifier used in a circuit to convert AC voltage to half-wave DC voltage.

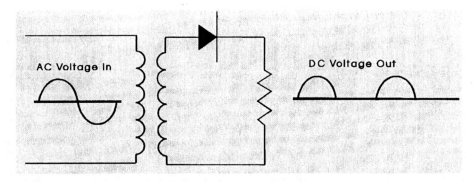

forward-bias condition occurs when the AC voltage sine wave provides a positive voltage to the anode and a negative voltage to the cathode. During this part of the AC cycle, the diode is forward biased, and it has very low resistance so current can flow. When the other half of the AC sine wave occurs, the diode becomes reverse biased, with negative voltage applied to the anode and positive voltage applied to the cathode. During the time the diode is reverse biased, a high-resistance junction is created, and no current flows through it.

Rectification is the process of changing AC voltage to DC voltage. One of the main jobs of the diode is to convert AC voltage to DC voltage. Most electronic circuits used in industrial circuits need some DC voltage to operate. Because the equipment is connected to AC voltage, a power supply is required to provide regulated DC voltage for the solid-state circuits, and the diode is part of the power supply that rectifies the AC voltage to DC.

19.9
HALF-WAVE AND FULL-WAVE RECTIFIERS

When one diode is used in a circuit to convert AC voltage to DC voltage, it is called a *half-wave rectifier*, because only the positive half of the AC voltage is allowed to pass through the diode; the negative half is blocked. The rectifier shown in Fig. 19–10 is a half-wave rectifier. The half-wave rectifier is not very efficient, because half of the AC sine wave is wasted.

If four diodes are used in the circuit, they can convert both the positive half-wave and the negative half-wave of the AC sine wave. Figure 19–11 shows a circuit with four diodes used to rectify AC voltage to DC voltage. Because the four diodes can convert both the positive half and the negative half of the AC sine wave, this type of rectifier is called a *full-wave rectifier*.

19.10
THREE-PHASE RECTIFIERS

Larger three-phase AC motors, pumps, and fans used in industrial systems can have their speed changed to run more efficiently by changing the frequency of the voltage supplied to them. In these applications, six diodes are used to convert three-phase AC voltage to DC voltage, and then a microprocessor-controlled circuit converts the DC voltage back to three-phase AC voltage; the frequency of this voltage can be adjusted to change the speed of the motors. Figure 19–12 shows a six-diode, three-phase rectifier. Notice that the supply voltage to the diodes is three-phase AC voltage, and the output voltage from the rectifier consists of six positive half-waves.

19.11
TESTING DIODES

One of the tasks that you must perform as a technician is to test diodes to see if they are operating correctly. One way to do this task is to apply AC voltage to the input of the diode circuit and test for DC voltage at the output of the diode circuit. If the amount of DC voltage at the power supply is half of what it is rated for in a four-diode bridge rectifier circuit, you can suspect one of the diode pairs has one or both diodes opened. If this occurs, you can use an ohmmeter to test each diode to determine which one is faulty.

When you are testing the diodes with an ohmmeter, it is important that all power to the diode circuit be turned off. You should remember from earlier chapters that the ohmmeter uses a battery as a DC voltage source. Because you know the diode can be tested for forward bias and reverse bias with a DC voltage source, you can use the ohmmeter as the voltage source and the meter to test for high resistance and low resistance through the diode junction. Figure 19–13a shows an example of putting the positive

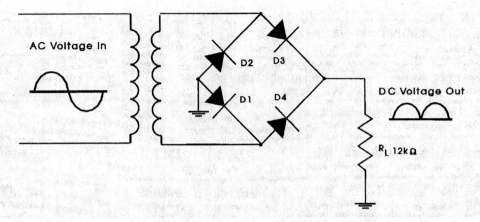

Figure 19–11 Four-diode, full-wave rectifier. This type of rectifier is often called a full-wave bridge rectifier, because the diodes are connected in a bridge circuit.

Figure 19–12 A six-diode bridge used to rectify three-phase AC voltage to DC voltage.

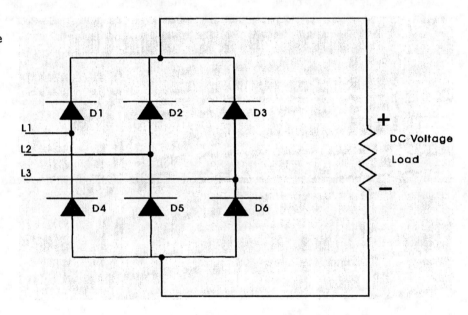

Figure 19–13 (a) Using the battery in an ohmmeter to forward-bias a diode. The diode should have low resistance during this test. (b) Using the battery in the ohmmeter to reverse-bias a diode. The diode should have high resistance during this test.

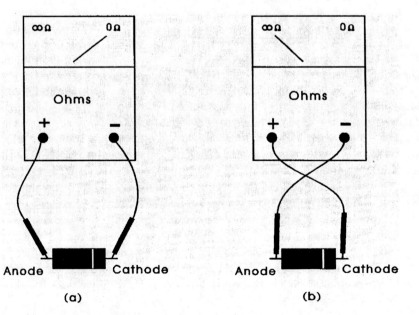

ohmmeter terminal on the anode of the diode and the negative ohmmeter terminal on the cathode of the diode to cause the diode to go into forward bias. During this test the diode is forward biased, and the ohmmeter should measure low resistance. When the ohmmeter leads are reversed, as in Fig. 19–13b, so that the negative meter lead is connected to the anode of the diode and the positive meter lead is connected to the cathode of the diode, the diode is reverse biased. When the diode is reverse biased, the ohmmeter should measure infinite (∞) resistance. If the diode indicates high and low resistance, it is good. If the diode indicates low resistance during the forward-bias test and the reverse-bias test, the diode is shorted. If the diode shows high resistance during both tests, it is opened.

19.12

IDENTIFYING DIODE TERMINALS WITH AN OHMMETER

Because the ohmmeter can be used to determine if a diode is good or faulty, the same test can be used to determine which lead of a diode is the anode and which lead is the cathode. When you use the ohmmeter to test the diode for forward bias and reverse bias, you should notice that the ohmmeter will indicate high resistance when the diode is reverse biased and low resistance when the diode is forward biased. When the meter indicates low resistance, you know the diode is forward biased, so the positive lead is touching the anode and the negative lead is touching the cathode. This method works when you are testing any diode. If the diode has markings, you can identify the cathode end of the diode because it has a strip around it. Figure 19–14 shows two types of diodes; the anode is identified in each.

19.13

LIGHT-EMITTING DIODES

A light-emitting diode (LED) is a special diode that is used as an indicator because it gives off light when current flows through it. Figure 19–15 shows a typical LED and its symbol. From this figure you can see that the LED looks like a small indicator lamp. You will probably encounter LEDs on various controls, such as thermostats. The major difference between an LED and an incandescent lamp is that the LED does not have a filament, so it can provide thousands of hours of operation without failure.

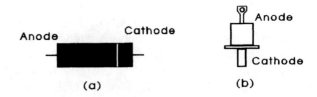

Figure 19–14   (a) Typical diode with anode and cathode identified; (b) power diode with anode and cathode identified.

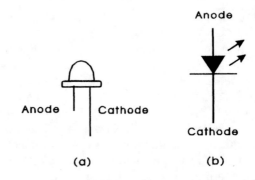

Figure 19–15   (a) A typical light-emitting diode (LED). Notice that the LED looks similar to a small indicator lamp. (b) The electrical symbol for a LED.

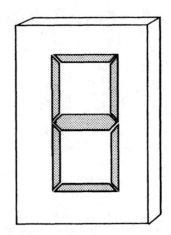

Figure 19–16   LEDs used in seven-segment displays. The seven-segment display can display the numbers 0–9.

An LED must be connected in a circuit in forward bias. Because the typical LED requires approximately 20 mA to illuminate, it is usually connected in series with a 600- to 800-Ω resistor, which limits the current. Figure 19–16 shows a set of seven LEDs that are connected to provide a seven-segment display. The seven-segment display can display all numbers 0–9. Seven-segment displays are used to display numbers on thermostats and other electronic devices.

LEDs are also used in optoisolation circuits, where larger field voltages are isolated from smaller computer signals. The LED is encapsulated with a phototransistor. When the input signal is generated, it causes current to flow through the LED; and light from the LED shines on the phototransistor, which goes into conduction and passes the signal on to the computer.

19.14

PNP AND NPN TRANSISTORS

Two pieces of N-type material can be joined with a single piece of P-type material to form an NPN transistor. A PNP transistor can be formed by joining two pieces of P-type material with a single piece of N-type material. Figure 19–17 shows the electronic symbol and the material for both a PNP and NPN transistor. The terminals of the transistor are identified as the emitter, collector, and base. The base is the middle terminal, and the emitter is the terminal identified by the arrowhead.

19.15

OPERATION OF A TRANSISTOR

A transistor can be connected in a circuit to perform a wide variety of functions. The simplest function for a transistor to provide is the function of an electronic switch. Figure 19–18 shows a transistor as an electronic switch. In this type of application, the base terminal of the transistor provides a function like the coil of a relay, and the emitter-collector circuit provides a function like the contacts of a relay. When the proper amount and polarity of DC voltage is applied to the base of the transistor, the resistance between the collector and emitter is relatively low, which allows the maximum amount of circuit current to flow through the emitter-collector circuit. The transistor at this time acts like a relay that has its coil energized.

When the polarity of the voltage on the base of the transistor is reversed, the emitter-collector circuit is changed to a high-resistance circuit, which acts like the relay when the coil is de-energized. The major advantage of the transistor is that a very small amount of voltage or current on the base can switch the transistor from high resistance to low resistance. Because the base current is very small and the current flowing through the collector is very large, the transistor is called an *amplifier*. Transistors are used in a variety of applications, including thermostats, compressor-protection circuits, and variable-frequency motor drives.

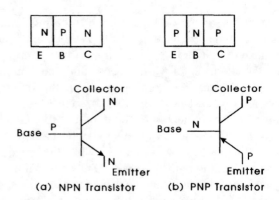

Figure 19–17 Electronic symbol and diagram of (a) NPN and (b) PNP material joined to form transistors.

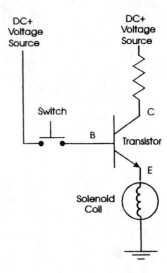

Figure 19–18 A transistor used as an electronic switch.

Figure 19–19 shows a PNP transistor and an NPN transistor as two PN diode circuits. The equivalent diode circuits are shown with each transistor to give you the idea of how the two junctions work together inside each transistor. Because each transistor is made from two PN junctions, each junction can be tested, just as with the single-junction diode, for forward bias (low resistance) and reverse bias (high resistance). If you must work on a number of systems that have electronic circuits, you can purchase a commercial-type transistor tester that allows you to test the transistor while it is in the circuit or when it is out of the circuit.

19.16

TYPICAL TRANSISTORS

You can identify transistors by their shapes. Small transistors are used for switching control circuits, and

Figure 19–19 PNP and NPN transistors shown as their equivalent PN junctions. Each PN junction can be tested for forward bias and reverse bias.

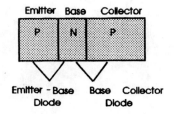

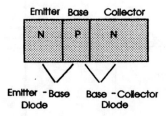

larger transistors are mounted to heat sinks so that they can easily transfer heat. Figure 19–20 shows examples of two types of transistors.

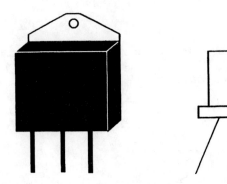

Figure 19–20 Typical transistors that are used for power control and switching.

19.17
TROUBLESHOOTING TRANSISTORS

Transistors can be tested by checking each P and N junction for front-to-back resistance. Figure 19–21 shows these tests. You can see each time the battery in the ohmmeter forward-biases a PN junction, the resistance is low; and when the battery reverse-biases the junction, the meter indicates high resistance. You can test a transistor in this manner if it has been removed from the circuit. You can also test transistors while they are connected in circuit with a commercial-type transistor tester. The transistor tester performs a similar front-to-back resistance test across each junction.

19.18
UNIJUNCTION TRANSISTORS

Approximately two dozen different types of electronic devices are made by combining a number of different sections of P and N material together. Some of these devices are designed specifically for switching larger voltages and currents, and other devices, such as the unijunction transistor, are designed to produce a small pulse of voltage that can be used to turn on or bias other switching devices. The unijunction transistor (UJT) is used to produce a pulse of voltage that is used to turn on silicon-controlled rectifiers (SCRs). SCRs are used to control large DC voltages and currents; they are introduced later in this chapter. As a maintenance technician you may be more familiar with current-switching devices such as SCRs or triacs, but you must remember they will not operate without devices such as UJTs, which are used to turn them on and off.

The unijunction transistor is made of a single PN junction. Figure 19–22 shows a diagram of the PN material and the electronic symbol for the unijunction transistor. The terminals for the unijunction transistor are called base 1, base 2, and emitter. From the diagram you can see that the base 2 and base 1 leads are mounted in the large section of N-type material. This means that if you measure the resistance between the base 2 and base 1 terminals, you would measure the same amount of resistance regardless of the polarity of the meter leads. When you measure the resistance between base 2 and the emitter, you would find it is like a diode PN junction, and the polarity of the ohmmeter battery would cause the PN junction to be forward biased one way and reversed biased when you reversed the leads. The base 1–to–emitter junction also acts like a diode junction.

19.19
OPERATION OF THE UJT

The simplest way to explain the operation of a UJT is to show it in a typical circuit. Figure 19–23 shows the UJT in a circuit with a resistor and capacitor connected to the emitter terminal and a resistor connected to each of the base terminals. When voltage is applied to the resistor and capacitor that is connected to the emitter, the capacitor begins to charge. The size of the resistor controls

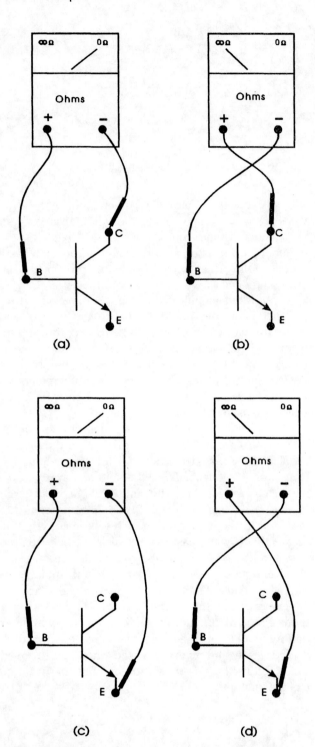

(a)   **(b)**

(c)   **(d)**

Figure 19–21 (a) Testing the base-collector junction of an NPN transistor for forward bias; (b) testing the base-collector junction of an NPN transistor for reverse bias; (c) testing the base-collector junction of a PNP transistor for forward bias; (d) testing the base-collector junction of a PNP transistor for reverse bias.

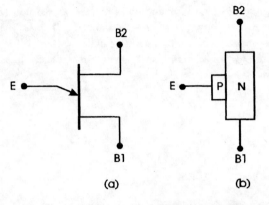

(a)   **(b)**

Figure 19–22 (a) Electronic symbol for the unijunction transistor. The terminals for the UJT are base 2, base 1, and emitter. (b) P and N material in the unijunction transistor.

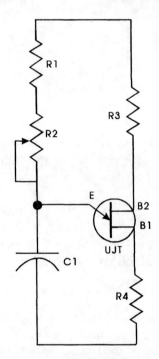

Figure 19–23 A UJT connected to an oscillator circuit to produce sharp output pulses. The size of the resistance and capacitance connected to the emitter controls the frequency of the pulses.

the time it takes for the capacitor to charge. This time is referred to as the *time constant* for the circuit.

The same circuit voltage is applied to the resistors that are connected to base 2 and base 1. The resistors connected to base 2 and base 1 create a voltage drop with the internal resistance of the UJT. When the charge in the capacitor grows to a value that is larger than the voltage drop across the UJT and the resistor connected to the base 1 lead, a current path is developed through the emitter and through the resistor that is connected to base 1. When current flows through this path, it creates a voltage pulse that increases to its maximum voltage very quickly. The waveform of this pulse is shown in the diagram, and you can see that it turns on and increases to its maximum value immediately. The duration of the pulse depends on the internal resistance of the UJT and the ratio of the size of the base 2 and base 1 resistors.

When the UJT is producing the pulse, the capacitor is discharging, and it begins to charge up immediately and repeats the process as its voltage charge increases to a point where it is larger than the voltage across the UJT. At this point the UJT produces another pulse. The time between pulses is determined by the size of the resistance and capacitance of the resistor and capacitor that are connected to the emitter. If the amount of resistance is increased, the time for the capacitor to charge increases and the time between pulses increases. If the amount of resistance at the emitter is reduced, the capacitor charges more quickly and creates pulses that are grouped closer together. This type of circuit is called an *oscillator*, and the number of pulses the oscillator produces in 1 s is called the *oscillator frequency*.

19.20
TESTING THE UJT

The unijunction transistor can be tested in the same way as a PN junction. The UJT must be isolated from its circuit, and you can test the PN junction between base 2 and emitter and then test the PN junction between base 1 and emitter. Use an ohmmeter and switch the polarity of its probes so that you forward-bias and reverse-bias each junction. Remember that when you forward-bias a PN junction, it should have low resistance, and when you reverse-bias a PN junction, it should have high resistance.

19.21
THE SILICON-CONTROLLED RECTIFIER

The silicon-controlled rectifier is called an SCR, and it is made by combining four PN sections of material. Figure 19–24a shows the electronic symbol for the SCR, and Fig. 19–24b shows the PN material for the SCR. The terminals on the SCR are identified as the anode, cathode, and gate. Because the SCR is basically a diode that is controlled by a gate, its symbol uses the arrow from the basic rectifier diode that you studied at the begin-

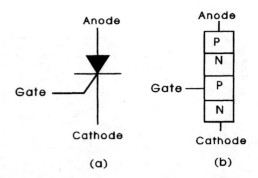

Figure 19–24 (a) Electronic symbol for the silicon-controlled rectifier. The terminals of the SCR are the anode, cathode, and gate. (b) P and N material in a silicon-controlled rectifier. The P and N materials are combined to make a PNPN junction.

ning of this chapter. When the SCR is turned on, it can conduct large amounts of DC voltage and current (more than 1,000 V and 1,000 A) through its anode and cathode. The major difference between the SCR and the junction diode you learned about in Section 19.5 is that the junction diode is always able to pass current in one direction when the diode is forward biased. The SCR is forward biased by applying positive voltage to its anode and negative voltage to its cathode. At this point the SCR still has high resistance between its anode-cathode junction. If a positive voltage pulse is applied to the SCR gate, the SCR's anode-cathode junction will have low resistance and the SCR will be in conduction. When the pulse is removed from the gate of the SCR, it will remain in conduction because the positive current that comes through the anode replaces the voltage the gate provided. The only way to turn the SCR off is to provide reverse-bias voltage to the anode-cathode or reduce the current flowing through the anode-cathode to zero. You should remember that the AC sine wave has zero voltage right before it provides the negative half of its waveform. This means that if the SCR is powered with AC voltage, the SCR is turned off when the AC waveform goes through 0 V and then to its negative half-cycle. When the AC voltage waveform goes positive again, a gate pulse can be provided, and the SCR can go into conduction again. The gate is used to provide a pulse that is used to cause the SCR to go into conduction.

19.22
OPERATION OF A SILICON-CONTROLLED RECTIFIER

Figure 19–25 shows an SCR connected in a circuit to control voltage to a DC load. The source voltage for this

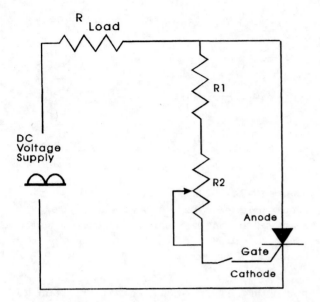

Figure 19–25 An SCR shown in a circuit controlling DC voltage and current to its load.

circuit is AC voltage. The main advantage of the SCR is that it will not go into conduction until it receives a pulse of voltage to its gate. The timing of the pulse can be controlled so that it can be delivered any time during the half-cycle that controls the amount of time the SCR is in conduction. The amount of time the SCR is in conduction controls the amount of current that flows through the SCR to its load. If the SCR is turned on immediately during each half-cycle, it conducts all the half-wave DC voltage just like a normal diode rectifier. If the gate delays the point at which the SCR turns on and goes into conduction at the 45° point of the half-wave, the amount of voltage and current the SCR conducts is 50% of the full applied voltage.

The other important feature of the SCR is that it will go into conduction only when its anode and cathode are forward biased. This means that if the supply voltage is AC, the SCR can go into conduction only during the positive half-cycle of the AC voltage. When the negative half-cycle occurs, the anode and cathode are reversed biased, and no current flows. This means the SCR is automatically turned off when the negative half of the AC sine wave occurs. Because the positive half of the AC sine wave occurs for 180°, the SCR can provide control of only 0° to 180° of the total 360° AC sine wave.

It is also important to understand that because the turn-off point of the SCR is fixed to the point where the sine wave begins to go negative, the SCR can be controlled only by adjusting the point where it is turned on. The point where the SCR is turned on and goes into conduction is called the *firing angle*. If the SCR is turned on at the 10° point in the AC sine wave, its firing angle is 10°. If the SCR is turned on at the 45° point, its firing

angle is 45°. The number of degrees the SCR remains in conduction is called the *conduction angle*. If the SCR is turned on at the 10° point, its conduction angle is 170°, which is the remainder of the 180° of the positive half of the AC sine wave.

19.23
CONTROLLING AN SCR

Figure 19–26 shows an SCR in a circuit with a UJT connected to its gate. The load in this circuit is a DC motor. This type of DC motor is often used as a damper-control motor. The circuit is powered by AC voltage, and the variable resistor in the oscillator (capacitor-resistor) circuit sets the timing for the pulse that is used to energize the gate of the SCR. Notice that a diode rectifier provides pulsing DC voltage for the capacitor, which charges to set the timing for the pulse that comes from the UJT. Because this DC voltage comes from the original AC supply voltage, it has the same timing relationship of the original sine wave. This makes it easier to adjust the pulse from the UJT to turn on the SCR gate at just the right time to control the firing angle of the SCR from 0° to 180°. In reality the firing angle is usually controlled from 0° to 90°, which gives sufficient range of control to adjust the output DC voltage that is sent to the DC motor. You should remember that the speed of the DC motor can be controlled by adjusting the voltage sent to the armature and field. This diagram shows the waveform for the voltage at each point in this circuit. The load in this circuit could also be any other DC-powered load.

19.24
TESTING AN SCR

You will need to test an SCR to determine if it is faulty. Because the SCR is made of PN junctions, you can use forward-bias and reverse-bias tests to determine if it is faulty. In this test you should put the positive probe on the anode and the negative probe on the cathode. At this point the ohmmeter will still indicate the SCR has high resistance. If you use a jumper wire and connect positive voltage from the anode to the gate, you should notice the SCR will go into conduction and have low resistance. The SCR remains in conduction until the voltage applied to its anode and cathode is reverse biased or until the voltage applied to the anode and cathode is reduced to zero. This means that you can turn the ohmmeter polarity switch to the opposite setting or you can

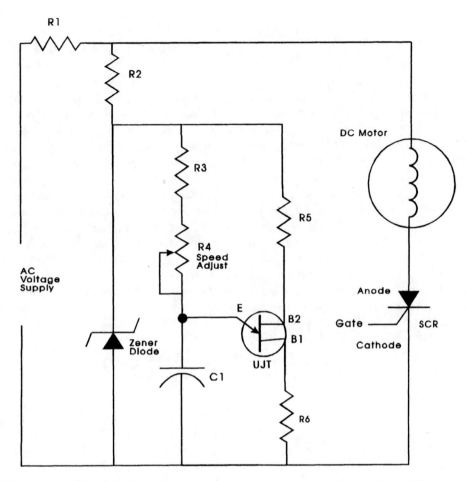

Figure 19–26 An SCR used to control a DC motor. A UJT is connected to the gate of the SCR to control its firing angle.

remove one of the probes, and the SCR will stop conducting. It is important to understand that the amount of current needed to keep the SCR in conduction is approximately 4 to 6 mA. This means that some high-impedance digital volt-ohmmeters will not have enough current when set to the ohms range to keep the SCR in conduction. If this is the case, you may need to test the SCR with an analog ohmmeter. The analog ohmmeter is a type of ohmmeter that has a needle and scale. You should also test the SCR for reverse bias to ensure that it has high resistance. Sometimes an SCR will not go into conduction because it has developed an open in its anode-cathode circuit. Other SCRs may stay in conduction at all times, which means the SCR is shorted.

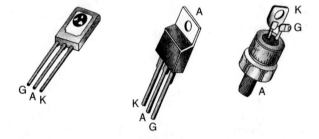

Figure 19–27 Typical SCRs shown in a variety of packages and case styles.

19.25

TYPICAL SCRS

SCRs are available in a variety of packages and case styles that include three terminals and larger types that

have threads so that they can be mounted directly into heat-sink material. The heat-sink material is made from metal and in some cases includes a fan that moves air over fins that are molded into the material to help remove the large amounts of heat that build up into the devices. Figure 19–27 shows examples of different types of SCRs.

19.26
DIACS

A unijunction transistor provides a positive pulse for the gate of the SCR, which allows the SCR to control large DC voltages and currents. A *diac* is an electronic component that provides a positive and negative pulse used as trigger signal for a device called a *triac*. A triac is similar to an SCR, except it can conduct voltage and current in both the positive and negative direction. Thus a triac provides a controlling function for AC voltage and current that is similar to the SCR. As a maintenance technician you may not notice the diac, because it is used as a device to produce a firing pulse for a triac. You learn later in this chapter how the triac is used in speed controls and other AC voltage-control circuits. Figure 19–28 shows the electronic symbol for a diac. Notice that the diac symbol consists of two arrows that show voltage pulses can be positive or negative. Because the diac symbol has two arrows, it can be shown in two different styles.

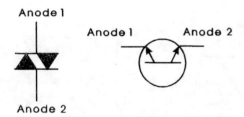

Figure 19–28 Electronic symbols for the diac. Notice the diac has two arrows, and it can be represented by either symbol.

19.27
OPERATION OF A DIAC

When voltage is applied to a diac in a circuit, its PN junction remains in a high-resistance state and blocks the voltage until the voltage level reaches the breakover level. Thus when AC sine-wave voltage is applied to the circuit, the sine-wave voltage will increase to its peak level, return to zero, and then increase to negative peak voltage. Each time the voltage increases toward its peak and exceeds the breakover level of the diac, the diac provides a sharp output pulse of the same polarity as the voltage that is supplied to it. For example, if the breakover voltage level is 18 V, the diac blocks voltage until the AC voltage exceeds the 18-V level. At this point, the diac produces a pulse that is approximately 18 V. Figure 19–29 shows the diac in its circuit and the pulse that it creates.

19.28
TRIACS

A *triac* is basically two SCRs that have been connected back to back in parallel so that one of the SCRs will conduct the positive part of an AC signal and the other will conduct the negative part of an AC signal. As you know, the SCR can control voltage and current in only one direction, which means it is limited to DC circuits when it is used by itself. Because the triac acts like two SCRs that are connected inverse parallel, one section of the triac can control the positive half of the AC voltage and the other section of the triac can control the negative half of AC voltage. Figure 19–30a shows the electronic symbol of the triac, and Fig. 19–30b shows the arrangement of its

Figure 19–29   (a) Diac in a circuit; (b) the pulse the diac produces.

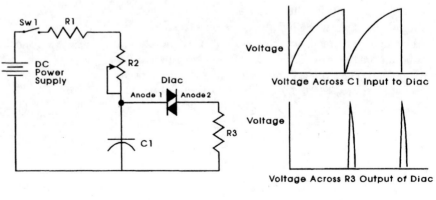

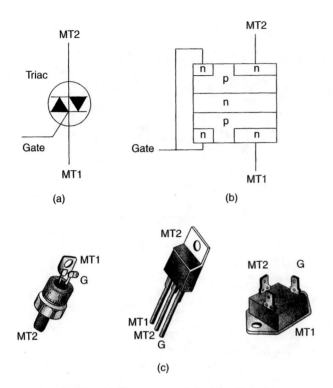

Figure 19–30 (a) Electronic symbol for the triac; (b) P-type and N-type material for the triac; (c) typical triac semiconductor devices.

P-type and N-type materials. The terminals of the triac are called *main terminal* 1 (MT1), *main terminal* 2 (MT2), and *gate*. Because the triac is basically two SCRs that are connected in an inverse parallel configuration, MT1 and MT2 do not have any particular polarity.

19.29
USING A TRIAC AS A SWITCH

A triac can be used in an industrial circuit as a simple on-off-type switch. In this type of application, the MT1 and MT2 terminals are connected in series with the AC load. When the gate gets a positive pulse signal, the triac turns on for the positive half of the AC cycle. When the AC voltage waveform returns from its positive peak to 0 V, the triac turns off. Next the negative half-cycle of the AC voltage waveform reaches the triac, and it receives a negative pulse on its gate and goes into conduction again.

This means that the triac will look as though it turns on and stays on when AC voltage is applied. The load that is connected to the triac will receive the full AC sine wave, just as if it were connected to a simple single-pole

switch. The major difference is the triac switch can be used for millions of on-off switching cycles. The other advantage of the triac switch is that the gate pulse can be a very small amount of voltage and current. This allows the triac to be used in temperature control where the temperature-sensing part of a thermostat can be a small solid-state sensing element called a *thermistor.* The sensing element can also be a very narrow strip of mercury in a glass bulb that is very accurate but can carry only a small amount of voltage or current.

Another useful switching application for a triac is the solid-state relay. Section 19.32 explains AC and DC solid-state relays. Figure 19–31 shows a triac in a thermostat connected to a 24-VAC gas valve in a furnace. The temperature-sensing element is connected to the gate terminal of the triac. When the temperature increases, it causes the mercury in the temperature-sensing element to expand and make contact between the two metal terminals. The small amount of voltage flowing through the sensing element is sufficient to make the triac go into conduction and provide voltage to the gas valve.

19.30
USING A TRIAC FOR VARIABLE VOLTAGE CONTROL

A triac can also be used in variable voltage control circuits, because it can be turned on any time during the positive or negative half-cycle, similar to the way the SCR is controlled for DC circuit applications. In this type of application, a diac is used to provide a positive and negative pulse that can be delayed from 0° to 180° to control the amount of current flowing through the triac. This type of circuit can be used to control the amount of current and voltage supplied to electric heating elements. This allows the amount of current and voltage to be controlled from zero to maximum, which in turn allows the temperature to be controlled very accurately. Figure 19–32 shows a diagram of a triac used to control an electric heating element that is powered by an AC voltage source. Notice that a variable resistor is connected with a capacitor to provide an oscillator pulse to the diac. Because the resistor and capacitor are connected to an AC voltage source, the pulse will be both positive and negative as the AC sine wave changes polarity. The triac is connected to the same AC voltage source, so the timing of the pulse from the diac to its gate will always be synchronized with the polarity of the voltage arriving at the main terminals of the triac.

Figure 19–31 A triac uses a switch to turn on voltage to a gas valve for a furnace. The triac receives its gate signal from a small amount of voltage that moves through the temperature-sensing element.

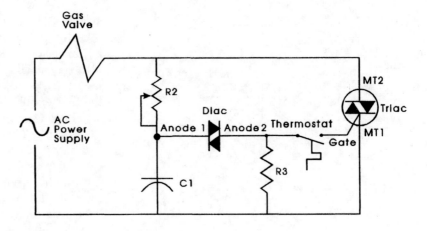

Figure 19–32 A triac used to control variable voltage to an AC electric heating element. Notice a diac is used to provide a positive and negative pulse to the triac gate.

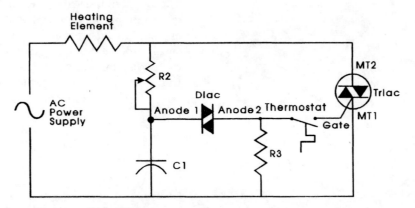

19.31

TESTING A TRIAC

Because the triac is made from P-type material and N-type material, it can be tested like other junction devices. The only point to remember is that because the triac is essentially two SCRs mounted inversely parallel to each other, some of the ohmmeter tests are not affected by the polarity of the ohmmeter leads. In the first test for a triac an ohmmeter should be used to test the continuity between MT1 and MT2. When no gate pulse is present, the resistance between these terminals should be infinite regardless of which ohmmeter probe is placed on each terminal. Figure 19–33 shows how the ohmmeter should be connected to the triac. In Fig. 19–33a you can see that the positive ohmmeter probe is connected to MT1 and the negative probe is connected to MT2. Because no voltage is applied to the gate, the resistance should be infinite. When voltage from MT2 is jumped to the gate, the triac will go into conduction and the ohmmeter will indicate low resistance.

In Fig. 19–33b you can see that the positive ohmmeter probe is connected to MT2, and the negative probe is connected to MT1. Because no voltage is applied to the gate, the triac is not in conduction, and the ohmmeter will indicate the resistance at this time is high. When voltage from MT2 is jumped to the gate, the triac will go into conduction, and the ohmmeter will indicate the resistance is low. It is important to remember that the triac will stay in conduction only while the gate signal is applied from the same voltage source as MT2. As soon as the voltage source is removed, the triac will turn off.

19.32

OPERATIONAL AMPLIFIERS

Operational amplifiers have become a common component in heating, air-conditioning, and refrigeration systems because they allow very small signal voltages from sensors to be amplified to a usable voltage to turn compressors and other loads on and off. The *operational*

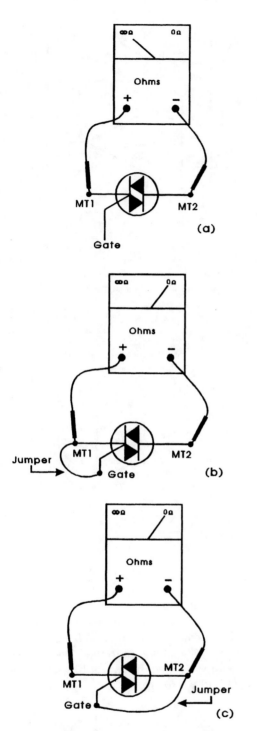

amplifier, also called an *op amp*, is made of dozens of PN-junction devices such as transistors that are manufactured on a single integrated circuit (IC). You will find them in most of the newer solid-state controls such as thermostats, voltage monitors, and other protection and control devices.

The op amp provides two basic functions. First, it can take a very small signal and amplify it. For example, a sensor can send a 100-mV signal to an op amp, and the op amp can multiply the signal strength by 10 times ($\times 10$) so that the output signal will be 1 V, or the op amp can multiply the signal by 100 times ($\times 100$) so that the output signal will be 10 V. Second, the op amp can change the polarity of the voltage of the input signal, or it can keep it at the same polarity. For example, if a negative sensor voltage is applied to the inverting input terminal of an op amp, the output signal will be positive. If the negative input signal is applied to the noninverting input, the output will remain negative. Each op amp input can also receive a signal that spans -0.5 to $+0.5$ V and produce an output voltage that spans -12 to $+12$ V. Figure 19–34 shows the electronic symbol for an operational amplifier with all its terminals identified.

In Fig. 19–34b you can see that the op amp has two input terminals identified as terminals 2 and 3 and an output terminal identified as terminal 6. Terminal 2 is identified by a negative sign ($-$), which indicates that this is the *inverting input*. If an op amp receives a signal on the inverting input, the voltage polarity will be inverted when the signal reaches the output. Any signal the op amp receives on its noninverting input will remain the same polarity when it reaches the output. The op amp also has two terminals where power supply voltage is applied. It is important to understand that when the op amp amplifies as signal, the output signal cannot be any larger than the supply voltage that is connected to the op amp at its V+ (terminal 7) and V− (terminal 4) terminal. Typical power supply voltage can be 12, 15, or 18 VDC. You should also notice that the op amp is mounted on an eight-pin integrated circuit, which means that the chip will have eight pins. The chip has an identifier mark at the top of the chip to help identify the number 1 terminal.

Figure 19–33 (a) Testing a triac by placing the ohmmeter's positive probe on MT1 and its negative probe on MT2. Since there is no gate voltage, the triac does not go into conduction, and the meter reads high resistance between MT1 and MT2. (b) Testing a triac by placing the ohmmeter's positive probe on MT1 and its negative probe on MT2. When positive gate voltage is applied from the MT1 probe, the triac goes into conduction. (c) Testing a triac by placing the ohmmeter's positive probe on MT1 and its negative probe on MT2. When negative gate voltage is applied from the MT2 probe, the triac goes into conduction.

19.33
THE OP AMP AS AN AMPLIFIER

Like a transistor, the op amp has the ability to amplify an input signal. The amount of amplification is called *gain*. The major difference between the transistor and the op amp is that the amount of amplification (gain) for the transistor is designed into the transistor when it is manufactured. In contrast, the amount of gain for the op amp

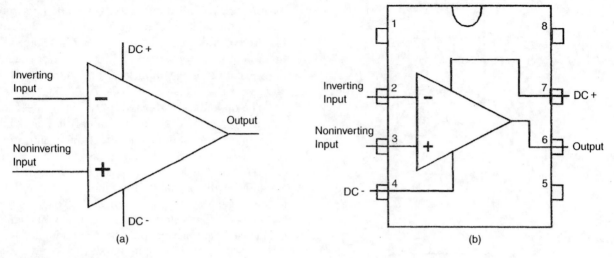

Figure 19–34   (a) Electronic symbol for an operational amplifier (op amp); (b) pin outline of an integrated circuit for an op amp.

Figure 19–35   An op amp with a resistor connected to the inverting input and a feedback resistor connected between output and inverting input.

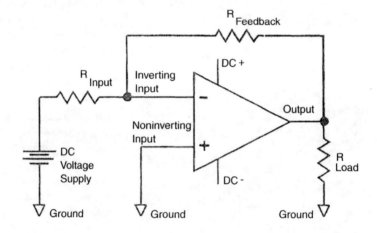

is determined by the external resistors connected to it. This means that it is possible to design a circuit where an op amp can have any amount of amplification. The general values of amplification are between 10 to 100.

Figure 19–35 shows a typical op amp circuit. The resistor that is connected to the noninverting input is called the input resistor (R_{in}), and the resistor at the top of the op amp that is connected between the output terminal and the noninverting input is called the feedback resistor (R_F). When the input signal is connected to the inverting input, the op amp is called an inverting input. The amount of gain for the inverting op amp is determined by the formula

$$Gain = \frac{R_F}{R_{in}}$$

The formula for the noninverting op amp is

$$Gain = 1 + \frac{R_F}{R_{in}}$$

In an example circuit the resistor connected to the noninverting op amp is 10 kΩ and the feedback resistor is 100 kΩ, so the gain is 10 for this circuit. If the input resistor is changed to 1 kΩ, the gain for the circuit will be 100. If a 0.1-V signal is sent to the input, the output will produce a 1-V signal if the gain is 10, and the output signal will be 10 V if the gain is 100.

19.34
CONTROLLING A RELAY WITH AN OP AMP

One circuit you will find in industrial electricity systems is a relay controlled by an op amp. One type of circuit has a temperature sensor (that produces a millivolt signal) to control a 10-hp, three-phase fan motor. Figure 19–36 shows an example of this circuit. In this circuit

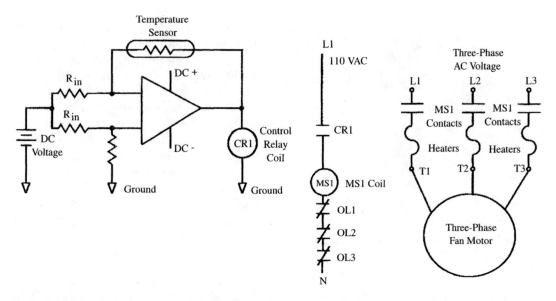

Figure 19–36 A millivolt signal from a temperature sensor which is used to control the coil of a small relay, and the relay controls the coil of a motor starter and 10-hp three-phase fan motor.

the op amp receives (CR1) a small signal from a temperature sensor. The coil of a small relay is connected to the output of the op amp. The contacts of the small relay control voltage to the coil of a motor starter (MS1) and the three-phase motor is connected to the contacts of the three-phase motor starter.

When the temperature sensor sends a 0.1-V signal to the input of the op amp, the op amp increases the signal to 10 V at the output, which sends voltage to the coil of the small relay. When the coil of the small relay is energized, its contacts close and provide voltage for the coil of the motor starter, and the contacts of the motor starter close and provide 240 VAC to the 10-hp fan motor. When the temperature drops, the voltage from the temperature sensor drops below 0.1 V, and the small relay coil drops out. When this occurs, the motor-starter coil becomes de-energized, and the fan motor turns off.

19.35

CONTROLLING THE SPEED OF A THREE-PHASE MOTOR WITH A VARIABLE-FREQUENCY DRIVE

The speed of a single-phase or three-phase AC induction motor can be controlled by adjusting the frequency of the voltage that is sent to it. In modern air-conditioning and refrigeration systems, it is important to be able to vary

the speed of motors in the system to provide more efficient use of electricity and to provide a system that is able to control temperatures more precisely.

In the 1980s, manufacturing technology for transistors and other solid-state devices made it possible for transistors to be able to control larger voltages and currents. Prior to that development, triacs and SCRs had been the devices used for controlling large voltages and currents. With the advent of larger transistors, the frequency of larger voltages could be manipulated to a point where the speed of three-phase and single-phase motors could be controlled by changing the frequency of their supply voltage.

Figure 19–37 shows a typical frequency drive. The drive in the picture costs approximately $450 and can control compressors or fans up to 1 hp. Larger drives can control motors up to 100 hp. In some air-conditioning systems the frequency control board is built into the control cabinet with the other air-conditioning controls.

19.36

OPERATION OF THE VARIABLE-FREQUENCY DRIVE CIRCUIT

Figure 19–38 shows the circuit diagram for a typical variable-frequency drive. You should notice that the circuit consists of three sections. The first section is the *rectifier section*, where a three-phase diode bridge

Figure 19–37   A typical 1-hp variable-frequency drive amplifier. (Courtesy of Rockwell Automation's Allen-Bradley Business)

rectifier changes the three-phase AC voltage to pulsing DC voltage. The second section is the *filter section*, where the pulsing DC voltage is smoothed to pure DC. The third section is the *transistor switching section*, which produces the three-phase AC voltage at the desired frequency.

In the rectifier section you can see that six diodes are connected in a bridge circuit to convert the three-phase 60-Hz voltage to DC voltage. The actual rectifier section is mounted on a single module and can be exchanged by removing the three input voltage wires and the two DC bus connections. Each diode in the module can still be tested for front-to-back resistance ratio with an ohmmeter just as if it were an individual diode. The output of the rectifier section is 12 half-wave pulses 60° apart. The output of the bridge rectifier is connected to two large copper conductors that are called the *DC bus*.

The filter section of the circuit consists of several capacitors that are connected in parallel with the DC bus

and a large inductor that is connected in series with the DC bus. The capacitors charge and discharge in synchronization with the input voltage. This causes the half-wave signal to be converted to pure DC. The capacitors are used to filter the voltage part of the waveform, and the inductor is used to filter the current part of the waveform.

The transistor section consists of six transistors, two for each output phase. You should notice that one transistor of each phase is connected to the positive DC bus, and the second transistor is connected to the negative DC bus. Each transistor is controlled by a base firing circuit that is controlled by a microprocessor chip (small computer). At the appropriate time, each transistor is turned on in six distinct steps. Figure 19–39 shows an example of the waveform that results from turning the transistor on in six steps. You can see that the transistor that is connected to the positive DC bus conducts and produces the positive part of the output AC waveform, and the transistor that is connected to the negative DC bus produces the negative part of the output AC waveform. The first set transistors produces the A phase for the three-phase output waveform, and the second set produces the B phase. The third set of transistors produces the C phase of the AC output waveform. You can see that the output frequency can be adjusted by changing the timing of the base firing circuit for each set of transistors. Typical control for a variable-frequency drive (VFD) is from 0 to 120 Hz, which allows the motor to be controlled between 0 and 200% of its rated rpm. In reality the motor is generally adjusted from 60 to 130% of rated rpm, which provides sufficient control for the system being adjusted. It is important to note that the VFD can only control the motor for a short period of time at rpms above 100% because the motor will overheat unless the motor is rated specifically for a drive.

It is important to be able to adjust the speed of a conveyor motor or large fan to provide variable air flow. The

Figure 19–38   Circuit diagram of a variable frequency drive.

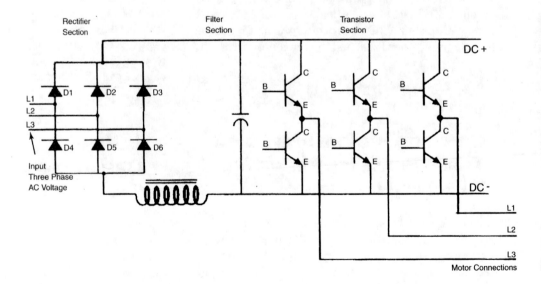

VFD can also control the speed of a pump or compressor motor. If the motor needs to continually run above or below rated speeds gear drives should be considered rather than use the variable frequency drive.

19.37
SOLID-STATE RELAYS

Another solid-state device that you will see frequently used in industrial control systems is called a *solid-state relay (SSR)*. The solid state relay uses DC voltages for the control part of the circuit, which acts like the coil of a regular relay, and it can switch AC voltages through the part of the circuit that acts like the contacts. The

solid-state relay uses SCRs, transistors, or triacs to control various loads. It operates very much like an electromechanical relay in that it is energized by a small voltage that controls a larger voltage and current.

Figure 19–40 shows a picture of a typical solid-state relay and provides a block diagram of the operation of the relay. Figure 19–41 shows the electrical diagram for a solid-state relay that uses an LED and a photo transistor to accept the DC input and a triac to control the AC voltage and load in the output. The LED and phototransistor are called an *optoisolation device*, or an *optocoupler*. When a DC input signal is received at the SSR, it is converted to light through an LED. The LED and phototransistor are manufactured in an integrated circuit so that the light from the LED shines directly on a phototransistor. The integrated circuit is imbedded in the solid state relay. When the light strikes the phototransistor, its collector-emitter circuit goes to low resistance, and the collector current flows through it to the base of the triac. The AC load is controlled directly by the triac. Notice that the power source for the load must be isolated from the input signal.

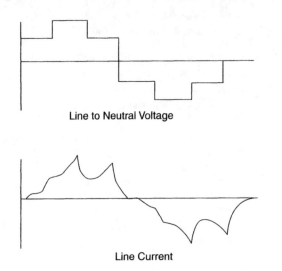

Line to Neutral Voltage

Line Current

Figure 19–39   The six-step waveform of the output transistors.

Figure 19–40   A solid-state relay rated for DC input and 120-VAC output.

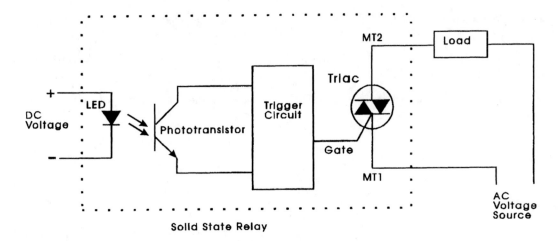

Solid State Relay

Figure 19–41   An electronic diagram of a solid-state relay that has a DC input signal and 120-VAC output.

QUESTIONS

Short Answer

1. Explain P-type and N-type material.

2. Explain the operation of a diode (PN) junction and show the input AC waveform and the output DC waveform.

3. Identify the terminals of a transistor and explain its operation.

4. Explain the operation of the SCR and explain the type of circuit where you would find one.

5. Explain the function of the unijunction transistor when it is used with an SCR.

True or False

1. The function of the triac is to provide switching similar an SCR, except the triac operates in an AC circuit.

2. The light-emitting diode (LED) is similar to a DC lightbulb in that it has a very tiny filament.

3. The main function of diodes and SCRs is to convert AC voltage to DC voltage.

4. The full-wave bridge rectifier uses one diode.

5. The UJT and diac are important solid-state devices that provide pulse signals to other components to use as a firing signal.

Multiple Choice

1. A PN junction is forward biased when _____.
   a. positive voltage is applied to the N-material and negative voltage is applied to the P-material
   b. negative voltage is applied to the N-material and negative voltage is applied to the P-material
   c. its junction has high resistance

2. The rectifier _____.
   a. converts DC voltage to AC voltage
   b. converts AC voltage to DC voltage
   c. can be used as a seven-segment display for numbers

3. A circuit that has one SCR in it can _____.
   a. control AC voltage
   b. control DC voltage
   c. control both AC and DC voltage

4. The transistor operates similar to a relay in that _____.
   a. its emitter is like the coil and its base and collector are like a set of contacts
   b. its collector is like the coil and its base and emitter are like a set of contacts
   c. its base is like a coil and its emitter and collector are like a set of contacts

5. The solid-state relay _____.
   a. uses an LED to receive a signal and a transistor to switch current
   b. uses a capacitor to receive a signal and an IC to switch current
   c. uses a transistor to receive a signal and an IC to switch current

PROBLEMS

1. Draw the symbol for a diode and identify the anode and cathode.

2. Draw the symbol for an NPN and a PNP transistor and identify the emitter, collector, and base.

3. Draw the symbol for a silicon-controlled rectifier (SCR) and identify the anode, cathode, and gate.

4. Draw the electrical symbol for a unijunction transistor and identify base 1, base 2, and the emitter.

5. Draw the electrical symbol for a triac and identify MT1, MT2, and the gate.

6. Draw the block diagram of the variable frequency drive.

CHAPTER 20

Lockout, Tag-out

OBJECTIVES

After reading this chapter you will be able to:

1. Identify the lockout, tag-out procedure.

2. Identify sources of electrical energy.

3. Identify sources of other types of energy.

4. Identify sources of potential energy and kinetic energy on a machine.

20.0

INTRODUCTION

A wide variety of safety precautions are used in factories today to ensure that workers are not endangered or injured while they are working. One program in use is called *lockout, tag-out*. This program is designed to ensure that all power to the machine is de-energized and locked in the off position prior to working on the system and that tags are attached to the disconnect switches to indicate who is working on the machine.

The lockout, tag-out program is mandated to all factories by the Occupational Safety and Health Administration (OSHA). The exact standard is OSHA Standard 1910.147. This program explains techniques that are to be used to ensure that uncontrolled release of energy is locked in the off position prior to working on the machine. The program is designed to prevent injury that may be caused by uncontrolled machine motion, electrical shock, or electrical burns or by coming into contact with hazardous materials.

The National Institute for Occupational Safety and Health (NIOSH) has maintained data that show a wide number of factory accidents that resulted in serious injury or death. These accidents include personnel being crushed between moving parts on a machine, such as the clamps on a molding machine, having arms and legs caught in moving conveyor equipment, and amputation of arms or legs from blades. These accidents occurred because coworkers did not pay attention when workers were working inside equipment and machines, or because workers did not understand the dangers of working around machinery that still has power applied to it.

The lockout, tag-out procedure is devised to allow maintenance and other personnel to safely work on and around machinery where unexpected start-up of the equipment or the unexpected release of stored hydraulic, pneumatic, or other type of energy may cause injury or death. The procedures should be used any time when equipment is being set up, inspected, or adjusted or any time when maintenance is being conducted.

These procedures are particularly important when guards must be removed to set up the machine or change molds and dies. During this time workers are especially vulnerable to conditions where body parts such as arms and legs are exposed to moving parts. In these conditions workers may sustain serious injury from a crushing accident, or they may have their clothing snagged, which can pull them deeper into moving parts in the machinery.

The procedure should also cover the disconnection and lockout of electrical power when necessary. There are times when machines need to be lubricated or cleaned by personnel, and it is important that all power is turned off. This will prevent personnel from being shocked if they come into contact with electrical terminals or if the cleaning fluids and equipment they are using come into contact with electrical terminals. All hydraulic or pneumatic power should also be disabled at this time by bleeding their pressure to zero. This will prevent accidental activation of a hydraulic or pneumatic actuator, which could cause a cylinder to extend and crush someone.

Fuels must also be controlled. This includes exposure to fuel intakes and outlets and exposure to extreme heat such as from steam pipes or other exposed heating elements.

20.1

A TYPICAL LOCKOUT, TAG-OUT POLICY

Each factory must create a policy for lockout, tag-out and ensure that the policy is being enforced. The following illustration may be used as an example, but it does not necessarily include all the facets of your shop.

The policy ensures that electrical disconnect for all equipment shall be locked out to the de-energized state where possible while maintenance work is being completed on said equipment. If it is not possible to lock out the electrical disconnect, equipment must be protected against possible operation by unauthorized personnel during maintenance procedures. In all cases, all the power sources for the equipment, such as electrical, hydraulic, pneumatic, and gravity, must be disabled and made inoperable so as to prevent injury to personnel. Employees must never try to operate or reapply any of the power sources that have been locked out or tagged out by another person. The person who applies the lockout or tag-out equipment is the only authorized person to remove these safety devices, unless the procedure continues past the work shift of the person. If the work continues past the work shift of the person applying the lockout device, the person should remove his or her lockout and allow the new shift workers to place their device on the equipment. If the device must be removed by personnel on a different shift they must follow the procedure for such work.

20.2

DETERMINING THE LOCKOUT, TAG-OUT POLICY AND PROCEDURE

The lockout and tag-out procedures should be determined and enforced by each individual company. There are many sources for examples of what should be included in these procedures. This chapter provides a basic version of a typical procedure. It is limited to a few of the things that you should do during the lockout, tag-out procedure, and it is not intended to be a complete procedure. Each work site has specific machines and equipment that may require additional procedures.

When a company is designing lockout and tag-out procedures, it must keep in mind the different types of

energy that the system uses and the way the machine is engineered to operate, such as using gravity for part of the machine operation. While writing the procedure, the company must also have a plan to evaluate how well the procedure is working to prevent accidents, and finally it must also design a means to enforce its procedures to ensure they are being used at all times by all personnel. The evaluation and enforcement part of the procedure should also include continuing education. This will ensure that all personnel remain aware of the potential dangers that are ever present and that they will learn to perform the procedure automatically at all times.

20.3

IDENTIFYING SOURCES OF KINETIC ENERGY AND POTENTIAL ENERGY

Energy in and around a machine may take the form of potential or kinetic energy. *Kinetic energy* is energy that is in motion. This is the energy that is easy to see and understand when you are around a machine. For example, the energy that a plastic injection molding machine uses to move its mold (clamp) open and closed is easily observed when the clamp is moving. *Potential energy* is energy that is stored and is available for use after the power source is removed. Its dangers are not always so apparent. For example, a hydraulic system may be designed to store hydraulic fluid in a pressurized accumulator where it is waiting to be released when a hydraulic valve is activated and the stored pressure moves a hydraulic cylinder. What is dangerous about this form of energy is that it continues to be present even after the hydraulic pump is turned off and locked out.

Other forms of potential energy include parts of the machine that utilize gravity for part of its operation. For example, during each cycle, a metal stamping press may raise a die several feet above the parts that are being stamped from the die. If the machine stops with the die in the top-dead-center position above the die, it has the potential of falling from gravity when power is turned off. In this type of application, any part of the machine that could move from gravity must be supported in such a way that it cannot move when power is removed. Another problem arises when the machine or the die must be checked or tested while it is running.

Another example of potential energy is the energy that is stored in rotating parts of a machine, such as a rotary saw blade or rotary grinding wheel. In these applications, the machine uses an electric motor or similar power source to provide large amounts of energy to get the saw blade or grinding wheel up to top speed. When the blade or wheel is spinning at top speed, it has a large amount of energy stored in its mass, which will cause it to continue to spin even after the power to the motor is turned off. In most conditions, the heavier the blade or wheel, and the higher the speed, the more energy it will store and the longer it will continue to spin when power is turned off. In these applications, the company must evaluate the potential for danger and provide braking to stop the blades or wheels as quickly as possible. Another method of protection is to provide sufficient guards so that personnel cannot come into contact with the moving parts. In some cases the guards are interlocked with the moving parts to cause them to stop moving immediately any time the guard is moved or removed.

In some machines, springs are used to provide an additional source of energy. Any time a spring is under compression or extension it has the potential of causing physical harm. This type of energy is not always apparent since some springs are not visible inside spring-return-type pneumatic cylinders. When the ends of the cylinder are removed during maintenance, the spring may release its energy and cause harm.

20.4
DESIGNING SAFETY INTO THE MACHINE

The personnel who are developing the lockout, tag-out procedure must be aware of all these types of energy sources and the potential problems they present. They should evaluate the machine and provide guards and stops where necessary and build in other safety equipment to minimize the dangers to personnel.

20.5
DOCUMENTATION AND TRAINING

Documentation and training are essential parts of the lockout, tag-out program. Documentation will ensure that each procedure is explained in detail so that personnel are continually reminded of using lockout and tag-out procedures at the appropriate times when they are on the job. Training is essential to ensure that the employees understand the proper procedure and perform it correctly. Training helps new employees to understand the dangers of working around equipment and reminds them that they each have a personal responsibility in ensuring that lockout and tag-out procedures are used at all times.

20.6
OSHA INSPECTIONS

The Occupational Safety and Health Administration may become involved if accidents occur that are caused by not following procedures or if incorrect lockout, and tag-out procedures have been used. The problem becomes extreme if OSHA finds that the employer does not have lockout and tag-out procedures or if these procedures are not being enforced.

20.7
WHAT IS LOCKOUT?

Lockout procedures include manually disconnecting the electrical and hydraulic power and placing a locking device (usually a padlock) on the disconnect so that it cannot be reconnected without removing the padlock. Figure 20–1 shows a typical lockout safety center that contains all the safety equipment needed for a lockout procedure. If more than one employee is working on the equipment, each should place his or her padlock on the disconnect. A special device that allows multiple padlocks to be connected on the same disconnect is shown in Fig. 20–2.

When the clasp is closed, up to six padlocks can be installed at one time. The clasp is secured around the disconnect, and it cannot be opened until the last padlock is removed.

The lockout, tag-out procedure also requires that each padlock and lockout device be identified with a picture and the name of the employee who placed the lockout on the disconnect. Lockout devices must be supplied by the employer and are to be maintained by the employee.

In some cases, the work on a machine involves multiple personnel, and multiple padlocks will be applied to the disconnect. When all work is completed and it is time to return power to the machine, all the locks must be removed and all personnel must be accounted for so that they are not in danger.

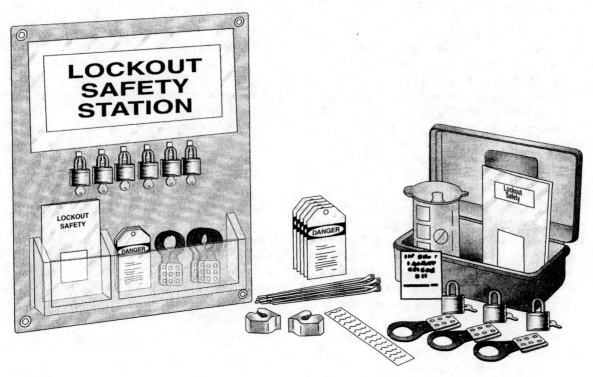

Figure 20–1 A typical lockout safety center and safety kit that provides all the safety devices needed for electrical lockout.

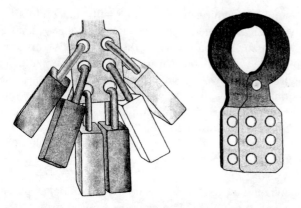

Figure 20–2 (a) A typical lockout device that can accept multiple padlocks; (b) a typical lockout device.

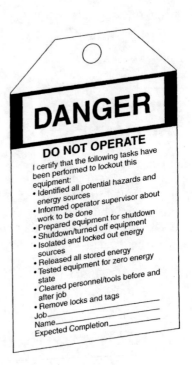

DANGER

DO NOT OPERATE

I certify that the following tasks have been performed to lockout this equipment:
- Identified all potential hazards and energy sources
- Informed operator supervisor about work to be done
- Prepared equipment for shutdown
- Shutdown/turned off equipment
- Isolated and locked out energy sources
- Released all stored energy
- Tested equipment for zero energy state
- Cleared personnel/tools before and after job
- Remove locks and tags

Job_____
Name_____
Expected Completion_____

Figure 20–3 Typical lockout tag.

FRONT   BACK

Figure 20–4   Additional types of lockout tags.

20.8
WHAT IS TAG-OUT?

Figure 20–3 shows a typical tag that is used for tag-out procedures. The tag has a written warning that explains why the machine has been tagged. This warning also explains that the power should not be applied to the machine until the person working on the machine has completed the work and is ready for power to be reapplied.

All tags used in the lockout, tag-out procedure should be uniform and standardized so that they are easily recognized by all employees. Figure 20–4 shows additional types of lockout tags.

20.9
DESIGNING A LOCKOUT, TAG-OUT PROCEDURE FOR YOUR MACHINES

The first step in a typical lockout, tag-out procedure is to evaluate each machine and determine all dangerous conditions that are present when the machine is running and when it is locked out for maintenance. This includes identifying all possible sources of power to the machine and listing them by type, such as electrical, hydraulic, and pneumatic. It also includes identifying any hazardous material, chemicals, sources of radiation, or exposure to sources of thermal (heat) energy.

The next step includes identifying all employees who may work on or around the machine (operators, inspectors, supervisors, and others). These employees should always be notified when the machine is going to be turned off and locked out and tagged out. There should also be a means of notifying affected personnel who work on other shifts. The notification should include who shut the machine down and what is being worked on. Other information could include the expected time when the machine will be in operation again.

The next step is the actual shutdown of the equipment. This includes pressing the stop button, moving any other switches to the off position, and making sure that all motion on the machine is stopped. All sources of energy should be secured in the off condition and all forms of potential energy should be neutralized so that they do not present a hazard. After you have identified all sources of energy and have ensured they are in the off position, you can apply lockout and tag-out hardware to ensure that they remain in the off position until you are ready to turn them back on.

It is important at this time to identify several methods of disrupting power that are unacceptable because they can easily be bypassed and pose a danger to workers. For example, if you simply remove the fuses to a machine and do not lock out the fuse box, someone could replace the fuses and apply power to the machine while you are still working on it. If you turn off a pneumatic hand valve but do not lock the valve in the off position by a chain or other blocking device, it may be turned on again and cause serious injury. Some equipment is controlled by a circuit breaker rather than a disconnect-type switch. Figure 20–5 shows lockout devices for circuit breakers. You can attach the lockout device to ensure that power to the circuit is turned off and remains off. If the equipment is powered by an extension cord, you must secure the end of the extension cord to ensure that the power is turned off to the equipment. Figure 20–6 shows typical lockout equipment for an electrical extension cord.

If the equipment has air pressure or steam pressure that must be turned off and secured, you will need to use a safety donut for locking out these valves. Figure 20–7 shows a lockout donut for a globe valve, and Fig. 20–8 shows a lockout for a ball valve.

It is also important to identify any interlocking power sources or auxiliary power sources to each machine. For example, if a machine uses a robot to load and unload it, it is possible for some electrical signals from the robot cabinet to be sent to the electrical cabinet of the machine. If the machine is locked out, but the robot is not, you could have problems with the robot moving into the machine space, or you could have problems with electrical wires from the robot that terminate in the machine cabinet still being powered. Since many of these signals are 110 VAC, they can pose an electrical shock hazard.

After all power is shut down and locked in the off position, you should make a number of tests to ensure that all the power is in fact turned off. You should also test all blocks that are in place to ensure that they are secure and holding the machine in a safe condition. After all tests are made you can begin your work.

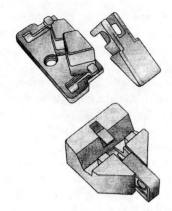

Figure 20–5 Circuit breaker lockouts.

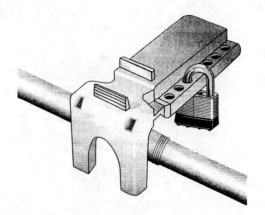

Figure 20–8 Ball valve lockout.

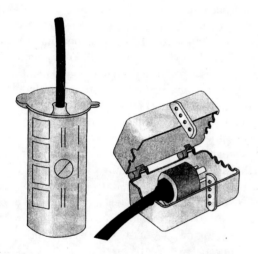

Figure 20–6 Lockout for electrical plugs and extension cords.

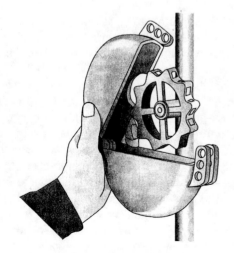

Figure 20–7 Safety donut for locking out air and steam valves.

20.10

REMOVING THE LOCKOUT, TAG-OUT DEVICES AND RETURNING POWER TO THE MACHINE

After all work has been completed on the machine, you are ready to remove the lockout and tag-out devices and return power to the machine. You must make several checks prior to removing the lockout and tag-out devices. First, ensure that all work is complete and that all tools and any test equipment have been removed. All guards and safety doors should be returned to the normal condition and tested to ensure that they operate correctly. Be sure that all parts of the machine that have been disassembled are reassembled properly.

When you have completed these tasks, you must physically account for all personnel who are in the area to ensure that they know and understand that you are ready to turn on power to the machine again. This is probably the most important part of the procedure, especially if more than one person is working on the machine, or if several employees routinely work on the machine when it is in normal operation.

When you are ready to apply power again, be sure you follow the operational checklist, so that power is returned to the machine in the proper order. For example, some machines must have circulating pumps running before steam valves are returned to their open position.

20.10.1 Removing a Coworker's Lockout and Tag-out Equipment

The general rule is that the person who placed the lockout on the machine must be the one who removes it. But there will be times when work is started on the day shift and completed 14 hours later on another shift. In this case the person who placed the lockout, tag-out equipment on the machine is not present to remove it. When this occurs, it is permissible for a supervisor or person of responsibility to remove the lock. The personnel on each shift who may remove a lock should be identified when the lockout, tag-out procedure is written. The supervisor may have a master key or permission to cut the lock off with bolt cutters. The person who removes the lock should always file a report to indicate why the lock was removed and who performed the action.

20.11

REVIEW OF LOCKOUT, TAG-OUT PROCEDURES

The lockout and tag-out procedure is designed to ensure that all sources of energy are locked out and tagged out whenever maintenance procedures require that power must be removed from the machine. The

lock provides security to ensure that the disconnect switch remains in the off position, and the tag provides a means to identify who placed the lock on the disconnect. When the procedure is created, a comprehensive review of all safety aspects of a machine should be accounted for. The lockout, tag-out practices should include training and education to ensure that all employees understand the safety hazards that are present with each machine and that they continually use correct safety procedures when working on and around the machine. The procedures should identify all sources of kinetic and potential power as well as sources of hazardous material, thermal energy (steam), and radiation.

Lockout and tag-out procedures are mandated and enforced by OSHA, and all procedures, equipment, and training are to be provided by the employer. Each employee is responsible for utilizing proper lockout and tag-out procedures and for maintaining the equipment that has been issued to him or her. Be sure to use the correct lockout and tag-out procedure any time you are working around equipment.

QUESTIONS

Short Answer

1. Explain what a lockout, tag-out procedure is.

2. What do the initials OSHA stand for?

3. What is kinetic energy?

4. What is potential energy?

5. Who is allowed to remove a padlock from a lockout?

True and False

1. A vertical press that is in the open position so that its platen could fall is an example of potential energy.

2. A fluid accumulator that is under pressure for a hydraulic press is an example of kinetic energy.

3. Lockout devices are available that allow more than one padlock to be connected to the same lockout.

4. The lockout device should have a picture of the owner on it in case the lockout is lost.

5. It is permissible to have electrical power applied to a machine if you need to make voltage readings as part of your troubleshooting procedure.

Multiple Choice

1. Who is responsible for creating the lockout, tag-out procedure in a factory?
   a. OSHA
   b. Each worker should create his or her own procedure.
   c. Each factory should create its own procedure.

2. The forms of energy that you should be aware of during a lockout procedure include _____.
   a. hydraulic and electrical power
   b. hydraulic, electrical, and potential energy sources such as gravity
   c. only electrical power sources

3. If an employee on the day shift places a padlock on a lockout device, and a second-shift employee completes the repair, _____.
   a. the second-shift supervisor can remove the first-shift padlock
   b. the second-shift employee can remove the first-shift padlock
   c. no one on the second shift can remove the padlock, it cannot be removed until the employee that placed the padlock on the lockout comes back to work

4. Prior to removing a padlock from a lockout device and restarting a machine, you should _____.
   a. replace all guards and turn on the power
   b. alert all employees in the area who work on this machine that you are turning on the power and starting the machine
   c. call the operator to start the machine

5. The major problem with potential energy sources during a lockout procedure is that _____.
   a. the hazard they present may not be readily apparent such as a machine part that is supported by hydraulic pressure that will drop when the pressure is released
   b. potential energy is much more dangerous than electricity
   c. potential energy is not a real threat, like kinetic energy, so it can be ignored

PROBLEMS

1. You are asked to create a lockout, tag-out procedure for a plastic injection molding machine that has both electrical power and hydraulic power. List all the things you should be aware of for the lockout procedure.

2. List four sources of potential energy you should be aware of when you devise a lockout procedure.

3. List the things you should do prior to restarting a machine when you are ready to remove the padlock from the lockout device.

4. You are asked to help implement a lockout, tag-out procedure for your factory. Explain what you would consider as part of the education and training of all employees.

5. You are asked to help develop an inspection procedure for the lockout, tag-out procedures in your factory. Explain what you would include in creating the inspection part of the lockout, tag-out program.

CHAPTER 21

Troubleshooting

OBJECTIVES

After reading this chapter you will be able to:

1. Explain the fundamentals of troubleshooting.

2. Explain the difference between a symptom and a problem.

3. Develop a wiring diagram from a ladder diagram.

4. Develop a ladder diagram from a wiring diagram.

5. Develop a procedure for locating a loss of voltage in a circuit.

21.0
INTRODUCTION

As a technician or maintenance person, you will be requested to troubleshoot equipment that has stopped running or is not operating correctly. Many people fear this task because it looks so complex. This chapter will help you gain the skills that make troubleshooting much easier and more predictive. Technicians who have been in the field for many years have developed tests to verify the condition of every component that is found in an electrical control or load circuit. You must believe in the value of these tests and strictly follow their procedures. You can review these tests and their procedures in the previous chapters of this book. After reading this chapter you will have the skills to find problems in any machine that you are requested to troubleshoot. You will also learn troubleshooting processes that can be used every time you troubleshoot something, and you will be able to verify the condition of the components you suspect are working correctly or the components you suspect are faulty.

Troubleshooting can be broken into multiple scenarios. You may be asked to troubleshoot a machine that is already in operation and has a wiring diagram as well as a ladder diagram that explains the sequence of operation of the machine. In this case, you would use all the diagrams to test the operation of the machine and determine which parts are inoperative.

In some cases you may be asked to try and troubleshoot a machine that has a wiring diagram but does not have a ladder diagram. In this case, you must convert the wiring diagram to a ladder diagram to determine the sequence of operation for the machine. If the machine has only a ladder diagram but does not have a wiring diagram, you must convert the ladder diagram into a wiring diagram so that you can determine the location of the components. In some cases you may be able to open the cabinet door and identify where the components are located.

This chapter explains how to troubleshoot a machine when you have both a wiring diagram and a ladder diagram. It will also explain how to convert a wiring diagram into ladder diagram and how to convert a ladder diagram into wiring diagram. Both of these skills are very useful when you are working on the job to troubleshoot machines.

21.0.1
Working Safely

The most important point to remember when you are troubleshooting is to work safely and follow all safety procedures provided in the previous chapters in this text. You should strictly follow lockout and tag-out procedures to ensure that the electrical, hydraulic, pneumatic, and mechanical energy sources are safely turned off and secured when you are working on the machine. Some tests require that these systems be fully energized to complete the test. In these cases be sure that you are following all safety rules and are fully aware of the potential for injury when you are working around live circuits. You should also work as though each circuit and power system is live and active any time you are working around a machine.

21.1
WHAT IS TROUBLESHOOTING?

Many technicians believe that troubleshooting is a process to determine what is broken on a machine. A little known secret that the best troubleshooting technicians have learned is that it is actually easier to identify what is working correctly on a machine and, through a process of elimination, determine what is broken. This chapter explains the basic things you should do when you are trying to troubleshoot a machine. You should develop the steps into a procedure that you use each time that you troubleshoot a machine. Many technicians do not use a consistent procedure each time. Instead they try different things each time they troubleshoot a different machine, and sometimes they try to remember what was broken on the machine the last time and work from there, as though the problem is simply repeating itself. You will learn to develop a procedure that you can use every time you troubleshoot a machine.

21.2
THE DIFFERENCE BETWEEN SYMPTOMS AND PROBLEMS

When you are first learning to troubleshoot a machine, you may find what appears to be a problem, but instead is only a symptom of the problem. For example you may find a blown fuse on a machine and think you have found the problem. In fact, if you replace the fuse, the machine will begin to operate again. What is hard to understand is that the machine has a motor that is overheating and overworking because it has a bad bearing, and this is what is causing the fuse to blow. Each time you replace the fuse, the motor will start and run again until it draws enough overcurrent and blows the fuse again. If you continue to replace fuses, the motor will eventually overheat to the point at which it is totally destroyed. In this example, the blown fuse is only a symptom; the bad bearing is the problem. You will learn in this chapter how to determine what are problems and what are symptoms. You may need to revisit some of the early chapters in this text to determine how components, such as relays, transformers, and motors, operate.

21.3
STARTING THE TROUBLESHOOTING PROCEDURE

The first step in troubleshooting a machine should be to determine what the machine was doing prior to stopping. Sometimes this can be done by simply talking to the machine operator or a worker who is assigned to the machine. Sometimes this is the person who was loading and unloading the machine, or it may be a person who was setting up and operating the machine. You can ask a series of questions to determine if the machine was operating correctly prior to this malfunction. For example, you may ask the operator to try and operate different parts of the machine in manual mode. Sometimes you will find that the machine can operate partially or that one part of the machine can operate. If you have not observed this machine operating before, you may need to look around in the general area to see if there are other machines that are similar to the machine that is malfunctioning. You may be able to determine the operation of the machine by observing a similar machine that is close by. If you cannot observe a machine, you may have to ask the operator the basic sequence of operation for the machine.

21.3.1
Observing the Machine and Using Your Senses

When you first begin to troubleshoot a machine, it looks like a very large, complex job. As you work with more experienced technicians, you will find that they use a number of methods to determine what is wrong with the machine. For example, they may spend a few minutes just looking at the machine and trying to get an idea of what components are running or turned on. They may also try to identify what power systems the machine has, such as the type of voltage, if there is hydraulics, or if there is a pneumatic system. You should also try to identify the moving parts on the machine and determine how they operate. Do they use gravity? Do they use belts? Do they use hydraulic cylinders? Another important part of this process is to identify what types of motors the machine has and how many there are. Do the motors have variable-frequency drives (VFDs) or other types of acceleration and speed controls, or are they connected directly to a motor starter? Finally, you may take a minute to determine whether the machine uses regular hard-wired controls or a programmable logic controller (PLC).

A second to step in the process of troubleshooting includes using your other senses beyond vision. For example, you may want to listen for problems, such as motors that are under extreme load, hydraulic pumps that are fully loaded, or belts and pulleys that are squeaking which indicates they are running improperly and causing friction. These clues will help you later when you are trying to determine what is wrong with the machine. Another sense that is usually overlooked is the sense of smell. In some cases you will be able to smell a motor that is overheating before you are able to determine from some other test that it has a problem. You may also be able to smell a machine that has overheated hydraulic oil or that has rubber seals that are becoming faulty. After years of experience, you will learn what a machine smells like when it is operating correctly and what it smells like when it is faulty.

Some technicians have become very good at using their hands or a screwdriver in their hands to determine if there is any abnormal vibration on the machine. Vibrations that are extreme can be felt and will help to determine if the machine is malfunctioning. There is also sensitive test equipment that can determine if a machine has abnormal vibrations.

21.3.2
Swapping Parts

When you have tested your machine and have identified a faulty component, you may want to swap a part to ensure that the part you suspect is faulty is actually bad. This procedure has several drawbacks. First you are putting a second machine out of commission to remove the part to do the test. This may create additional problems and loss of more production. The second problem this may create is that some technicians utilize the component swap method too soon prior to thoroughly testing the part, and basically guess which part is bad and waste a lot of time changing parts. The advantage of swapping a part is that if the part is expensive or has a two- or three-day shipping time, you can ensure that you have actually located the faulty part before you order a new one and wait for it to be shipped.

The basic rule of thumb to follow when swapping parts is to ensure that you have completely tested the part while power is applied and you have completed an alternative test with power off. If you have completed these tests and are sure the part is faulty, you can safely change or swap the part to verify your suspicions.

21.3.3
When to Use a Voltmeter to Troubleshoot

The voltmeter is the meter of choice for most technicians. The voltmeter allows the technician to make tests in the circuit with power applied. It also gives the advantage of being able to make a large number of preliminary tests without disconnecting any wiring or components. If you are called to troubleshoot a machine that has stopped completely, the chance of it having an open circuit, where voltage is interrupted, is very large. The first test that you should make with the voltmeter is to measure the source voltage. You should always set the meter to the highest voltage range until you have determined the actual voltage. After measuring source voltage, you will need to use the ladder diagram for the machine and measure voltage throughout the circuit wherever you suspect the voltage has stopped. When you find the loss of voltage, you can turn power off to the machine lock it out, disconnect the wire or component that you suspect is faulty, and test it with an ohmmeter.

Some problems that you might encounter when measuring voltage will provide you with inaccurate information. The first problem is voltage feeding backward through a circuit. This usually occurs in three-phase circuits that have one blown fuse. If you have one blown fuse in a three-phase circuit, it is possible for voltage from one of the remaining supply lines that has a good fuse to feed through a transformer winding or motor winding and flow backward up to the bottom of the blown fuse. When you test the output side of each fuse, your meter will measure voltage, and it will appear that all the fuses are good. If you are aware this can happen, you can use a fuse puller to pull all three of the fuses from their holders and test each fuse for continuity with an ohmmeter.

The other problem that is beginning to appear to troubleshooters is odd values of voltage in a circuit. If a circuit has a variable-frequency drive, the drive can produce voltages that are above or below 60 Hz. Since most digital and analog voltmeters are designed for 60 Hz. The voltage reading on the meter will be slightly more or less than it actually is, and the technician may feel that the circuit is faulty when in fact it is operating correctly. You can use a true rms reading voltmeter to get an accurate reading in these types of circuits.

21.3.4
When to Use an Ohmmeter to Troubleshoot

It is important to understand when to use an ohmmeter. All power to a circuit must be turned off before you use an ohmmeter to take measurements since the ohmmeter presents very low impedance to the circuit. When impedance is low, it will create a short circuit path if you place the meter leads across the terminals that have voltage present. This will cause the large current to create sparks and possibly cause molten metal to explode into your face, and it will certainly damage the meter. In most cases, you will use the ohmmeter to make a continuity test with power off, after you have used a voltmeter to locate a faulty switch or wire. When you are making the continuity test, you should zero the meter and use the lowest ohm setting. This setting will detect the slightest amount of resistance in the switch or wire. If the resistance is very high, it is called infinity, and it indicates that an open circuit is present in the contacts or the wire. If you are testing a wire, this test indicates the wire is faulty and must be replaced. If you are testing a switch, be sure you are on the correct terminals that make a contact set, and then change the switch to its other positions. If the meter reading remains at infinity, you have determined the switch contacts are faulty, and the switch must be replaced.

Another reason to use an ohmmeter is to test the actual number of ohms in the windings of a motor, transformer, or coil. When you are testing a single-phase motor, you should measure the resistance in each winding accurately so that you can determine that the winding with the highest resistance reading is the start winding and the winding with the lowest resistance reading is the run winding.

If you are troubleshooting electronic circuits, you will also need to measure the amount of resistance accurately. For example when you are trying to identify the terminals of a transistor, SCR, or triac, you will need to measure the resistance and the polarity of the meter probes. Be sure to switch the meter to the setting that will provide the most accurate resistance reading.

21.3.5
When to Use a Clamp-on Ammeter to Troubleshoot

A clamp-on ammeter is a valuable troubleshooting instrument that can indicate the presence of current in a circuit. For example, if you suspect that a motor with three-phase supply voltage has one blown fuse, you can quickly determine which leg of the three phases has the blown fuse by measuring the current in each of the three legs with the clamp-on meter. The leg that has the blown fuse will indicate 0 A, and the two legs where the fuses are good will indicate current that is higher than rated for the circuit, because the motor will be drawing locked-rotor current since it will not be able to start. This test is more effective than the voltage-drop test, which is subject to feedback voltages.

The clamp-on ammeter can also be used to evaluate the current load of any circuit wire. This is especially helpful in determining how close a circuit is to overloading the circuit wire which will cause a fuse to blow. You can also tell if a motor is beginning to wear out and have bad bearings or lacks lubrication. If the motor current is higher than its rating, you can begin to suspect that it is overloaded. If the load is within tolerance, you can suspect that the motor needs lubrication or the bearing may be bad.

The clamp-on ammeter can also be used to adjust belts on belt drive systems. You can apply the clamp-on ammeter and adjust the belt tension while watching the meter. If the current increases above the motor rating, the belt is too tight. If the current rating is too low, the belt is not tight enough.

21.3.6
Dividing the Machine's Electrical System into the Load Circuit and the Control Circuit to Make Troubleshooting Easier

The part of the electrical system that has the motors in it is called the load circuit. Other electrical loads may include lights and electric heaters. The control circuit includes all the switches and devices that control voltage to the coil of motor starters, relays, or solenoids. When you are troubleshooting a circuit that has pushbutton switches controlling a motor-starter coil, you can quickly eliminate parts of the circuit that are functioning and the parts that are not by seeing if the coil of a motor starter is energized. If the coil of the motor starter is energized, it indicates that all the switches and the coil are operating correctly, and you should shift your test to the load circuit where the motor is connected to the motor starter.

Another way to understand this concept is to think about troubleshooting a machine the way you would try to locate a specific card in a deck of playing cards. Suppose I pull one card from the deck and then ask you to guess what the card is. As you know, there are 52 cards in a standard deck of cards that has hearts and diamonds that are red cards and has clubs and spades that are black cards. So if you just start guessing, without a method, it may take you 52 guesses to get the correct one. This section shows you a method that you can use later when you are troubleshooting a machine.

For this example let's say the card I have pulled from the deck is the 9 of hearts. Using this process, the first question would be to ask if the card is a black card. The answer is no, so we can deduce that the card must be red. This one question eliminates 50 percent of the deck. We now know the card must be a heart or diamond, and we can stop thinking about all the spades and clubs in the deck.

The next question will be, is the card a heart? (It does not matter whether I ask if the card is a heart or if I ask if it is a diamond, because the answer will indicate which one it is.) In this case the answer is yes, the card is a heart, so I can stop thinking about all the diamonds. With two questions, we have eliminated 75 percent of the cards.

Now that I know the card is a heart. I will continue to ask questions that eliminate the remaining cards 50 percent at a time. I know there are 13 hearts and they range from a high of ace and a low of 2, so I will next ask if the card is lower than an 8. The answer is no, so I have eliminated the 2, 3, 4, 5, 6, and 7. The remaining cards are 8, 9, 10, jack, queen, king, and ace. Since the queen or jack is the midpoint. I would next ask if the card is below a queen. The answer is yes, so now. I know the card must be a jack, 10, or 9. Then the next question is, is the card lower than a 10? Since the answer is yes, I now know the card is the 9 of hearts.

You may be wondering what this card example has to do with troubleshooting. The four suits of cards—diamonds, spades, hearts and clubs—represent the four power systems on most machines: electrical, hydraulic, pneumatic and mechanical. When you come upon a broken machine, your first questions should determine which of these systems are working and which are inoperative. You need to develop questions and tests that will eliminate the most possibilities. Let's say that you have found that the air pressure and the mechanical parts of the system operate correctly, but the hydraulic and electrical systems will not energize. The next test would be to see if you can get the electrical system energized. You can see that this process cuts down the number of parts of the machine that could possibly be faulty.

When you are testing the electrical system, you would check the load and control system that controls the hydraulic pump. The load circuit consists of the circuit that shows the three-phase supply voltage connected to the motor starter and the hydraulic pump connected to the motor starter. The control circuit includes the master stop push button, master start push button and the coil of the motor starter for the hydraulic pump. The process of selecting tests that will eliminate large portions of the machine will logically help you find the problem on any machine every time you troubleshoot.

21.3.7
Why It Is Important to Test a Circuit with Power Applied

You may wonder why it is important to test a circuit with voltage applied to the circuit. You may even think that this is a dangerous practice. You will find that it is a very safe way to test a circuit, and it is also the most

reliable. The problem with testing a circuit for continuity with the power off is that it is a very slow process because all the wires must be disconnected before they can be reliably tested and it is very easy to have faulty readings from parallel circuits that you are not aware of. If you expect to become a qualified electrical technician, you must be able to repeat the voltage-loss test explained in the following section without making mistakes. You must also be able to make the test quickly and accurately.

21.3.8
Troubleshooting the Push-Button Switches in the Control Circuit

Testing the control circuit in a system is a simple procedure. You should remember that the control circuit may contain several push-button switches that may be faulty or not operate correctly. Figure 21–1 shows such a circuit. The problem could be any one of several things such as the loss of the power supply, a loose or broken wire, a faulty pilot switch, or a faulty load (motor starter or relay coil). This means that you must troubleshoot the complete circuit instead of picking on individual components and testing them at random. The test that will be outlined in the following procedure is called the *voltage-loss* test, and you should learn to use this test whenever you are trying to locate a faulty component or wire in an inoperable circuit. We will use the control circuit diagram shown in Figure 21–2.

Start the procedure by testing for voltage across the load terminals (the two terminals of the motor-starter coil). If voltage is present at the terminals of the motor-starter coil, you can be sure that the power supply and all the pilot switches and interconnecting wires are working correctly, and the remainder of the troubleshooting test should focus on the coil.

If no voltage is present at the terminals of the coil, the test should focus on the loss of voltage somewhere

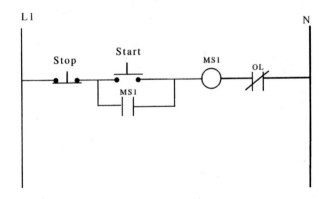

Figure 21–1 Control circuit diagram for troubleshooting.

between the source and where it travels through the pilot switches to the coil. If voltage is not present at the power supply, be sure to check for a blown fuse, open disconnect, or other power supply problems. If voltage is present, it may be 24, 120, or 240 V or higher, and it may also be AC or DC. At this point in the test, all that you are concerned with is that the supply voltage is the same voltage that is specified on the electric control diagram.

After you have determined that the proper amount of supply voltage is available, the remainder of the tests should focus on the pilot switches and interconnecting wires. The fastest and most accurate test for locating an open wire, loose terminal, or faulty pilot switch is to test for a voltage drop (loss). This test requires that power remain applied to the circuits while you touch the meter probes to each test point. *You must be aware of where you are placing your hands, tools, and meter probes at all times because severe electrical shock could result if you come in contact with the electrical circuit.* Even though leaving power applied to a circuit while you test it presents an electrical safety condition, it is necessary because control circuits tend to be complex and you may not be able to find the problem with the power off.

Figure 21–2 also shows the test points where you should place the voltmeter probes to execute the test. This test can be used to locate problems in any circuit, regardless of the number and types of pilot devices. The test points are identified as A through J. The first test should be made with the voltmeter probes touching points A and J. which is across the power supply. Use this test to make sure that your meter is on the right range. From this point, leave one of the voltmeter probes on point J and move the other probe to point B. If no voltage is available at point B, and you had voltage at point A, this test indicates the wire between the stop

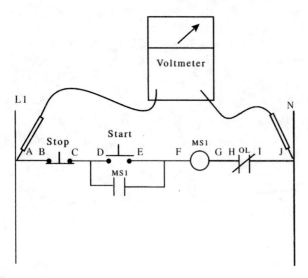

Figure 21–2 Wiring diagram with test points for the voltage-loss test.

push button and the power supply is open. You can turn off the power and test the wire for continuity, and you should be able to verify that the wire has an open. A continuity test uses the resistance range on the VOM meter. If the wire has an open, the meter will indicate infinite resistance; and if the wire is good, it is said to have continuity and the meter will indicate low resistance. The wire must be isolated (one end removed from its terminals) so that you do not measure resistance from a parallel circuit that may be connected to the wire you are testing. If your continuity test indicates that the wire has an open, you should remove and replace the wire to put the circuit back in service.

If you measured voltage when the meter probe is on point B and point J, the test indicates that the circuit is good to point B. You can move the meter terminal probe to point C. If you do not have voltage at point C and you had voltage at point B, the test indicates that the stop push button is open. Remember that the stop push button is a normally closed push button, so you should have voltage at point C. Check to be sure that the stop button is not depressed. Remember that it may be a maintained type of switch and require someone to return it to its normal condition. If the switch is in its normal condition and no power is present at point C, the switch is faulty and must be replaced.

If you are testing for a stop switch that will not de-energize the relay coil, be sure to test point C and ensure that voltage is not present there when the stop switch is depressed. If voltage is still present at point C even when the stop button is depressed, you can assume that its contacts are welded or stuck together and that it must be replaced.

To continue the test for a loss of voltage in this circuit, we will resume from the last point where voltage is present, at point C, and continue the test by moving the probe on to point D. If voltage is present at point C and not at point D, your test indicates that there is an open in the wire that connects the stop button and the start button. Since the voltage test indicates a loss of voltage at point D, you can turn off power to the circuit and test this wire for continuity; and if this test indicates an open, you should change the wire.

If you have voltage at point D, you have determined that the circuit is operating correctly to this point and you can move to point E. Your test should indicate that you do not have voltage at point E, because the start push button is a normally open switch. You will need to depress the start push button and check for voltage at point E. If you do not have voltage at point E when the switch is depressed, you should turn off the power and test the switch for continuity. If the switch is faulty during the continuity tests, you should replace it.

If you have voltage at point E, your next test should be at point F. If you do not have voltage at point F, you can predict that the wire between the start push button and the coil of the motor starter has an open in it. You can turn off the power and test this wire for continuity.

If you have voltage at point F, you should move the meter lead from point J to point G. Meter probes will now be touching both sides of the motor-starter coil. Since the coil is a load, it should have full voltage applied to if it is supposed to become energized. If you have voltage at points F and G, you can predict that the circuit wire and switches are all operating correctly and providing voltage to the coil. This is perhaps the most important point in troubleshooting. The main function of all of the wire and switches in every circuit is to provide the proper voltage to the load. In this circuit the load is the motor-starter coil; and if you measure voltage at its terminals and the contacts are not pulled in, you can predict the circuit wires and switches are operating correctly, and you should suspect the motor-starter coil is faulty. You can turn off power to the circuit and test the motor-starter coil for continuity. If the coil is good, the test should indicate resistance in the range of 20 to 2,000 Ω depending on how much wire is in the coil. If the test indicates infinite resistance, the coil is open and it should be replaced. Remember, do not apply voltage to a coil that is not installed in a relay.

If you had voltage when the meter leads were on points F and J, but did not have voltage at terminals F to G, your test has indicated that the circuit has an open between point G and point J. The final two tests to locate the problem in the circuit are perhaps the most difficult to understand, because it appears as though the circuit is operating correctly when your test determined that voltage is present at point F to point J. In reality, your tests have proved that the wire and switches on the left side of the motor-starter coil are operating correctly. Since the coil will not energize, you can then predict that the problem is in the wire between points G and J or the overload contacts are open.

The best way to locate the fault in this part of the circuit is to move your left meter lead to point F and leave it there for the reminder of the tests, and resume the test by moving the right meter lead to test point J, then to I, and then to H. The point where you lose voltage is the place where there is an open in the circuit. For example, if you have voltage from point F to point J, but you do not have voltage when you move the right meter probe to point I, it indicates that the wire between J and I has an open. Turn off the power and test the wire for continuity. If your test indicates that you have voltage at point F to point I, you should move the right meter lead to point H. If you do not have voltage at point H to point F, it indicates that the overload contacts are open. You can verify this by turning off the power and testing the over-load contacts for continuity. If you have voltage from point F to point H, you can predict that the last wire to be tested, wire G to H, has an open. You can turn off the power and test the wire for continuity. Remove and replace the wire if it is faulty.

21.4

DEVELOPING A LADDER DIAGRAM FROM A WIRING DIAGRAM

Many times when you troubleshoot a system, a wiring diagram of the electrical circuit will be provided, but a ladder diagram will not be available. You can develop your own ladder diagram by tracing the circuits in the wiring diagram. The material in this section provides an example of converting a wiring diagram to a ladder diagram. Figure 21–3 shows a wiring diagram of a typical motor control system that is connected to 230 VAC. You can see in this diagram that the major components in this system are the power supply, transformer, start and stop push buttons, motor starter, control relays and hydraulic pump motor. One trick that will help you with this process is to take a colored pen and mark over each wire in the wiring diagram as you add it to the ladder diagram. When you have all the wires in the wiring diagram colored, you have completed the ladder diagram.

We will draw the load circuit separate from the control circuit. Since the load circuit has fewer components

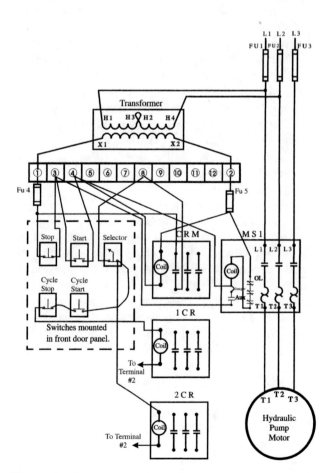

Figure 21–3 Wiring diagram for conversion to a ladder diagram.

Figure 21–4 Load circuit of the ladder diagram.

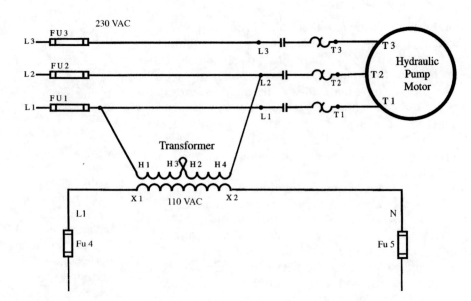

and is simpler, we will start with it. When you are converting a wiring diagram to a ladder diagram, you can start by locating the loads (motors) and the motor starter controlling each of them. In the example wiring diagram, you can see the motor is at the bottom right and the motor starter is down the right side. The voltage for the load circuit comes from the three fuses at the top right side of the diagram. When we convert the load part of the wiring diagram to the ladder diagram, you will notice that it looks very similar. Figure 21–4 shows the load circuit across the top of the ladder diagram. The load voltage also provides the primary voltage for the control transformer at terminals H1 and H4, which are connected to L1 and L2. The secondary terminals of the transformer X1 and X2 provide the 110 VAC for the control circuit, so we will also show the transformer and fuses for the control circuit and identify them as L1 and N, which creates the two power rails for the entire control portion of the ladder diagram.

You can always start the ladder diagram of the control circuit with two power rails. The transformer primary winding will be connected to the power supply, and the secondary will be connected to L1 and N at the top of the power rails for the ladder diagram. You can start developing the ladder diagram by placing the motor starter coil between the two rails of the ladder diagram. This is shown in Figure 21–5. (You can place the coils of relays and motor starters in any rung of the ladder diagram you are building at this time. The types of switches controlling the coil will determine the actual logic of the circuit. You will learn with experience how to locate the rungs in the ladder diagram.)

The next step in the procedure is to find the motor starter and its coil. From the top of the coil you can trace power to the N side of the circuit. From the wiring diagram you can see that power must travel through the three sets of normally closed overloads and then to fuse

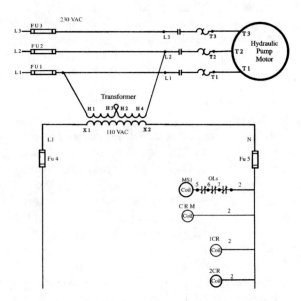

Figure 21–5 Ladder diagram with coils shown in position.

5, which is connected to terminal 2, which is connected to N (neutral line of the circuit). The next process is to trace the wires back to L1 (terminal 1). The wire out of the bottom of the motor-starter coil runs to a terminal (black dot). From the terminal, one of the wires goes to terminal 4 on the terminal board, and the other wire goes to the top of the auxiliary contacts. To minimize confusion, at this time you should follow only the wire from the motor-starter coil to the start push button and disregard the other wires that are connected to the coil and to terminal 4. Be sure to color the wires in the wiring diagram that you have added to the ladder diagram (see the ladder diagram in Fig. 21–6).

The next step is to trace the wire from terminal 4 to the right side of the start push button. From the left side

Figure 21–6 Ladder diagram with first line filled in.

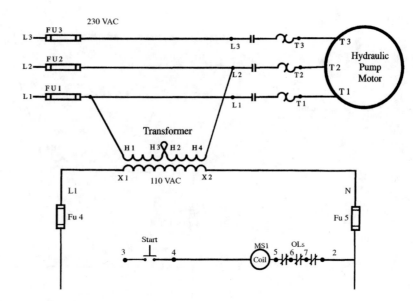

Figure 21–7 Start of ladder diagram with coils in place.

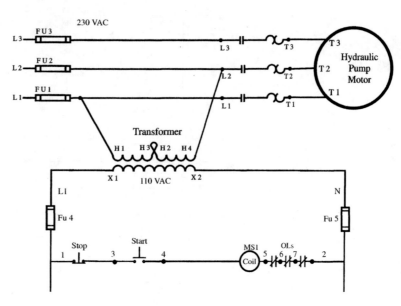

of the start push button, the wire runs back to terminal 3 on the terminal board. From terminal 3, a wire runs back to the right side of the stop push button; and from the left side of the stop push button, the wire runs back to the bottom of fuse 4, which is connected to terminal 1 and L1 (see the ladder diagram in Fig. 21–7).

At this point you may get confused by the other wires that are connected to each of the terminals. For example, the auxiliary contacts are shown connected to terminals 3 and 4. You should add the auxiliary contacts and wires connecting the contacts to terminal 3 and terminal 4 at this time.

The final component in this circuit is the coil of the control relay master. You have already connected the right side of this coil to terminal 2. The wire from the left side of this coil is connected to terminal 4. When you have added these components and wires to the ladder diagram, it will look like Fig. 21–8. You should notice

that the coil of the CRM is connected in parallel with the coil of MS1. Be sure to color these components and wires on the wiring diagram as you add them to the ladder diagram.

You are now ready to draw the next line of the ladder diagram. You have already drawn the MS coil and the CRM coil, so you are ready to add the next coil, which is the coil of control relay 1CR. The first wire to add to the coil is the wire that runs from the bottom of the 1CR coil to terminal 2 through fuse 5. The wires that run from the top of the 1CR coil connect the coil to the terminal on the left side of the cycle stop switch. From the right side of the cycle stop switch, a wire runs to the cycle start switch. Note that the wire can be a jumper, since the cycle start and cycle stop switches are mounted on the door of the electrical panel and are some distance from the terminal board. In some circuits, the wires would go all the way back to the terminal strip.

Figure 21–8 Ladder diagram with MS coil and CRM coil.

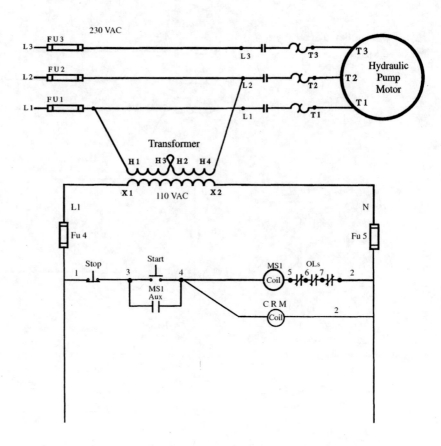

Figure 21–9 Ladder diagram with 1CR drawn in the third rung.

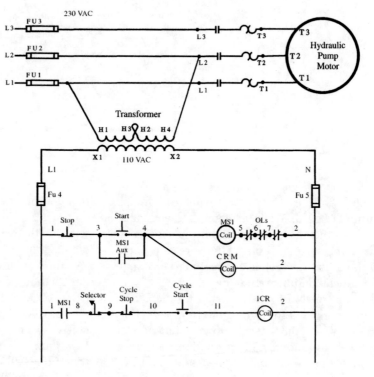

The next place the wires run are from the right side of the cycle start push button to the right side of the selector switch. From the left side of the selector switch, the wire runs to terminal 8. From terminal 8 the wire runs back to the bottom of the normally open contacts of control relay CRM. Out of the top of these contacts,

the wire runs back to the bottom of fuse 4, which is connected to terminal 1. See Fig. 21–9 to see this part of the ladder diagram. Be sure to color the wires in the wiring diagram as you show them in the ladder diagram.

The final part of the ladder diagram is the coil of control relay 2CR. The wire from the right side of the coil

Figure 21–10 Completed ladder diagram.

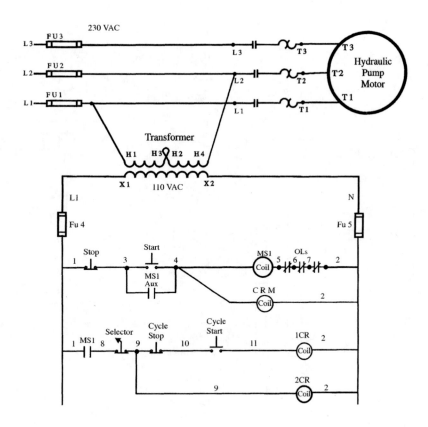

is connected to terminal 2, which is connected to N. The wire from the left side of the 2CR coil goes to the left side of the selector switch, which makes its way back to open contacts of control relay CRM and then back to terminal 1 for L1. The coil of 2CR is connected in parallel with the coil of 1CR. This completes the process of converting the wiring diagram to a ladder diagram. (Fig. 21–10).

21.5
CONVERTING A WIRING DIAGRAM FROM A LADDER DIAGRAM

At times you may have a ladder diagram (also called an elementary diagram) that provides the sequence of operation for the machine, but you do not have a wiring diagram that shows information such as the physical location of the terminals and the number of contacts on relays. This usually occurs when a machine is being rewired or when a new circuit is added to the electrical control circuit. When you are wiring the new components into the circuit, you will use many of the same skills that you would use to draw a wiring diagram from the ladder diagram. With practice, you can create a wiring diagram from a ladder diagram in a very short time.

The first step in this process is to make a visual inspection of the electrical cabinet and sketch the components that you see mounted inside the cabinet or components such as push-button and selector switches that are mounted in the door of the electrical cabinet. The sketch should be accurate with respect to the location and relative position of each component. It is also important to identify where terminal connections are on a component such as the terminals where power comes into the device and goes out. The simplest way to identify these contacts is to use identifiers as top of the component and bottom of the component or as left-side terminal and right-side terminal. We will use these identifiers as we develop the diagram.

This section provides an exercise that shows how to develop a wiring diagram from a ladder diagram. In this exercise we will use the picture of the inside of the electrical cabinet and the components mounted in the cabinet door as shown in Fig. 21–11. From this picture we will develop the sketch of the components which will become the skeleton for the wiring diagram. This sketch of the components is also shown in Fig. 21–11. We can use the ladder diagram provided in Fig.21–10 to begin making the wiring diagram. We will take advantage of the wire numbers provided in the ladder diagram. We will complete only the first two lines of the ladder diagram for this exercise. In reality, you would need to convert only the part of the ladder diagram where you

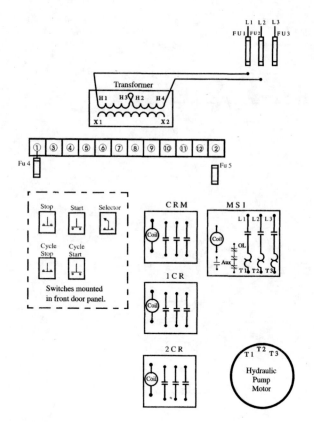

Figure 21-11 Layout of all components in the electrical cabinet to start conversion of the ladder diagram to a wiring diagram.

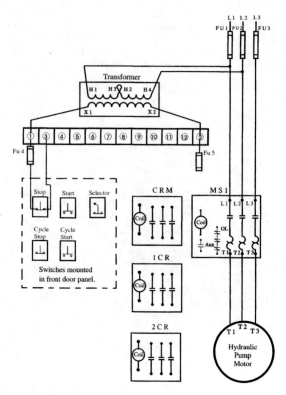

Figure 21-12 Wiring diagram with wires for the transformer and motor load drawn in.

suspect the problem exists. You would convert the entire diagram only if your company wanted a completed wiring diagram to go with the ladder diagram.

We start the process by copying the wires where voltage enters the ladder diagram. This will include the two wires connected to the primary winding of the transformer the two wires that run from the secondary terminals at the bottom of the transformer to the terminals at the top of the fuses, the wires that run from the bottom of the fuse on the left that runs to terminal 1, and the wire from the fuse on the right side of the diagram that runs to the #2 terminals on the terminal board. Figure 21-11 also shows this part of the diagram.

Figure 21-12 shows the wires that run from terminal 1 to the terminals on the left side of the master stop push button and from the terminals on the right side of the stop push button to terminal 3 on the terminal board. The diagram in Fig. 21-13 shows that the next wire runs from terminal 3 on the terminal board to the terminal on the left side of the master start push button and from the terminal on the right side of the start push button to terminal 4 on the terminal board.

Figure 21-13 also shows the wires that run from terminal 3 to the terminal on the left side of the normally open auxiliary contacts that are mounted on the motor starter and from the terminal on the right side of the aux-

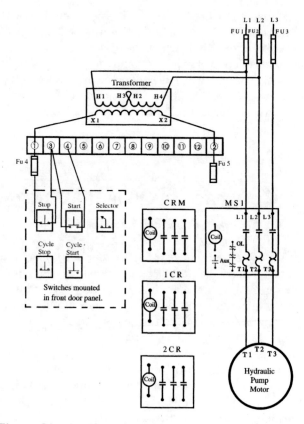

Figure 21-13 Wiring diagram with wires connected between stop and start push buttons.

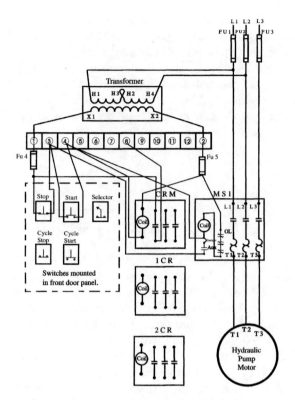

Figure 21–14 Wiring diagram with auxiliary contacts wired in and coils connected to terminal 2.

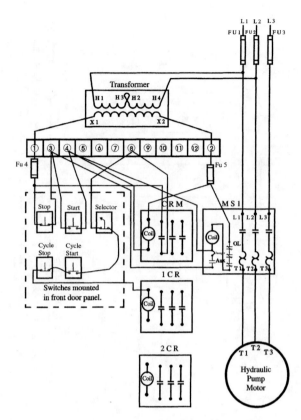

Figure 21–15 Wiring diagram with wires connected to cycle start and cycle stop and to the coil of 1CR.

iliary contacts back to terminal 4. Figure 21–14 shows the diagram where the next wire connects the overloads of the motor starter to terminal 4 and terminal 7. Note that the terminal on the left side of the first overload is connected to terminal 4 and the terminal on the right side of the third overload is connected to terminal 7. The first overload contact is connected directly to the second overload contact, and the second overload contact is connected directly to the third overload contact. (Since the overload contacts and the auxiliary contact are all mounted on the motor starter, some manufactures may connect the #4 wire of the overloads directly to the auxiliary contacts on the motor starter rather than run the wire all the way back to the terminal board. This practice will save wire since the distance between the auxiliary contacts and the overloads on the motor starter is relatively short. The #5 wire and the #6 wire that run between the three sets of overload contacts are not sent back to the terminal board. These wires are relatively short since the distance between the overloads is short.)

The final wire, #7, that comes from the right side of the third overload contact connects the last overload to the coil of the motor starter. This wire may also be run directly between the overload and the coil. The coil is also mounted on the motor starter, and the distance between the coil and the overload contracts is rather short. This practice will save a substantial amount of wire, compared with running the wire back to the terminal board.

The next wire in this part of the diagram runs between the coil of the motor starter and terminal 2. Another wire runs between terminal 7 and the coil of the master control relay (CRM) and from the coil to terminal 2.

The next section of the wiring diagram includes the wires for the second rung of the ladder diagram (Fig. 21–15). You can see that the first wire runs between terminal 1 and the top terminal of the first set of normally open contacts that belong to the CRM relay. One end of wire 8 is connected to the bottom terminal of the contacts, and the other end is connected to terminal board terminal 8. From terminal 8 on the terminal board the next wire runs to the terminal on the left side of the selector switch. The next wire, #9, is connected to the terminal on the right side of the selector switch on one end, and the other end is connected to terminal board terminal 9. From terminal 9 on the terminal board the next wire runs to the terminal on the left side of the cycle stop push button. One end of the #10 wire is connected to the terminal on the right side of the cycle stop push button, and the other end is connected to terminal board terminal 10. From terminal 10 on the terminal board the next wire runs to the left side of the cycle start push button. One end of wire 11 is connected to the right side of the start cycle push button, and the other end is connected to terminal

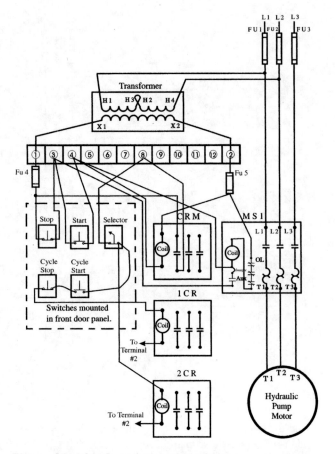

Figure 21–16   Completed wiring diagram.

board terminal 11. From terminal 11 on the terminal board the next wire runs to the terminal on the left side of the 1CR coil. The final wire in this rung has one end connected to the left side of the 1CR coil and the other end connected to terminal 2.

Notice in the ladder diagram that a second control relay, 2CR, is wired in parallel to the coil of 1CR. This part of the circuit starts with wire 9, which is connected to terminal 9 in the terminal board on one end and to the terminal on the left side of the 2CR coil. The final wire in this circuit connects the terminal on the right side of 2CR coil to terminal 2 (Fig. 21–16).

21.6
AUTO MODE AND MANUAL MODE

When you are troubleshooting a machine, you may find that it can operate in automatic mode and manual mode. A selector switch allows the operator to switch the machine between automatic mode and manual mode. A good procedure to follow is to try to operate

each section and component of the machine in manual mode. If the components operate in manual mode, you know that they are functioning correctly. If the machine does not operate in automatic mode, after testing it in manual mode, you can suspect that the components and wiring in the control circuitry are causing your problems when the system is in automatic mode.

21.7
REVIEW OF TROUBLESHOOTING

Now that you have seen all the features of troubleshooting, you should remember several important points. First, always work safely and use lockout and tag-out procedures when you are troubleshooting. You should also remember that troubleshooting is the process of determining what systems are working correctly on a machine, so that you can eliminate them from your process and focus on the part of the machine that is malfunctioning. Tests have been presented in this text that will allow you to test any component or control of the electrical system and verify if the part is good or bad. These include in-circuit tests and tests for when a part is removed from a system. If possible, you should test components while they are in circuit. If you determine that a part is faulty, then remove it and test it again when it is out of circuit to verify your previous findings.

When you are troubleshooting, you should use electrical ladder diagrams and wiring diagrams whenever possible. If you have a wiring diagram and need a ladder diagram, you can convert the diagram using the procedures shown in this chapter. You can also convert a ladder diagram to a wiring diagram when you are installing components or need the wiring diagram for other tests.

The final point to remember about troubleshooting is to use your senses—vision, hearing, smell, and touch—to evaluate a machine that is malfunctioning. Also be sure to ask personnel who work around the machine how the machine operates under normal conditions. Check other similar machines, if they are available, to verify operation. Try to operate all the sections or components of the machine in manual mode before you test them in automatic mode.

You must learn to develop a strategy that strictly follows procedures. You should select tests that will eliminate the largest portions of the machine, like the example that showed the best procedure for identifying a card in a deck of cards. Finally, you should keep very accurate and detailed records of your troubleshooting and maintenance functions.

QUESTIONS

Short Answer

1. Explain why you should use the voltage-loss method to detect an open in a circuit.

2. Explain the difference between a symptom and a problem.

3. Explain when it would be appropriate to swap a part that you thought was faulty.

4. Explain why it is important to use the same procedure every time you troubleshoot a machine.

5. Explain when you should use an ammeter to make a test in a circuit.

True or False

1. When you test a component with an ohmmeter, you should always ensure the power is turned off.

2. When you test a component with a voltmeter, you should always ensure the power is turned off.

3. When you are called to troubleshoot a machine, you should use your senses to determine what parts of the machine are operating correctly.

4. You could use your "sense of feel" to determine if a machine is vibrating excessively.

5. A clamp-on ammeter could be used to determine which line of the three lines that supply voltage to a motor has a blown fuse.

Multiple Choice

1. A wiring diagram _____.
   a. shows the sequence of operation for the circuit
   b. shows the location of each component in the circuit
   c. shows how each component operates
   d. shows how a machine should be troubleshot

2. A ladder diagram _____.
   a. shows the sequence of operation for the circuit
   b. shows the location of each component in the circuit
   c. shows how each component operates
   d. shows how a machine should be troubleshot

3. Troubleshooting may best be described as _____.
   a. finding the part of the machine that is broken
   b. identifying the parts of a machine that are working correctly and through a process of elimination determining what is faulty
   c. swapping parts until the machine starts running again
   d. removing parts from a machine and testing each one with an ohmmeter

4. The control circuit _____.
   a. includes the motor and other loads in a circuit
   b. includes the motor and other controls in a circuit
   c. includes switches and other controls in a circuit
   d. all of the above

5. The load circuit _____.
   a. includes the motor and other loads in a circuit
   b. includes the motor and other controls in a circuit
   c. includes switches and other controls in a circuit
   d. all of the above

GLOSSARY

Alternating Current (AC)—The flow of electricity in a circuit that constantly switches between positive and negative charges.

Ampere—A unit of current flow. One amp is equal to 6.24×10^{18} electrons flowing past a point in 1 second. The steady current in two parallel conductors separated by one meter in the meter-kilogram-second system which exerts a strength of 2×10^{-7} newtons per meter length

Ambient Temperature—The temperature of the environment surrounding an object.

American Wire Gauge—The standard used to determine the size of conductors.

ANSI—American National Standards Institute.

Apparent Power—Current caused by capacitive reactance and inductive reactance. Apparent power does not take into account the phase shift caused by the capacitor or inductor. When converting electrical energy to a nonelectrical form, this is the voltage-ampere it takes the instrument being used to convert it.

Armature—Bundles of wires wrapped around an iron core in a magnetic field, which when rotated produces an electrical current.

Auxiliary Contacts—Contacts connected in parallel with the start push button. These contacts serve as a seal-in circuit after the motor-starter coil is energized.

Back-EMF—The back voltage (electromotive force) created by windings when a motor begins to run. It is large enough to cause the coil of the potential relay to energize and pull its contacts open, which de-energizes the start winding.

Bimetal strip—A strip made of two metals, one of brass and the other of steel. When the bimetal strip is heated over a Bunsen burner, the strip will curve toward the steel side. When the strip is cooled in liquid nitrogen, it curves toward the brass side. Bimetal strips are useful for detecting and measuring temperature changes.

Capacitor—A device that has one or more pairs of conductors. The conductors are separated by insulators that are used to store an electrical charge. A capacitor helps improve electricity flow through distribution lines by reducing energy losses.

Centrifugal switch—Also known as the end switch in a set of electrical contacts mounted on spring tension. The steel keeps the contacts in the open position.

Circuit breaker—A circuit is a path through which energy travels. A circuit breaker "breaks" or stops that path of electricity. If there is a chance of overloading, a circuit breaker is designed to break the electrical circuit. Circuit breakers sense over-current and excess heat.

Contactor—A larger version of a relay. It has a coil and contacts that are a connection for the passage of a current. The contacts of the contactors touch together in a manner that allows electrical current to flow. Depending on which side it touches, the contactor will either open or close a circuit.

Continuity—A term used to identify a good wire or zero ohms.

Counter-EMF—The generated voltage that is out of phase with the applied voltage.

Current Relay—A special relay designed to energize the start winding of a single-phase motor. The current relay is typically used with sealed compressor motors.

Delta Connection—Windings connected in the shape of a triangle. Used in a motor or three-phase transformer. The Greek letter "D" is called Delta and has the shape of a triangle.

Delta Voltage—The amount of voltage that you would measure between terminals from a delta-connected transformer. Typical delta voltage is 240 or 480 volts.

Diac—A bidirectional trigger diode (device) that specifically triggers a triac or an SCR. The diac does not conduct current until breakover voltage is reached.

Diode—An electronic device that restricts the flow of currents to one direction (one-way valve). This semiconductor has only two terminals.

Direct Current (DC)—An electrical current flowing in one direction. [Flow of electrons from the negative $(-)$ terminal to the positive $(+)$ terminal; goes in a complete circuit.]

Dual-Element Fuse—A two-element fuse that provides short circuit and slow overcurrent protection. The first element is a time-delayed element that allows the motor to receive a larger starting current for a short period of time while the motor starts. The second element is a short circuit element that protects the motor and circuit against a very large short circuit current.

Duty Cycle—Duration of transmitting time; also the percentage of time a motor runs.

Electromagnetic Induction—Movement between the wire and the magnetic field that produces the current.

Electromotive Force (EMF)—Voltage energy per unit charge, which is reversibly converted—from chemical, mechanical, or other form of energy—into electrical energy in a device such as a battery or dynamo.

Enclosure—A box that encloses an electrical part in one particular vicinity.

End Switch—Also called a centrifugal switch. A switch mounted in the end plate, electrically connected to the start winding that is located in the stator of a single-phase motor.

Flux Lines—The lines of force that eminate from the poles of a permanent or electro magnet.

Frame Size—Specific physical dimensions of motors, including critical mounting dimensions. Regulated by NEMA.

Frequency—The number of cycles, repetitions or vibrations, in a certain amount of time in an alternating electrical current. Number of cycles per second.

Full-Wave Rectifier—A rectifier that uses four diodes to convert the full AC wave, while dealing with the positive and negative half-cycles, to DC voltage.

Fuse—A device placed in a circuit in order to protect the circuit from overcurrent. Inside the device, there is a piece of wire (made of easily melted metal). When the current becomes too strong, it causes the metal to melt, which in turn causes the circuit to be broken.

Ground-Fault Interrupter—A type of circuit breaker that protects generating equipment from damage by shutting off power when it senses currents caused by ground faults.

Half-Wave Rectifier—A diode rectifier that eliminates either negative or positive alternation of the input AC cycles, resulting in the conversion of AC waves to pulsating DC waves.

Hertz—A unit of measure of frequency equal to one cycle per second.

High Leg Delta Voltage—Voltage that is derived from L3-N of a delta-connected transformer. Should not be used to power 208-volt components.

Horsepower (hp)—The unit used to measure a motor or engine's power. One horsepower equals 30,000 foot-pounds per minute.

Impedance—In an alternating electrical circuit, this is the apparent resistance caused by a capacitor, inductor, or resistor in an AC circuit. The formula to find impedance of a charge is $z = \sqrt{R^2 + X^2}$.

Inductive Reactance—The opposition caused by the inductor in circuits.

Infinite—Boundless, without limits, extending beyond measure or comprehension, endless; having no limit in space, extent, number, or time.

Inrush Current—Also known as the pull-in current. It is typically three to five times as large as the hold-in current.

Ladder Diagram—A diagram of a control circuit and a load circuit. The overall shape of the diagram looks like a wooden ladder and shows the sequence of operation.

Light-Emitting diode (LED)—A special diode that is used as an indicator because it gives off light when current flows through it. An LED does not have a filament, and so it can provide thousands of operations without failure.

Load Circuit—A part of a circuit where the motor is located.

Locked Rotor Amperage (LRA)—A large amount of current that is drawn when voltage is first applied to the motor. The amount of current is large because the amount of resistance in the windings is very small and the rotor has not started to rotate.

Lockout, Tag-out—A program mandated by OSHA that is designed to make sure all power to a machine is de-energized and locked prior to working on the system. Tags are attached to the disconnect switches to indicate who is working on the machine.

Magnet—Material that has an attraction to iron or steel. The atoms within are grouped in regions known as dipoles.

M (mega)—Prefix meaning large, representing one million.

Micro—Prefix meaning minute, representing one-millionth of a specified unit.

Milli—Prefix denoting a factor of one-thousandth.

NEC—National Electric Code.

NEMA—National Electrical Manufacturers Association.

Normally Closed Contacts—Contacts that remain in the closed position when there is no voltage applied to the circuit. When they are used as overload contacts, they open to protect the motor from system overloads.

Normally Open Contacts—Contacts that remain in the open position when there is no voltage applied to the circuit. When they are used as auxiliary contacts, they close when the relay coil is energized.

Ohm's Law—$E = IR$; $I = E/R$; $R = E/I$; where E = voltage impressed on a circuit, I = current flowing in a circuit, and R = circuit resistance. Ohm's law is used for calculating voltage drop, fault current, resistance, and other characteristics of an electrical circuit.

Over-current—Any current in excess of the related current of equipment or the ampacity of a conductor. It may result from an overload, a short circuit, or a ground default.

Parallel Circuit—A circuit in which all the loads have the same voltage supplied. The circuit allows current along more than one path to return to the power supply.

Period—The amount of time in between two accruing functions. The period of AC voltage is calculated as the time it takes for one cycle to occur.

Potential Relay—The potential relay uses a high-impedance coil with a set of normally closed contacts to energize a start-capacitor, capacitor-run (CSCR) single-phase compressor motor. When the motor reaches approximately 75% full RPM the CEMF will become large enough to energize the potential relay coil and cause the contacts to open. When the contacts open the start winding becomes de-energized.

Power—The magnification factor of an optical device such as in an alternating current circuit. The current multiplied by the voltage gives you the power.

Power Factor—The amount of actual power in comparison to the apparent power. To find the power factor, divide watts by voltamperes.

Power Supply—Source of voltage for a circuit. The major part of a programmable logic controller.

Pressure Switch—A type of switch that activates a circuit when pressure falls or rises. It can be used to turn on a compressor when the pressure inside the system falls, to make sure that it sustains proper pressure.

Reactance—The inductance and capacitance in a circuit causing opposition to the flow of alternating electrical current.

Rectifier—A device that changes alternating current to direct current.

Resistor—An element in an electrical circuit that provides resistance.

Root-Mean-Square (RMS)—The square root of the average of the squares of a set of numbers.

Rotor—A rotating shaft in a motor, generator, or alternator. Coils of wire mounted on the alternator's rotating part.

Series Circuit—A simple circuit that includes a power source, a switch for control, and a motor as the load. There is only one path in the circuit for current to travel to the load and back to the power source.

Service Factor (SF)—A rating that indicates how much a motor can be safely overloaded.

Silicon-Controlled Rectifier–(SCR)—A four-layer device with three terminals—the anode, the cathode, and the gate. The SCR is commonly used to control high-power AC circuits such as electric motors.

Slip—The difference in the rated speed of a motor and the actual running speed of the motor.

Star Connection (Wye Connection)—The winding connection for a three-phrase transformer or motor that is in the shape of the letter Y.

Stator—Stationary part of the motor.

Synchronous Speed—The rated speed of a motor that is proportional to the frequency of the electrical current that operates it.

Torque—The amount of rotating force that the motor shaft has. This force is needed to pump the piston of the compressor and turn the fan blades for a condenser fan.

Triac—Two SCRs that have been connected back-to-back in parallel so that one of the SCRs will conduct the positive part of an AC signal and the other will conduct the negative part of the AC signal.

True Power—Current caused by a resistor in a circuit.

Variable-Speed Drive—A controller that sends a specific frequency of voltage to a motor to change its speed.

Voltampere (VA Rating)—The rating for a transformer calculated by multiplying the primary voltage and current or the secondary voltage and the secondary current.

VOM Meter—(Volt-ohm-milliampere meter) A meter that can read volts, ohms, or milliamps. The ammeter is actually designed to read any small amounts of current, such as 1/1,000 amps. The quantity 1/1,000 amps is called a milliamp (mA), which can also be written 0.001A.

Wild Leg Voltage (high leg voltage)—The high leg delta voltage of 208 volts. It will occur between L2-N if the neutral point is produced by a center tap of the L1-L3 winding.

Wye Voltage—L1-N or L2-N voltage for a wye-connected transformer. It will always be more than half the voltage of L1-L2.

INDEX